Third Generation Wireless Systems,

Volume 1

Post-Shannon Signal Architectures

For a complete listing of the *Artech House Mobile Communications Library*, turn to the back of this book.

Third Generation Wireless Systems,

Volume 1

Post-Shannon Signal Architectures

George M. Calhoun

Artech House
Boston • London
www.artechhouse.com

Library of Congress Cataloging-in-Publication Data

A catalog record for this book is available from the Library of Congress.

British Library Cataloguing in Publication Data

A catalog record for this book is available from the British Library.

Cover design by Igor Valdman

© 2003 ARTECH HOUSE, INC.
685 Canton Street
Norwood, MA 02062

10 9 8 7 6 5 4 3 2 1

CONTENTS

PREFACE

This is a book about the technological foundations of the next generation of wireless networks, including the so-called third generation, or 3G, for the mobile Internet. It is a vast subject, encompassing a variety of new services, applications, and air interfaces. In this book I have focused on the lowest layers of the system: how the wireless information-bearing signal is constructed, and the choice of physical-layer transmission techniques and coding strategies capable of withstanding the extreme environment of the wireless channel.

I have tried to emphasize the general lines of technical development and core principles, rather than focusing on specific air interface standards. Underneath the wide variety of prospective technical solutions, there are a few common engineering challenges. The third generation and other new wireless standards must operate at much higher bit rates, and must deliver a much cleaner signal. Conventional approaches to managing the physical channel will not suffice. We are entering a new historical phase of intensive innovation, in which I believe that wireless technology will again become the leading edge for communications engineering in general. This book is planned as the first of three or four related volumes on this next generation wireless. Subsequent volumes will deal with higher-level architectures designed to support multiple access in wireless networks, as well as the emerging suite of wireless applications (beyond voice) that will soon transform these networks into an infrastructure for true multimedia services.

This is also a book about an intellectual revolution, the outlines of which are just now becoming visible. The source of illumination is the challenge of optimizing the wireless channel (although its effects will undoubtedly extend to all corners of communications and information science). Today we are beginning to realize that many of our received ideas no longer entirely fit the physical and engineering realities of the systems we are building. As has often happened in the history of science, engineering solutions are outpacing the standard models. Our ways of thinking about communications problems are in flux, and our explanations are far less coherent than we like to imagine.

The historical reference point—our ancien regime—is a body of thought we may call the Shannon theory.[1] For more than 50 years, the field of communications engineering has oriented itself around the work and insights of Claude Elwood Shannon (1916–2001). His seminal 1948 article on the mathematical theory of communication packaged a set of powerful ideas about information, coding, and the communications process that from the very beginning has seemed both unusually cogent and nicely self-contained, or complete, at least in principle. Although the Shannon theory was for a long time something of an intellectual curiosity (rather than a practical engineering program), it held sway over us as a kind of ontological proof (an "existence proof") that laid down a metaphysical foundation for what has become the core of our "information age," a technological infrastructure based on the manipulation of digital signals. It seemed to draw clean limits and boundaries, to define the game board of communications—just as a chess game is defined by 64 squares and a few simple rules of movement for the chess pieces. The Shannon theory might not "solve" the engineering equations of interest (any more than the basic rules of chess disclose the winning strategy), but it created the right sort of finiteness for taming a complex subject, and a deterministic framework to prove that a solution does exist. A decade or so after Gödel had shown that arithmetic and other axiomatic systems in mathematics were hopelessly open-ended, Shannon began the perpetual search for the grail of perfect computability all over again.

There is often a mundane problem at the fulcrum of the paradigm shift. Copernican astronomy was motivated (it is said) by the problems of calendar construction. The practical spur for Newtonian mechanics is

---

1. "Shannon theory" is admittedly a loose shorthand for a set of ideas and techniques that encompass a variety of disciplines and may be referred to variously as information theory, communications theory, or coding theory, although all of these terms have been defined with different scopes as well.

sometimes said to have been the need to forecast the tides [1]. The Shannon problem is capacity. Roughly, this means: How fast can we send digital signals in a given channel? Is there indeed a limit, or is the effective signaling speed only a function of our hardware? If our communications systems were unconstrained in terms of capacity, there would be no need to seek efficient coding schemes. If there were not a mysterious trade-off at work between signal power and capacity (or bandwidth), noise would not really be a problem—we could just crank up the power and shout louder. Indeed, the pre-Shannon engineering world view was defined by such assumptions. The idea of a signal-to-noise ratio (SNR) existed, but the implicit solution to a poor SNR was to increase the power of the signal. In fact, within very broad limits, the wireline telecommunications world is unconstrained with respect to capacity. With a few exceptions, it is always possible to add capacity to a wireline network, just by running new wires. It is also—with a few exceptions—relatively unconstrained by noise, at least in principle. If we are willing to spend the money, we can boost the bit rate by using better quality wireline facilities, more repeaters, and so forth. Most wireline protocols until recently have assumed an essentially error-free transmission environment.

The wireless channel is clearly another matter. For one thing, capacity is ultimately hard-limited by the finiteness of the radio spectrum. For another, it soon became apparent that shouting louder does not increase realizable system capacity. Wireless capacity is not merely a function of the willingness to invest in transmitter power. As wireless communications belatedly achieved mainstream status in the 1980s and 1990s, Shannon's ideas about how to construct a signal to optimize for channel capacity finally found fertile soil and began to produce a new suite of engineering solutions. Techniques such as source compression of analog sources and intensive error-correction coding began to find widespread application in second-generation wireless systems. The era of "classical" Shannon engineering was ushered in and soon extended to other sorts of "constrained" communications applications—notably the problem of how to transmit high-speed data signals over existing local loop wireline links.

Inevitably, a theory applied is a theory revealed (or perhaps, exposed). What we have come to understand in the last 20 years or so is how and where Shannon's ideas are undergirded by certain critical assumptions (which may not have been clearly understood at first). These assumptions are often (as we might expect) simplifications. Some are implicit limitations on the scope of the theory. Some are more provisional and questionable, based upon a perhaps imperfect conceptual analysis of certain aspects of the communications process. Some are apparently wrong. As well, we see more and more examples of countertheoretical results and techniques coming into use, as engineers have simply stumbled onto tricks that work without checking first for orthodoxy.

The instantiation of these statements is a main theme of this book, requiring us to go deep into specifics, and I will not attempt a detailed resumé of the matter here, but I feel certain that, by the end, the reader will at least be prepared to agree that there are a great many new and interesting developments in the wireless communications technology that do not easily fit into the classical Shannon model. This is the substance, then, of the admittedly provocative phrase "post-Shannon communications." Whether the response of the theory (and the theorists) to new data is to amend, to discard, or to supersede the Shannon theory, it will remain a landmark in the history of science. In any case, it is clear that large new conceptual extensions are required, which will bring in results from many fields lying outside the traditional bounds of the Shannon chessboard. It may be some time before we see either an amendment or a new theory that will cohere as well as Shannon's ideas have done for the past 50 years.

Here is the plan of the book: Chapter 1 is an introduction to the capacity problem in wireless, from a historical and a market perspective, to set the stage for the sudden relevance of the Shannon theory for working engineers. Chapter 2 is a synopsis of Shannon's life and core ideas. Chapter 3 offers a terminological framework for the rest of the book, and sets out the key engineering challenges that define the wireless technology agenda. Chapter 4 is a synopsis of the main technology strategies: signal hardening, signal shaping, and signal reconstruction or recovery. It reviews in a shorter space the material contained in Chapters 5 through 7.

Chapter 5 is devoted to signal hardening strategies, especially channel coding. It also deals with diversity techniques. Chapter 6 deals with signal shaping strategies, especially source compression. It also addresses baseband and RF shaping and adaptive system architectures. Chapter 7 deals with receiver designs that are intended to recover or reconstruct a damaged signal. A key topic is the understanding of the channel transfer function and how it adds information to the received signal. Technologies addressed include equalization and RAKE receiver designs.

Chapter 8 focuses on three approaches to constructing a more robust signal by forcing it to spread or expand in space, time, and/or frequency. Signal architectures such as direct sequence spread spectrum, OFDM and its variants, and space-time coding are presented. Chapter 9 is an epilogue devoted to a reconsideration of the role of the Shannon theory in this new era.

Overall, I find that even as only the first of several volumes, this book is overly ambitious and inevitably falls short of my own expectations. There are many topics that I have had to skim, a few to consciously neglect, and many of the conclusions I would have liked to have driven home decisively remain, at best, half-illuminated. Even the central theme of capacity optimization is only partially developed, since the real payoff of capacity enhancement is realized at the next higher level in the architecture—the multiple access scheme whereby many users contrive to share the finite channel resource. The discussion of multiple access technologies will be the focus of a forth-coming volume, and I may be able to close a few of the open loops at that point.

REFERENCE

[1]   Hall, A. R., *From Galileo to Newton,* New York: Dover, 1963.

1

THE GOLDEN AGE OF WIRELESS

1.1 The First Golden Age, 1890–1940

Radio was hot, once.

The first golden age of wireless lasted about 50 years—from roughly 1890 until 1940—and while it lasted, it was the paradigm of what we now call *high-tech*. It combined technical wizardry, greed, and glamour. It attracted the best engineers. It seduced the money men. One would have heard the word *Radio* whispered urgently all along Wall Street back in 1929, as the brokers' slang for RCA, a hot company and a hot stock, and an epitome of the bull market. The golden age was dominated by volcanic personalities: entrepreneurs like Marconi, De Forest, Sarnoff, and Armstrong, who fought each other bitterly for the priority of their ideas and their commercial interests. They sued each other over patents. They embroiled the courts in antitrust litigation. They urged Congress to create or dismantle monopolies. They lobbied and spent and bought and sold in the influence markets, and hired journalists to write their perfected versions of events. The public knew their names and counted their wives.

However, radio was something even more, a more distinct departure from its historical context than the modern term *high-tech* connotes. The consumer society at the turn of the twentieth century was still largely horse powered, and *technology* was mostly about steel and coal and gears and the transmission of mechanical force. The industrial revolution, reaching its climax during this same period, was based on well-known technological themes: the standardization of products, the division of labor, the centralization of the means of production, a transportation revolution (mainly railroads), and the substitution of steam and electric power for human muscle. It was the age of the dynamo, of force, of hardware, and the mechanical arts.

Radio was a totally new kind of technology. For one thing, it was not mechanical. No one could look at a radio receiver and think of it as a machine in the conventional sense. Even the telephone network was arguably a great mechanical contrivance, built around switching centers that underwent intense automation during this period, replacing human labor (switchboard operators) with complex mechanical switches. However, radio was based on different principles than those governing gears and levers and steam pistons. There was "hardware" to be sure, although it was of a strange, delicate sort—crystals and glass vacuum tubes. Nor did radio seem to have much to do with mere mechanical force. There was something else, which wasn't hardware or force at all: *the signal*. Telephones had signals, too, of course, generated in a deceptively straightforward way by the acoustic energy of the voice itself, translated directly into an electrical waveform, but radio signals had to be constructed in very precise and complex stages. A carrier frequency had to be synthesized and modulated with the desired information. The frequency of the modulated carrier might be stepped up or down, and eventually tuned to a precise frequency for actual transmission. Engineers gradually came to see that building a radio system was mostly about signal construction.

Physically speaking, those signals were *deeply* mysterious. The idea of instantaneous communication across vast oceans, with signals that somehow penetrated through solid matter, yet might reflect off of strange layers in the atmosphere, was beyond true comprehension. Radio was a window opening into a parallel universe, where even the physics were different. The engineers led the way, in the way that engineers do, by making things that worked, but De Forest no more understood the quantum principles of his vacuum tube than I do. Marconi was a shrewd businessman who understood the importance of patents and monopoly power, but his radio-telegraphy system was a brute-force affair. The so-called spark transmitter worked by creating electromagnetic explosions—like miniature lightning bolts—that splattered energy all over the radio spectrum. These explosions could be sequenced in a Morse code pattern, which *is* one way to communicate—just as burning down the henhouse is one way to roast a chicken.

Even Armstrong, the inventor of FM, and the most subtle-minded of the three great inventors, approached his problems from a strictly practical perspective. His demonstration of FM transmission in 1936 in New York was the last technology milestone in the golden age. He showed how to construct a new kind of signal that could be transmitted with great precision and efficiency, and he glimpsed the underlying principle—that he was trading what we would now call *bandwidth* for power, expanding the bandwidth in a controlled way to reduce the power required for successful transmission, while at the same time, strangely, increasing the robustness of the signal. His frequency modulation created a weaker signal that was somehow also a stronger signal. This paradox was a pregnant seed, but lay in a semidormant state.

In any case, by 1940, the golden age was over. Marconi had died in 1937 and in tribute the world had observed the very last moment of radio silence that there will ever be as long as the human species prospers on this planet [1]. De Forest, Fessenden, Alexander, Pupin, and all the rest were also gone, or retired, or in decline. Armstrong donated his patent on FM to the U.S. army in support of the war effort, which probably helped to delay its commercialization after the war (by eliminating the chance for a new monopoly). He struggled with Sarnoff and RCA to try to advance the cause of FM as a superior transmission system—which it clearly was—and grew frustrated and bitter over the resistance he met. One day in 1953, he walked out onto the balcony of his apartment in New York City and stepped off the edge.

The principal legacy of the golden age was the creation of *broadcast* radio and television. Technically, this involved the slow evolution of practical transmitters and receivers, capable of being manufactured at a low cost and sold as a consumer product. The progress during the 50 years of radio's development was tremendous; we have only to gauge the difference between the receiving stations built by the Marconi company at the turn of the century for transatlantic radio-telegraph service, which were the size of large buildings (several national monuments along the east coast of North America preserve these relics), to the compact radio receivers that had penetrated a large majority of American households by the late 1930s.

Commercially, the major milestone was the creation in 1919–1920 by the federal government—by force—of a patent pooling arrangement and a commercial entity to administer it, in order to consolidate what had become a highly fragmented and stalemated patent landscape for wireless technology. For the preceding two decades, the pace of technology development for both transmitter and receiver designs had been vigorous. A successful commercial system needed access to many branches of technology, because, of course, both transmitters and receivers had to be interoperable. Yet no single party held enough patent cards to do what AT&T had been able to do with the far simpler technology of the telephone: create a unified network. During the war, even the U.S. Navy had experienced difficulties in obtaining radio equipment because of patent problems. The prospects for commercial development were in jeopardy—as Armstrong later testified to the Federal Trade Commission: "It was absolutely impossible to manufacture any kind of workable [radio] apparatus without using practically all the inventions which were then known" [2, p. 99]. The Marconi interests could block the De Forest interests, and many other important patents were outside of the control of either. The result was that by the end of World War I, the technology of radio was ready to go to market, but no one could proceed to the business phase (except for very limited applications like Marconi's ship-to-shore service, which was already technologically obsolete). The U.S. government was determined to break through this stalemate in the national interest.

The solution was the creation of a new monopoly, a "radio trust"—the Radio Corporation of America (RCA). Between 1919 and 1923, RCA was able to pool the rights to more than 2,000 issued patents including "practically all the patents of importance in the radio science of that day" [2, p. 107]. The logic of the trust-builders proved correct. Radio broadcasting became an overnight success in the early 1920s. Radio technology penetrated the consumer market more rapidly than any other major new product before or since.[1] What had been a small and constricted business up until that point was energized by phenomenal injections of capital to build the new network infrastructure. Even greater than the economic impact was the new cultural power the networks began to wield. Steel and coal and other industrial commodities had supported the creation of huge business empires, but radio (and later TV) changed the world permanently. It is certainly possible to imagine the United States or any other country functioning quite well today without steel mills or coal mines of its own; it is impossible to conceive of the modern polity without the broadcast networks. This is the enduring legacy of the golden age.

1.2 A Quiet Interregnum, 1940–1990

The tremendous commercial success of radio did not accelerate the inventive process. Quite the contrary—after 1940, the field entered a period of

---

1. Penetration rates for various products can be mapped into a common time frame, and if this is done (as I did for radio, TV, telephony, and cellular radio in my first book, *Digital Cellular Radio*), broadcast radio was the fastest.

relative quiescence for the next 50 years. Technology froze around the new broadcast standards. Radio engineering focused on cost reduction and volume manufacturing. Probably the biggest new thing to happen, commercially and technologically, in this entire era was the advent of the "transistor radio"—a device-level innovation of great significance for achieving the production goals of manufacturers of radio and TV receivers, but not a fundamental advance in wireless technology itself.[2] The reduction in size and cost was impressive, but the performance and basic system capabilities were the same. I am old enough to remember the excitement of holding a transistor radio (Japanese-made, wonder of wonders!) in my hand in the late 1950s, and it was certainly exciting to be able to walk around with a radio that you did not have to plug in to the wall. But it was still "just a radio."

In fact, by the early 1980s, when I began my career in the wireless business, radio communications engineering had become a technological backwater. No *fundamental* new invention had broken through into the *commercial* sphere[3] within the active career span of anyone still working in the business. Indeed, I remember that in 1980, if you wanted to hire an RF engineer, you quite often found yourself talking to someone in his sixties or even his seventies. These were the survivors of the last wave of students who went into the "hot" field of radio at the end of the first golden age. By 1980, radio was so non-hot that few new engineering students chose to go into it. Why waste time in the slow lane? The technology hot zone of the late 1970s and early 1980s was the world of *digital* electronics. Integrated circuits were new and powerful, software was becoming the watchword for value creation, and the personal computer was expanding the commercial scope of the industry a hundred-fold. Anyone looking to do exciting engineering would have naturally gravitated to the digital world. Radio was not an option that most new students considered, which was why recruiting an engineering team was so hard. Our early business plans sometimes depended on keeping someone out of retirement.

Another factor in the relative stagnation was that the radio in *telecommunications* (as opposed to broadcasting) had indeed never really taken off. Of course, it was still (as in Marconi's day) the only good way to communicate with ships and airplanes, but the mainstream telecommunications industry in 1980 was almost completely wireless free. Point-to-point

---

2. I am slighting television, here, of course, although even TV was arguably invented and substantially perfected prior to 1940.

3. An important qualification, since there were interesting and even surprising things happening in the military field.

microwave systems were used in the long-distance network, and indeed were becoming the dominant long-haul technology, such that a lot of people were vaguely aware of microwave towers and their strange horn-shaped structures, but except for small groups inside Bell Labs and the AT&T Long Lines Division (the long-distance monopolist, it should be remembered), very few telecommunications engineers or managers had much exposure to any sort of wireless system or technology. It was true that someone at Bell Labs had written a prescient paper in 1948 describing a new idea for a wireless phone system, based on the idea of creating what the author called *cells* for reusing the limited frequencies many times in a confined geographical area. But even by 1980 there was still no such thing as "cellular radio." Moreover, the cellular idea was basically an innovation in the *switching* architecture, not in radio technology. The radios that were to be used for a cellular system were based on Edwin Armstrong's FM technology from the 1930s and were not fundamentally different (in terms of RF technology) from the mobile radios that had been used by police and some industrial users, and by a very, very limited number of mobile phone subscribers, for many years.

So, in the mid-1980s, when the FCC and various industry players launched a "revolutionary new service" called cellular radio, it was still—from an RF engineer's point of view—rather old news. The innovations in the cellular switching systems did not seem to depend much on the nature of the wireless transmission technology. In fact, faced with the option to consider newer types of transmission technologies (which we shall turn to in a moment), industry leaders Motorola and AT&T chose to go with FM, even though it was 50 years old and fully mature, and likely offered few opportunities for further fundamental improvements in economics or performance. As it turned out, this was a blunder, which became quite clear within a year or two, and which probably cost the industry fully 10 years in its commercial development cycle. (Unlike either the telephony business or the broadcasting business, most cellular carriers lost money for a considerable period of time.) The blunder was understandable, perhaps, given that "wireless" had played almost no role in the plans of any telecommunications company at that time. It was a peripheral field, expertise was scarce, and the potential was not fully appreciated.

It was a common view in the early and mid-1980s that cellular radio would be simply another value-added service, another noncore line of business, along with selling data modems and the odd satellite transponder. The idea that in anyone's lifetime there might be a *wireless* network that would surpass the wired telephone network in its subscriber base was completely unforeseen. I can remember attending a Supercomm trade show in Atlanta in 1985 or 1986 (Supercomm was the main industry trade show for the telecommunications industry) at which the main theme of the show was "Fiberworld"—huge companies were spending a lot of money to promote the idea that optical fiber would soon revolutionize the industry as it penetrated even into the home to bring new services, such as ISDN. There were dozens of technical and commercial presentations on the fiber revolution. Meanwhile, there was exactly one small technical session on wireless technology (which is why I was there), held in a small side room. We held our little conclave and got out from under foot. It was typical of the times. I remember making another presentation, probably in 1987, to a group of telephone executives on the potential of wireless technology to do more than just add a layer of high-end car-phone service for rich people—only to be interrupted vigorously by a fellow who started shouting at me, "Fiber, fiber—they're going to make fiber out of plastic. That's all we need. Plastic fiber." How could car phones compete with this vision? The executive management of the telephone industry was not wireless-ready.

1.3 The Digital Radio Revolution

What happened next was a true technology revolution. So many revolutions to endure, it sometimes seems—but this was a real one, even if it was essentially invisible to the consumer.

Cellular service was launched in the United States in 1983. *Within less than a year*, it became clear that there was a serious bottleneck in the system. It was not in the switch, but in the radio link—the *air interface*, as it began to be called—between the cell phone and the base station. FM was grossly inefficient. It used too much spectrum bandwidth to carry a call. (We will go through some calculations in Chapter 3 on this point.) Demand in some major markets soon began to put pressure on the system operators, and they quickly realized that they had built themselves into a corner. The Federal Communications Commission (FCC) tried to palliate the situation by allocating a bit of additional spectrum, but the real problem was much more fundamental. FM could not handle the demands that were being placed on it by the much broader service; this was apparent even in the very early days, before anyone had even thought about revising upward the estimated potential for ultimate growth in the cellular market. Even viewed simply as a car phone for rich people, cellular could not support the demand with FM as its air interface. Something had to be done.

The next part of the story is fairly familiar to those in the field. Beginning in 1987 (only 3 years after cellular was launched), the industry and the FCC began to evaluate technological alternatives for replacing FM as the air interface. In the end, two new technologies, *time-division multiple access* (TDMA) and *code-division multiple access* (CDMA), were approved. Both shared a common foundation: They converted the voice signal into *digital* form before encoding it for transmission over the radio channel. Once the voice signal has been converted to the binary language (1's and 0's) of computers and microprocessors, it can be manipulated with digital signal processors and powerful software algorithms. For example, a digitized voice signal can be compressed significantly—the inherent redundancy removed—without losing much quality at the receiving end.

The opportunity to implement voice compression turned out to be the main motivation for moving to a digital air interface. It was possible, with the compression techniques available in the late 1980s, to pack about three or four compressed digital voice signals into the same channel that can carry only one uncompressed analog FM voice signal. There were other benefits to be derived from digital transmission, but these second-generation air interface standards were focused on gaining spectral efficiency from the voice compression. Digitalization itself was seen as a necessary step (and indeed by some it was seen as a necessary evil) to facilitate voice compression. Other potential benefits were largely ignored. After all, these were voice-only systems; getting more voice calls into the available spectrum was seen as the definition of "spectrum efficiency" for the cellular industry. In fact, there was a considerable rear-guard fight by the proponents of analog FM who argued, correctly, that the same spectrum-efficiency gain of about a factor of 3 could be achieved by retooling the analog FM signal. Motorola pushed hard for its N-AMPS proposal, based on a narrowband FM signal that occupied only one-third the bandwidth of the standard (AMPS) FM signal. It was claimed that this approach might realize most of the voice efficiency gains without adding a lot of "digital overhead" that was certain to make the new digital phones much more costly (at least initially) than phones based on the mature FM technology. At that stage, the cost difference between analog and digital was being estimated to be at least two to one, if not more.

As late as 1990, it was not entirely clear that digital transmission would prevail. As long as its chief merit was as an enabler of voice compression, questions remained about why it was necessary to create an entirely new type of radio—a digital radio—just to support this one attractive function. It appeared to some, from the outside, that it was all a very roundabout and expensive way of gaining a relatively modest incremental improvement, which might instead be gained by tweaking the existing FM technology a little.

Digitalization of the voice signal forced a complete rethinking of the radio transceiver architecture. For one thing, the digitized voice signal had to be translated into an appropriate form to modulate a radio carrier. This meant that *digital modulation* was needed, to imprint the digitized voice signal on the radio carrier wave. Digital modulation, in turn, sometimes required a much more linear power amplifier. The digital signal required synchronization between the transmitter and the receiver. The more popular digital formats were based on TDMA, which meant that the transmission was no longer continuous.[4] Instead, the transmitter sent short bursts in a regular cycle. An FM system could perform its handshakes between transmitter and receiver once at the beginning of the call, but a TDMA system had to reacquire synchronization for each burst, many times a second. Digital signals proved to be rather brittle in some ways. Although FM may be robust enough to function well on a less-than-ideal channel, a digital signal is prone to sudden breakdown when its critical signal-to-noise threshold is crossed. Error correction techniques were applied to enable the receiver to detect and correct errors caused by poor channel conditions. Other compensation techniques were also needed, such as equalization (which is designed to combat the inevitable smearing together of the digital pulses caused by inherent characteristics of the wireless channel), and so on.

Digital signal processing (as it came to be called) changed radio technology fundamentally. FM radios had used digital controllers to talk to the switch, to handle call setup, handoff decisions, and the like, so the idea of having a microprocessor inside a cell phone was not per se a shock. But the computing power was now being brought to bear on the signal itself, to rearrange and create new kinds of signals with new physical properties. Even though the payoff was often seen as simply being the ability to enable voice compression, in fact, the entire radio had to be reengineered from top to bottom, from the filters and amplifiers, to the antennas and the acoustic microphones, and even the glue used to attach the microphones to the handset chassis.[5]

---

4. The U.S. TDMA standard was known as IS-54, and later IS-136. The Japanese TDMA standard was known as JDC. The European and global TDMA standard is known as GSM. Between them, the TDMA standards accounted for 80% to 90% of the digital market from the late 1980s until the end of the 1990s. Essentially, the second generation of cellular radio technology has been a TDMA generation.

The new digital radio architectures proved to be much more difficult to perfect than many originally expected. The complete development cycle for a new digital air interface has proved to be something between 4 and 7 years.[6] For reasons that will become clear in later chapters, wireless transmission brought engineers up against much more challenging physical problems than they had been accustomed to dealing with in transmission systems designed to work over wireline circuits. The technological backwater began to be flushed out, and the dynamic of innovation was revived. By 1990, it was clear that the interregnum was over. Radio engineering was suddenly an exciting field again. The entrepreneurs were back. Wireless was hot again.

1.4 The Capacity Crisis, 1995–2001

The capacity problems of the industry were only just beginning. In fact, the growth in traffic soon outstripped all the original "car phone" business models. The industry reached 25 million users in the United States by 1995 (about 10% penetration), breaking into the mainstream. But it was what happened next that created the true crisis:

- Between 1995 and 2001, the U.S. subscriber base *quadrupled* to reach 118 million subscribers by June 2001—a 40% penetration [3, pp. 23–24][7] (Figure 1.1).

- Minutes of use *per customer*, which had remained flat for years at around 120 minutes per month, suddenly in 1998 inflected upward,

---

5. The digital vocoders were so aggressive in eliminating redundancy from the voice signal that they became, in effect, less robust in some ways. Acoustic effects had to be managed much more carefully. The problem of controlling the acoustic properties of the adhesive glue used to secure the microphone was a real one that cropped up more than once in the implementation of second-generation architectures.

6. I would point to the development of TDMA as having begun in earnest in 1981–1982, by a team from IMM Corp (later InterDigital Communications) and M/A-COM Linkabit (whose key people later founded Qualcomm), with first commercial use in 1986 (in a wireless local loop application). CDMA for cellular was proposed in 1988–1989, and I would say that it did not reach the stage of unrestricted commercial deployment until about 1995. Motorola's iDEN system for digital SMR (a TDMA system) was proposed in 1989 or 1990, and again not fully debugged and commercialized until about 1995.

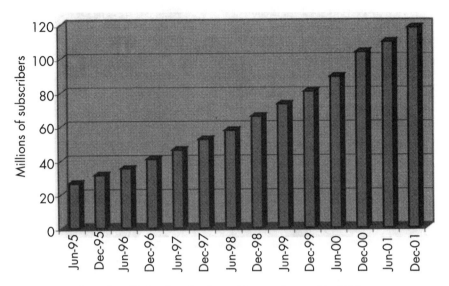

FIGURE 1.1 *Growth in U.S. subscriber base. (Source: [3, pp. 23–24].)*

nearly *tripling* by 2001 to more than 300 minutes per month[8] (Figure 1.2).

- Total minutes of use went up by about eight times between 1995 and 2001, from 38 billion minutes to 259 billion minutes [3, p. 167] (Figure 1.3).

By the end of the 1990s, there were a lot more users, with each user generating a lot more traffic—the industry was being forced to expand its infrastructure furiously. One dimension of the expansion was quantitative—more cell sites. Cellular operators in the United States had not reached 10,000 cell sites until 1992, 9 years after service began. The number of cell sites then doubled in the following 3 years. Since 1995,

---

7. The *New York Times* reported on February 14, 2002, that current subscribers in the United States have reached just short of 130 million people.

8. From *CTIA Report* [3, pp. 169–170]. The breakout in individual usage patterns was most likely a direct result of AT&T Wireless's flat-rate, 10 cents/minute billing plan, which changed the industry's pricing policy across the board and stimulated demand enormously.

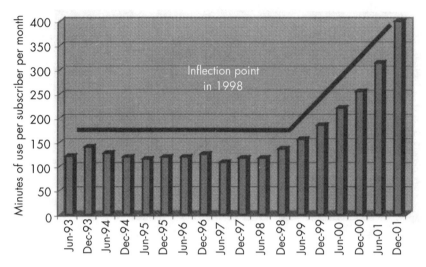

FIGURE 1.2 *Growth in minutes of use per customer in the United States. (Source: [3, pp. 169–170].)*

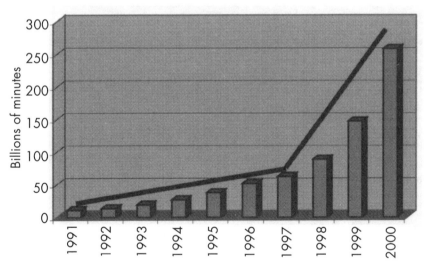

FIGURE 1.3 *Growth in total minutes of use in the United States. (Source: [3, p. 167].)*

however, the number of cell sites in the United States has *increased five and a half times,* to 115,000 cell sites by the end of 2001 [3, p. 139] (Figure 1.4).

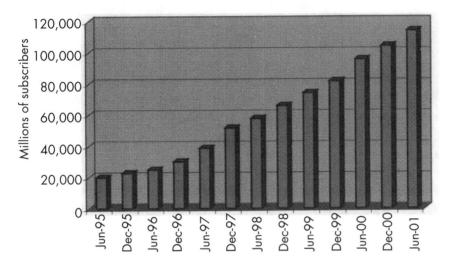

FIGURE 1.4 *Growth in number of cell sites in the United States. (Source: [3].)*

The other aspect of the expansion was qualitative. The capacity crisis finally drove the whole-hearted conversion of the cellular networks to the new digital air interfaces. The second-generation technology had been embraced tentatively at first, because of the considerably higher cost of digital equipment. Because the operators subsidized the cell phones to the end user they were initially hesitant to "push digital" too aggressively. Even as recently as 1995, analog phones still outnumbered digital phones by more than 20 to 1. The total number of digital subscribers in 1995 was only about 600,000 (out of 25 million total cellular users) [3, p. 38]. By 1997, the number of *new* digital subscribers being added surpassed new analog customers and, finally, at the beginning of 2000, digital subscribers exceeded analog subscribers as a percentage of the total. As of mid-2001, there were more than 85 million digital cellular phones—120 times what there had been just 6 years earlier (Figure 1.5).

This is only the first phase of the capacity crisis. The traffic on the network is still almost 100% voice telephony. It is narrowband traffic. Using the underlying digital vocoder rate as a benchmark, each second of use in today's second-generation systems generates about 10 Kb of user data to be transmitted. Based on the figures cited previously, the average user is today generating something like 180 Mb (or for those of us more used to computer metrics, around 22.5 MB) of data *per month*. In a voice-centric network,

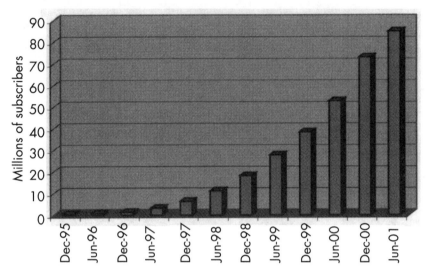

FIGURE 1.5 *Growth in digital versus analog phones in the United States. (Source: [3].)*

traffic growth can only come from greater penetration (more subscribers) and higher usage per subscriber. The penetration may double again (to 80%), but after that the market will reach saturation. The usage may grow somewhat, but there are natural limits on how much real-time talk-time a user can generate. Usage is already about 10 minutes per day, and wireline telephony statistics would indicate that this parameter, too, should flatten out before long. Based on the "saturation" trajectory, we might be prepared to see solid growth continue for a few years, then reach a plateau. It might well turn out that the big surge from about 1997 to 2002 will be a one-time event—comparable to other S-curves we have seen in the past, for products like television or personal computers. My family had one cell phone 4 years ago. Now we have four. It does not follow that by 2006 we will have 16. The second-generation digital systems should be able to support this slowing growth, perhaps until the plateau is reached. So why do we need a third-generation technology?

The answer can be given in a single word. It was not until about 1996—that is, well after the standards for all the current second-generation digital cellular air interfaces had been defined—that I first heard that word, that single word—*Internet*—in a discussion of wireless architectures.

It is worth recalling that by the mid-1990s a certain skepticism had developed in the wireless world regarding the demand for nonvoice services.

There had been a variety of wireless data offerings, none of which had achieved market success despite considerable support from regulators and leading commercial equipment suppliers.[9] Indeed, even the wireline networks had grown tired of promises involving data bearer services (i.e., those provided by the operator specifically to enable data communications). ISDN had been discussed for two decades and still had not arrived commercially. There were indeed many *private* data networks in the business sector, operating over short distances to connect computers and computer peripherals (mainly based on Ethernet-type protocols), but these *local-area networks* (LANs) were mostly isolated from the public telephone networks (except by dial-up modem access).

Data communications on the public network were being effected mostly by dial-up modem connections and by fax, two services that operated over the voiceband telephone network, mimicking voice telephony traffic. A lot of the pundits were concluding that data traffic would always adapt to ride on top of voice bearer services, and that would be that—a unified network of voice, fax, and voiceband modems, circuit-switched dial-up connections passing through standard telephone switching networks. True, the idea of broadband has been in the air for a very long time, without attaching itself to anything very specific. *Fiber-to-the-home* (FTTH) was one version of this idea that was quite clearly driven by technological possibilities (multimegabit transmission on fiber optics channels) rather than by a compelling service proposition. In a previous book, written in 1991, I examined the technical and economic arguments for FTTH and its variants, and I concluded that other than video (for which there were already several entrenched competitive delivery systems), there was no clear service proposition that would require and drive broadband in the access network [4]. There was no compelling reason, certainly, for *wireless* systems designers back then to anticipate a need for *wireless* broadband, especially given the inherent scarcity of the spectrum resource. In fact, the whole thrust of the second-generation architectures was *compression*. The overarching technical goal was to *narrow* the bandwidth of the signal, so that more signals could be crammed into the operator's finite spectrum allocation.

The Internet has changed all that. Notwithstanding its ups and downs in financial markets and public esteem, the Internet phenomenon of the last 5 years has shown us two things quite clearly: first, that circuit-switched bearer services will be supplemented, and even superseded for data

9. Both Motorola and Ericsson promoted wireless data fairly vigorously in the late 1980s and early 1990s.

communications, by packet-oriented, connectionless protocols that will drive new transmission technologies right down to the physical layer; and, second, that the consumer does indeed have an appetite for bandwidth that may approximate the underlying sense of the term *broadband.*

Actually, I think the demand for bandwidth presents two distinct manifestations. On the one hand, the Internet provides an *e-mail infrastructure*—to use the common term, although labeling it as such risks understating its significance. Indeed, viewed strictly as "e-mail" in the conventional sense of short text messages, there would be no need for broadband channels. This was, in fact, the way that wireless architects misled themselves in the mid-1990s when data services started to intrude. Text messages, short message services, and the like seemed to present no problem for second-generation air interfaces. At 10 Kbps, a TDMA or CDMA cell phone could handle a 1,000-character text message very easily. In fact, the problems of input/output (text entry and display) seemed to be far more problematic.

However, it is becoming clearer all the time that e-mail is morphing into a much bigger thing. In the business world, it has—quite suddenly—become routine to send documents of all sorts as e-mail attachments.[10] A 1-Kb text message is attached to a 1-Mb word-processing file, with graphics. And this e-mail may be sent to 10 people, generating at some point in the access network 10 times the original traffic. Other documents include e-mail-based fax transmissions (sent as packetized e-mail rather as separate circuit-switched transmissions), graphics materials, photographs, and even video clips. It is becoming common to submit reports, proposals, and presentations in "soft copy." I have not yet seen a quantitative treatment of this phenomenon, but it is clear that e-mail is a channel that is rapidly moving in the direction of broadband capacity needs. It is common to refer to e-mail as the "killer application" for the Internet, sometimes in a dismissive way ("just e-mail..."). If it is understood that it is becoming a way to transmit and receive not just short salutations, but increasingly large information payloads, we may be able to keep in mind the immense significance of this very new development.

The other manifestation of the Internet is as an enabled marketplace for information. *Enabled* means that it incorporates an open interface that does not require much more technical skill to access than dialing a phone or channel-surfing on a television. *Marketplace* implies several things. First, it is interactive. Consumers seek information from various sources (commercial

---

10. My own informal count is that about 25% of the e-mails I receive now include an attachment, ranging from a few kilobytes to several megabytes of data.

entities, libraries, government organizations), and those sources also try to push their own information back to the consumer. This entails, typically, the web-browsing experience as we have come to know it: A consumer request for information (in the form of a mouse click on a hyperlink button) produces a torrential download of information, graphics, advertising, sometimes audio and video, to create a virtual environment for the consumer to enter. Each new click leads to a reconfiguration of the environment with a new download. The relevant point for our discussion is that Web downloads produce large, often asymmetrical information flows. They are becoming richer all the time: more graphics, more information, more bits.

Another implication of the term *marketplace* is the economic dimension. Value transactions take place and proprietary information must be exchanged, encoded in ways that bulk up the information flows even further. A single transaction may generate a trail of subsequent information exchanges (confirmation of sale, notification of availability, or confirmation of shipment).

Consider the following calculations. First, as noted above, the average cellular voice user today appears to generate something like 180 Mb (22.5 MB) of digital data per month, based on 300 minutes of use per month. As we noted, this is only 10 minutes per day. This is a fair amount for voice telephony (at least viewed as an average), but it is clear that the newer forms of broadband usage that the Internet has created will drive higher holding times.[11]

Second, the bit rate will go up; 10 Kbps will be upgraded to much higher rates. The rich downloads, the e-mail attachments, video clips, video phones, and so on, will demand them.

Third—and most important of all—the quality of service criterion has to change. Today's cellular voice services can operate well with a bit error rate of 10^{-3}, even 10^{-2}. That is to say, they can tolerate a lot of channel errors, as they must tolerate in the noisy wireless channel. Data services cannot endure this level of errors. Internet protocols are designed around the assumption that the channel is essentially error free. Packet failures are implicitly diagnosed as indicators of network congestion, rather than a symptom of channel errors, and the protocols respond by (among other things) slowing down the rate. It is unlikely that IP protocols in their existing form could even sync up at the *bit error rate* (BER) levels for which today's cellular systems are engineered.

---

11. Albeit the idea of "holding time" is not strictly applicable to a packetized connectionless or quasi-connectionless service like the Internet.

So, let us assume that instead of 10 minutes a day, the data traffic generates the equivalent of 20 minutes a day. (Actual connect times for wireline Internet users are much higher.) Let us further assume that instead of 10 Kbps, the system must be designed to support 64 Kbps. Finally, let us assume that we need a BER of 10^{-8} instead of 10^{-3}.

The average subscriber, under these assumptions, will generate 2,300 Mb (288 MB) of data per month—more than 12 times as much traffic as his voice-only counterpart. If we tweak these assumptions to see what would happen if the service really became successful, we might raise the connect time per day to 30 minutes (not high for a business user), and look down the road to a 144-Kbps data rate. This scenario would generate 7,700 Mb (just under 1,000 MB) of digital data, or *42 times more traffic* from that subscriber node than the current level of voice traffic (Figure 1.6).

We might exacerbate the problem by factoring in the assumption of multiple recipients for heavy e-mail packages, or even user groups with mesh communications channels (for applications like Microsoft's NetMeeting software). We could run the model through the roof by looking at the higher bit rates that system architects want for "true" broadband—from 384 Kbps up to several megabits per second. Finally, if the holding times on wireless nodes ever begin to reflect the holding times for wireline Internet access, we will be measuring the offered traffic not in minutes but in *hours* per day.

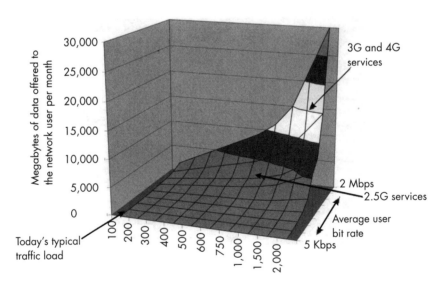

FIGURE 1.6 *Hypothetical growth in user traffic as a function of bit rate and usage levels.*

1.5 The New Golden Age

These services will not run successfully on wireless networks using second-generation RF technology. So, how do we get there from here?

One approach is to try to boost the transmission speeds on the existing systems by upgrading the modulation scheme (more bits per hertz) and by other means that make similar use of much of the existing infrastructure. The system operators have just begun to really use their second-generation networks, and they have a huge economic investment to work down. If there is a way to upgrade these networks, and preserve some of that investment, that would certainly be attractive. Because these upgrades are grafted onto existing second-generation (also called 2G) networks, they are often referred to as 2.5G solutions.

The other approach is to build entirely new networks, based on new RF technology designed from the start to support wireless broadband services. During the past several years, this approach has coalesced around the phrase *third-generation wireless,* or 3G. The market driver for 3G is absolutely clear: to support true Internet-type services. We probably do not need 3G for voice capacity needs. Voice growth will double and possibly double again, but it will be flattening out and various 2G and 2.5G solutions would probably meet that demand—if that were all. But the enormous new flood of wireless traffic that the nonvoice services are going to create cannot be met by 2G, and probably not by 2.5G in the long run. (We have learned that the "long run" may be rather short, in Internet time.)

There is still one last problem—the channel error problem, where information is irretrievably lost due to adverse physical conditions in the channel. Not only does a wireless broadband network need to be ready to support dozens of times more digital traffic per user than today, *it must deliver that traffic in ten thousand to one million times better condition.* It is difficult to factor this into the capacity calculation in a simple arithmetic manner. But we may get some idea of the impact if we assume that for a given modulation scheme, an improvement of five orders of magnitude in bit error rate performance might generally equate to something over 6 dB in the *signal-to-noise ratio* (SNR).[12] Six decibels equates to a factor of 4. This factor of 4 might show up in various calculations—as a reduction in range, or throughput (for a variable-bit-rate system), both of which trade off against

---

12. The actual SNR to BER relationship is unique for each modulation and coding scheme, and the curves are more steeply sloped at different points in the SNR range.

capacity in complex ways in a CDMA system (and 3G standards will be based on CDMA, it appears). It would be too simplistic to say that it simply imposes a 75% penalty on any 3G capacity calculation. But what *is* clear is that a 3G system is going to need a radically different type of radio technology in its air interface. It has to run faster, longer, and much cleaner than today's 2G systems.

The 2G challenge brought wireless back to the cutting edge in RF technology, and ended the 50-year interregnum of relative quiescence and stagnation in fundamental wireless development. The 3G challenge will energize the field in even more dramatic ways. Indeed, I would say that the 2G technology phase has for the most part involved importing techniques and solutions that had been previously developed in other fields (e.g., wireline telephony) into the wireless field, applying them, and then extending and refining them. A good example is the speech compression software, which was first explored in nonwireless contexts and then became a key part of the 2G technology package; 3G is going to push the field of radio technology much further, and reverse the direction of the flow of innovation. I predict that the wireless field is going to become the leading edge for telecommunications technology in general. Innovations in RF will find their way back into other streams of the high-tech economy.

There is a precedent for this. In the first golden age of radio, innovations in wireless technology led the way to the electronics revolution. The first true electronic device—the vacuum tube—was developed for wireless applications. Later it became the foundation for building the first computers. In the new golden age, which began in the 1990s and is still on its youthful upswing, I believe that once again we will see RF technology playing a seminal role in fertilizing many other fields with fundamental breakthroughs. The 2G designers largely found what they needed on the shelf, and adapted it for their needs. On the other hand, 3G technology is going to be driven to make those breakthroughs. The driver is the enormous appetite for capacity.

What *is* capacity? We have talked about *traffic*, so many bits per second or per month that a user can generate for possible transmission. Traffic can be measured empirically. But implicit in our discussion is the idea that a radio channel can somehow carry only so much traffic, only so many bits per second. It might seem that *capacity* could also be measured directly, by filling the channel until it cannot carry any more. This raises the question of what exactly are we filling it *with*? That is, what *is* information, and how can we quantify it? Maybe the rate of transmission of digital symbols (bits) is a proxy for the amount of information, although we sense that it may not be that simple, but even so—how do we know when the channel is full, when we reach its capacity? Perhaps it is full when we start to experience errors; but, no, errors occur even at very low transmission rates. Indeed, noise is ubiquitous, which suggests that some errors are inevitable at any transmission rate, or can errors be controlled?

We are getting slightly ahead of ourselves. But it is clear that capacity is a more abstract idea than raw bit rate. Signaling rates can be measured directly, but it is not as intuitively clear just how can we measure, or calculate, the *capacity* of a given channel, to determine whether that channel can or cannot support a given rate. In fact, defining the capacity of a signaling channel is *the* cornerstone problem of modern communications theory. Understanding capacity will lead to other important theoretical breakthroughs that underlie key modern concepts of *information, entropy*, and *bandwidth*. Historically speaking, just as "gravity" is intimately associated with Newton, and "uncertainty" with Heisenberg, the concept of channel capacity is closely connected with the name of Claude Shannon—whose acquaintance we must now make.

REFERENCES

[1] Jolly, W. P., *Marconi*, New York: Stein & Day, 1972, p. 271.

[2] MacLaurin, W. R., and R. J. Harman, *Invention and Innovation in the Radio Industry*, New York: Macmillan, 1949.

[3] Cellular Telephone Industry Association, *CTIA's Wireless Industry Indices: Semi-Annual Data Survey Results, 1985–2001*, December 2001 (referred to in text hereafter as *CTIA Report*).

[4] Calhoun, G., *Wireless Access and the Local Telephone Network*, Norwood, MA: Artech House, 1992.

2

SHANNON

2.1 "Our Shannon"

An otherwise rather sober 1991 technical paper on wireless modulation techniques, sponsored by the British National Space Centre, begins in a way that is not at all uncommon for even the driest of technical articles these days. The authors proclaim, in almost metaphysical tones, a *Quest*—defined as follows: *"The ultimate goal of all transmission systems is to approach the Shannon channel capacity limit"* [1]. The ultimate goal?

In Aldous Huxley's 1932 novel *Brave New World*, Huxley imagined a futuristic society based on mass production principles, in which it was customary for people to speak of "Our Ford" as a sort secularized divinity of the new order (in some contexts, strangely, the phrase tended to mutate into "Our Freud"). In our own world we appropriate other names—"Einstein" or "Heisenberg" or "Gödel"—as a shorthand to refer to the sometimes forbiddingly complex elements of our new understanding of the physical and logical universe. The names of our heroes provide us with a language to talk about things that we often do not fully understand.

"Our Shannon" is still less well known to the general public, although it is frequently predicted that one day his name will be as famous as Newton or Einstein. But for anyone involved in the field of communications or information technology, our Shannon is already a demigod. We have our Shannon limit, Shannon theory, and Shannon capacity. We have our *Collected Works of Shannon*. It is common, even after 50 years, for authors presenting the very latest technical developments in communications or information processing to begin with a genuflection toward "The Article" [2], which refers, of course, to the original 1948 article by Shannon entitled "A Mathematical Theory of Communications." The Article, and several others that followed shortly afterward, were the Big Bang of our field. They have spawned an enormous literature and a pervasive engineering methodology that is a part of everything digital. Shannon, in case you have not heard, discovered 1's and 0's: "Shannon was the person who saw that the binary digit was the fundamental element in all of communications. That was really his discovery, and from it the whole communications revolution has sprung" [3].

As our high-tech economy has caught up with Shannon's article, the relevance and significance of his ideas have grown in our estimation to the point where he seems an almost Promethean figure. It was Shannon, it is said, who came up with the "basic idea on which all modern computers are built" [3]. He is cited as a founding father for fields as diverse as telephone transmission, cryptography, and investment theory. It is sometimes implied that Shannon was the true inspirer of the audio CD: "Drag a knife point across the surface of a compact disc, and error correcting codes will mask the flaw, thanks to Shannon" [4]. (Since the CD was the first consumer product based on some of the ideas he sketched out in The Article, it is often assumed to be an easy "handle" by which the layperson can grasp Shannon's practical significance to our daily lives.)

This is the second wave of enthusiasm for "our Shannon." The first wave came and went in the 1950s, as all sorts of thinkers took up the ideas in The Article. There was no digital world back then. The appeal of Shannon's ideas was more philosophical. He had supposedly put his finger on a fundamental connection between information theory and the physics of thermodynamics. He disclosed, it was said, a startling equivalence between the properties of a coded signal (which he called *information*) and the second law of thermodynamics, which had so troubled Victorian metaphysicians with its alternative apocalypse (the "heat death" of a universe inevitably running down).

The significance of this conceptual linkage may have been puzzling for many, but it tapped into the intellectual strata where Big Thinkers like to hang out. For a decade or two after The Article, our Shannon was much more the property of philosophers (and even readers of *Life* magazine) than of the engineering community (who could not yet really figure out how to apply his ideas). The readiness to metaphorize his theory was unchecked. Finally, Shannon himself disavowed the philosophical enthusiasm for a theory that had been "oversold"—"Information theory has perhaps ballooned to an importance beyond its actual accomplishments" [5]—and with that, the vogue started to fade.

The new vogue among engineers is more substantive. Shannon's ideas are achieving practical relevance. The habit of reflexive citation is not just empty historicism[1]—it is an acknowledgment that the source is still vital. There is a sense that the original Shannon material is still unfolding its secrets even today. Insights that were tossed off, in some cases as mere asides, perhaps hardly noticed by the original readers, now assume enormous significance decades later. Unlike 99% of all the articles published in the engineering literature today, Shannon's works are actually readable—they *are* literature, in a sense. And they are read, I suspect, by many engineering students at one time or another, and probably reread by many researchers more than once, at different stations in our careers.

So many of Shannon's ideas are truly without precedent. His choice of problems, his vocabulary, his several ways of looking at the subject are all so novel that they seem to spring from an act of consummate genius rather than from a shared emerging intellectual tradition. Prior to Shannon, only a small handful of citations have any remaining relevance. After Shannon, everything that we have to say regarding the theoretical foundations of communications systems must accept his work as the point of departure. In short, his ideas are profound.

This combination of fecundity, clarity, and originality is a recipe for a sacred text, and quite often this is how Shannon's work has been handled by his successors. The language of intellectual absolutism seeps into the argument. It is claimed that Shannon has "proved" such and such, and set the "ultimate limit" on what we can attain. This "Shannon limit" is cited as though it were the speed of light, faster than which nothing may travel (except for those things that happen to do so[2]). In practical terms, Shannon's

---

1. *Webster's Third International Dictionary* defines historicism as "a strong or exaggerated concern with or respect for the institutions and traditions of the past."

2. It is very interesting to note that the definition of the speed of light as a physical limit is connected to the idea of information. There are, in fact, physical phenomena that propagate at rates that apparently far exceed the speed of light, but it is cur-

model is taken sometimes as a sort of nuclear solution for any and all problems of transmission engineering—as though if we could only unwrap it fully, it would illuminate every corner of the field and disclose the ultimate answers. As a "mathematical" theory, it implies that its posed problems may be exactly solved (some day, at least—since Shannon actually only proves the *existence* of certain solutions and does not say what they may be). It offers the hope that like many other mathematical problems we may one day possess absolutely certain knowledge in this field. The theory promises, in this guise, closure.

2.2 Claude Elwood Shannon (1916–2001)

The real Claude Shannon was born in 1916 in northern Michigan, and it is worth pausing to reflect just how distant from us, culturally and technologically, that world was. His father had been born during the American Civil War. World War I was now raging. There was a Russian czar, a German kaiser, and an Ottoman caliph. There was no radio, no TV, no movies with sound, no computer, no commercial airline industry, no home refrigeration or air conditioning, no penicillin. His family might well have owned a car (there is no record) and possibly a sewing machine or a Hoover. Telephones were rarities, although as a probate judge it would seem likely that his father might have been a subscriber.

When he died in 2001, at a nursing home in Massachusetts, Claude Shannon's life had spanned the many revolutions that have given rise to the postindustrial epoch. He was not a celebrity or a truly public man, and he had been retired for many years but he could not escape the halo of awe that his followers would place around him. Knowing his disdain for the inflated importance, as he saw it, of his own ideas, one can sense that this made him uncomfortable at times. He gave few interviews. By the time of his death, he had been burdened privately for years with Alzheimer's disease, and publicly with honors and accolades that verged on idolatry. They gave him degrees from Yale, Princeton, Michigan, Edinburgh, Northwestern, Oxford,

---

rently believed that they are incapable of transmitting *information*. An example is the strange interaction between separated pairs of particles with mutually dependent but indeterminate quantum states, whereby detection of the target property of one of the pair causes both particles to enter a determinate state, no matter how far apart they may be, at speeds that are clearly in excess of the light-speed barrier.

Carnegie-Mellon, and Penn; prizes from Israel and Japan; honors from all of the engineering societies; and the National Medal of Science, presented by Lyndon Johnson. They gave him honors, one suspects, to force this living legend to emerge, briefly, into the public eye.

His crucial work was done in the 1940s, and—before he turned his back on it in the 1950s—he produced about 10 articles that comprise the core of communications theory (and its subdiscipline, information theory). But he was really a crypto-polymath, who ranged into three different mega-domains—electronics, genetics, and communications—and brought his striking originality of thought to bear on each. He had arrived at MIT in 1936, and gravitated to work on what was then perhaps the most complex electronic device of its time, a differential analyzer, which was, in effect, an analog computer, in some sense analogous to a huge mechanical slide rule. A part of this system was a complex circuit built to control the operation of the differential analyzer itself. This controller circuit involved more than 100 switched relays, which worked by varying the "on" or "off" states of the circuitry to create different outcomes. Shannon, with an undergraduate degree in mathematics, came at this component of the system with a unique perspective. He saw that this complex of two-valued elements could be modeled by using a nineteenth-century logical technique called Boolean algebra.

George Boole was an English mathematician who, in 1854, published his masterwork—a treatise on what he called the "Laws of Thought" [6]. Essentially, this was a marriage of two classical fields, algebra and logic, and resulted in a powerful tool for thinking about the sort of systems with which Shannon was now concerned. Boolean algebra allowed Shannon to analyze and design complex circuits in a more deterministic process: "to change digital circuit design from an art to a science" [7]. He published his first papers to much acclaim, and developed the full concept for digital circuit analysis in his master's thesis ("one of the most important master's theses ever written" [8]).[3]

His next foray—into genetics—involved a similar attempt to apply algebra to the inheritability patterns of various genetic traits. In his Ph.D. thesis, published 12 years before Watson and Crick discovered DNA, Shannon tried to apply the same analytical strategies to understand Mendelian processes. Essentially, his approach was based on the abstract power of algebra freed from its original entanglement with arithmetic entities (i.e., numbers):

---

3. "You could use mathematics to calculate if a design was correct, instead of using trial and error."—Marvin Minksy, quoted in [4, p. xix].

To non-mathematicians we point out that it is a commonplace of modern algebra for symbols to represent concepts other than numbers, and
frequently therefore not to obey all the laws governing numbers.... In
the particular algebra we construct for genetics theory the symbols represent Mendelian populations.... Addition and multiplication are
defined to mean simple combination and cross-breeding respectively,
and it is shown that nearly all the laws of ordinary numerical algebra
hold here.... [9]

It rather sounds like a Ph.D. dissertation, more demonstration than
innovation. Had Shannon been working a few years later, or had more direct
interaction with biologists, this approach might have borne more interesting
fruit. Nevertheless, we must admit that there is a slightly supernatural suggestion of a deeper insight—that information science and biology had a
peculiar affinity, which the future would later reveal. Even when Shannon
failed to decisively penetrate his subject, there was inspiration in his choice
of problems to attack.

In the summer of 1940, Shannon was working in New York City for
the Bell System, which started him in the direction of his most important
work. The following academic year he spent at Princeton on a fellowship at
the Institute for Advanced Study, where he must have encountered some of
the great luminaries of his age like Einstein and Gödel, although there is no
record of it as such. He was there to work under the great "intuitionist"
mathematician Hermann Weyl. A somewhat neglected figure today, Weyl
possessed a genuine Old World philosophical mind, and had spent decades
developing a comprehensive critique of mainstream mathematical thought
based on a rejection of some of the artificial tricks and techniques that
had led to logical contradictions in the core of its structure [10]. He questioned the foundations of a mathematician's sense of certainty, and probed
deeply below the surface of the easy answers that many of his colleagues
produced.

Weyl was just then in the throes of dealing with the consequences of
Gödel's incompleteness theorems (which seemed to prove that in some
sense the ultimate goal of conventional mathematics—a complete system of
formal knowledge—was unattainable—"a shattering discovery" in Weyl's
words [11]). For Weyl, it was a decisive indication that mathematics needed
to retreat from arbitrariness, from the "there is" and the "all" language of pure
formalisms, and reroot itself in physical reality, "the one real world" as he
called it:

Gödel, with his basic trust in transcendental logic, likes to think that our logical optics is only slightly out of focus and hopes than after some minor correction of it we shall see *sharp*, and then everybody will agree that we see *right*. But he who does not share this trust will be disturbed by the high degree of arbitrariness of [formal mathematics]. How much more convincing and closer to the facts are the … arguments and systematic constructions of Einstein's general relativity theory.…

A truly realistic mathematics should be conceived, in line with physics, as a branch of the theoretical construction of the one real world, and should adopt the same sober and cautious attitude towards hypothetic extensions of its foundations as is exhibited by physics. [11, p. 235]

One can sense the active debate, pursued in-the-flesh perhaps in Princeton in the early 1940s with Einstein and Gödel and others. Into this intense, fermentive atmosphere our young Shannon fell. It would be intriguing to speculate about the nature of the interaction between the older man and the younger. Weyl was pushing for something much more allied with reality than traditional mathematics had become. The engineering orientation of a young man like Shannon would have rendered him sympathetic to Weyl's broader views, one would think. On the other hand, Weyl's methodological principles involved the rejection of many mathematical proof-tricks, notably the so-called "existence proof," whereby an entity is shown to exist by assuming the opposite and developing from that assumption a contradiction. (If I can show that, by assuming that unicorns do not exist, a contradiction results, I can claim to have proved that unicorns must exist.) The use of the existence proof is a bone of contention between different schools of mathematical philosophy, and Weyl was dead-set against it. In the event, Shannon's own proof of his coding theorem *was* an existence proof, as several generations of engineers have politely complained. Nevertheless, it would appear that it was at Princeton that Shannon turned finally to the set of topics that he later bound together in his "theory of communications."

The war was on by then, and Shannon was recruited to work at Bell Labs, where he spent the next 15 years. He apparently worked on a variety of projects, including the development of fire control systems for anti-aircraft weapons, cryptography, and various topics in telephony. For the next 7 years he worked while his magnum opus, The Article, gestated. The Article appeared in two parts in the *Bell System Technical Journal* in 1948. His ideas exploded "like a bomb" in the engineering community, brought him fame and honor, and even a quirky notoriety. He felt, undoubtedly, the impetus of what he had unleashed as it gained a life of its own, and over time it apparently made him uncomfortable. In 1956 he published a sort of open letter decrying the "bandwagon" that information theory had become [12]. He pushed the basic concepts a little further—and then seemed to lose intensity. By 1960, he was moving on, intellectually. "I just developed different interests," he said [4, p. xxviii].

Shannon's biographies today usually go on to speak about the odd streak of playfulness that everyone remembers him for, which manifested itself in all sorts of peculiar games and devices that he is credited with having invented. The eccentricity laid over this portrait seems almost too neat—Shannon is recalled as someone who liked to juggle, built machines to perform juggling tricks, rode a unicycle through the halls of Bell Labs, perhaps juggling while he went, built small mechanical clowns that could juggle five balls or more, tried out a mechanized pogo stick, built a mechanical mouse that could solve mazes and a machine to solve the Rubik's Cube, and so on.[4] On the other hand, he was a successful stock market investor who could cite off the top of his head his annualized rate of return on every stock in his portfolio [4, pp. xix–xxxiii]. He developed a system to beat the roulette table at Las Vegas. He contemplated how many ways a mirrored room could be constructed, each capable of producing a visual "infinity without contradiction" [4, p. xxxii]. (There were, he thought, seven such archetypal rooms, and he thought about trying to actually build them in his basement, but let it go.) He studied the logic of crossword puzzles, and calculated that the English language had enough redundancy to support only two-dimensional puzzles, not three-dimensional ones. He announced himself an atheist, but he meditated on the origins of life, its "gradual organization"—which he thought "the most incredible thing!" He thought of himself as "a natural device" and sided generally with the machines. "I can visualize sometime in the future when we will be to robots as dogs are to humans" [4, p. xxviii].

To reconstruct an organic personality from such fragments is difficult. Claude Shannon was a man who slipped almost too easily into our stereotype of the quirky inventor. Yet he wrote with a directness that belies his apparent eccentricity. There is nothing quirky or unfocused about any of his writings. There is instead—confidence: "I was confident I was correct, not only in an intuitive way, but in a rigorous way. I knew exactly what I was doing" [4, p. xxviii].

---

4. See, for example, the biographical summary provided in the preface to [8].

2.3 Shannon Theory

So, what exactly *was* he doing?

The best way to begin to grasp Shannon's theoretical work is to start with the problem with which he started: determining how fast a telegraph could operate. Telegraph speed had for many years been limited by the capabilities of the human operators to send or receive Morse code. A very good operator was able to send or receive 40 to 50 words of text per minute, and for ordinary telegraph links there seemed to be no problem supporting this rate. The first undersea telegraph cables, on the other hand, showed a strange deceleration and attenuation of the signal. The telegraph symbols—the dots and dashes—became stretched out into long slow waves of energy. Signaling at the landline rate of 40 words per minute caused these attenuated symbols to overlap with one another and blend together. To cope with this mysterious phenomenon, telegraph operators had to slow down their rate to only a few words per minute.[5] Clearly, the *capacity*—defined as the maximum signaling rate—was not the same for all telegraph channels.

Apart from undersea cables, however, it gradually became apparent that the normal (short) terrestrial telegraph lines could support much higher rates of transmission than human operators could physically manage. There *was* unused capacity. Eventually, the cost of copper wire facilities—which was the dominant cost of any landline network by a factor of probably 10 to 1—dictated that the telegraph companies needed to find a way to exploit this unused capacity. For several decades spanning the turn of the century, a variety of mechanical transmitting and receiving devices were developed to substitute for human operators. (e.g., the iconic "ticker tape" machine associated with the rise of Wall Street). By the 1920s signaling rates of as much as 500 words per minute were achievable on certain circuits.

What is the ultimate limit on the signaling rate of a telegraph system? Assuming that transmitter and receiver could be automated to operate at arbitrarily higher speeds, the question of channel capacity came to the fore. As the undersea cables had shown, the channel did possess certain physical characteristics that eventually would limit the signaling speed. That is, it appeared reasonable to speak of the capacity of the channel as a parameter that was subject to a physical limit of some sort.

---

5. The problem was a mystery for decades, probed by many telegraph inventors including Thomas Edison. It was only understood to be a function of the inductance of the extremely long undersea cable by engineers like William Thomson and Oliver Heaviside around the turn of the century.

2.3.1 Nyquist

The engineer whose name is most associated with the question of telegraph signaling speed is Harry Nyquist, who was later cited by Shannon as his only real direct intellectual antecedent for the development of the engineering framework underlying his theory. Nyquist was born in Sweden in 1889, eventually emigrated to North Dakota, and received his Ph.D. from Yale in 1917. He then joined AT&T where he remained until retirement in 1954. He was one of the senior luminaries at Bell Labs, and he worked on a range of problems in his career, accumulating more than 100 patents.

Two of Nyquist's papers are still read today.[6] In "Certain Factors Affecting Telegraph Speed" (1924), he took up the problem of telegraph capacity. In "Certain Topics in Telegraph Transmission Theory" (1928), he extended his ideas to provide a foundation for understanding digital signaling in general. Essentially, Nyquist's key contribution was to relate the maximum signaling rate (a proxy of sorts for "capacity") to the *frequency* bandwidth of the signal. That is, the wider the frequency band of the signal to be transmitted, the higher the signaling rate required to represent the information in that signal. Although this did not say anything directly about the capacity of the channel, it did offer a measure of the amount of information contained in a given signal. More specifically, and elegantly, he proved that the signaling rate needed to be exactly twice the bandwidth (in hertz) of the signal to be transmitted. To send a voice signal with a bandwidth of approximately 3,000 Hz (that is, with frequency components ranging between, say, 200 Hz up to around 3,200 Hz—which is normally defined as the *voiceband* in a telephone system), we need a signaling rate of at least 6,000 samples per second. In fact, the original standard for voiceband digitization was set at 8,000 samples per second, based on this sort of calculation. The derivation of Nyquist's result is presented in almost any textbook on digital transmission, and this signaling rate is referred to as the *Nyquist rate*.

His other important insight related to the way to pack the signaling pulses together. It was clear from experimentation that a transmitted pulse has a tendency to spread in time as it passes through the channel (Figure 2.1). This effect is universal, and is related in the first place to the length of the circuit. Undersea cable circuits are so long that the spreading becomes very great (which is why it was first observed there). But the same phenomenon occurs to some degree on any wireline telegraph channel, of any length.

---

6. Interestingly, his 1928 paper was recently republished in the *Proceedings of the IEEE* as a "Classic Paper." See the March 2002 issue.

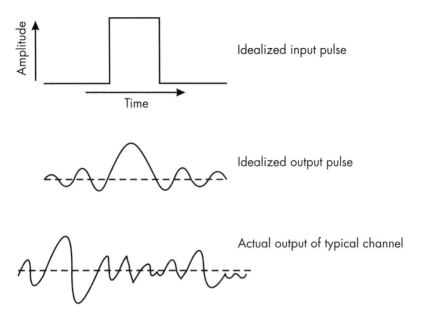

FIGURE 2.1 *Spreading pulse—channel impulse response.*

Nyquist saw that this could create an obstacle of a different sort to set a limit on the maximum signaling rate. If it was necessary to wait until one pulse had substantially died away before sending the next pulse, the signaling rate would be drastically affected.

Nyquist observed, however, that the spreading pulse oscillated between positive and negative voltages as it slowly damped down; it repeatedly crossed the zero voltage level. Nyquist realized that if you sampled the signal at the precise instant of such a zero-crossing, its voltage would be zero and the signal would effectively be absent from the channel at that precise moment. This suggested, finally, that if symbols could be properly and precisely spaced so that the peak of the following symbol occurs just at the instant of zero-crossing by the preceding symbol(s), then each symbol could be detected without any interference coming from the preceding symbols whatsoever (Figure 2.2). This remarkable result established the first benchmark for understanding the physical capacity of a signaling channel.

Nyquist's analysis was nevertheless somewhat artificial. First of all, it was obvious that even small errors in timing could result in displacing the sampling instant from the correct position, reintroducing interference. This led to a further insight: The shape of the signaling pulse at the transmitter

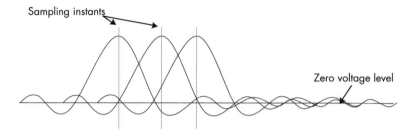

FIGURE 2.2 *Nyquist signaling: overlapping symbols with sampling at zero-crossing points.*

could determine how much energy would be contained in the oscillating "tails" of each received pulse, and how fast those tails would decay (Figure 2.3). Some pulse shapes were clearly superior in this regard to others, and

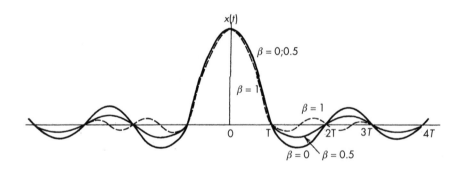

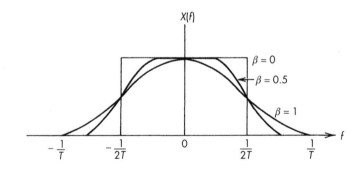

FIGURE 2.3 *Different Nyquist shapes based on different filter rolloff factors. (From: [13]. © 1989 McGraw-Hill Inc.)*

Nyquist shaping is still an important technique to minimize sensitivity of a digital signaling system to inevitable timing errors.

However, Nyquist did not really come to grips with what he called "departures from ideal conditions" [14]. He acknowledged that some type of interference is always present, discussed some of the sources of interference in telegraphy (such as "cross-fire and duplex unbalance") and he indeed took the important, if tentative, step of defining interference epistemologically as "the difference between the actual wave and the desired ideal wave." (Shannon would develop this perspective into an implicit element of a powerful new conceptual framework.) However, Nyquist did not really bring interference into his theoretical framework. He followed the classical scientific strategy of describing a simplified system precisely under idealized conditions, setting aside the problem of interference in the same way that a mechanical physicist might set aside friction from his idealized description of some physical device. Thus, in theoretical terms, Nyquist left the problem incompletely stated.

2.3.2 The Importance of Noise

When Claude Shannon came to the subject 20 years later, his first decisive innovation was the recognition that interference, or as he preferred to call it, *noise,* is an essential player in the communications game. The real problem, as he put it in the title of another article, is to analyze "communications in the presence of noise." He realized, in particular, that the analysis of channel capacity is incomplete without including noise in the calculation.[7] The Shannon model of the communications process takes shape in his famous diagram of a "general communications system" (Figure 2.4). This is the fundamental worldview of Shannon theory: a signal, making its way from transmitter to receiver, and mixing with noise in the channel. The implicit goal of the system is how to get the signal *through* the noise.

However, admitting noise into the game had troubling implications. If noise is always present, and if, as Shannon further assumed, it is by definition unpredictable and may assume *any* amplitude value at a given instant in time, then it would seem that error-free transmission is impossible in any channel. The number of errors may decrease as the SNR improves, but

---

7. In the opening paragraph of The Article, Shannon highlights the focus on noise: "In the present paper we will extend the theory [of Nyquist and Hartley] to include a number of new factors, *in particular the effect of noise* [emphasis added] in the channel ... " [2].

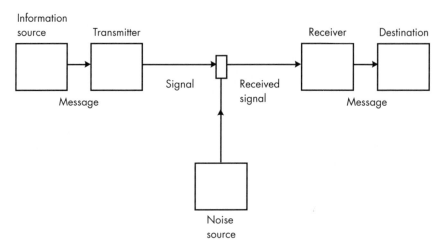

FIGURE 2.4 *Shannon's model of the communications process. (From: [15]. © 1993 IEEE. Reprinted with permission.)*

sometimes (no matter how infrequently) the noise will overwhelm the signal and cause an error. Perfect transmission would seem to be unachievable.

"On the contrary!" Shannon would argue. Error-free signaling *is* possible, even in a *very* noisy channel, as long as certain constraints are observed. The main constraint is that we must keep the signaling rate, the rate at which we are transmitting information, below the threshold of the channel capacity, C. If we do so, then perfect transmission is possible regardless of the noise.

What does this mean, exactly? We understand that analog signals, such as a voice telephony signal, blend with noise in the channel and cannot easily be unblended. Like the old-style vinyl records that cavemen used for recorded music, the scratches that accumulate with time and use are unavoidable. It would seem that noise must be a part of any signal that has passed through a real-world channel.

2.3.3 Discrete Sources and Entropy

Shannon made a crucial assumption: The signal with which he is concerned is not an analog signal. It is a *digital* signal, like the series of pulses transmitted by a telegraph: "The significant aspect [of the new theory] is that the actual message is one *selected from a set* of possible messages [emphasis in the original]" [2]. The message elements are drawn from a finite alphabet, like Morse code. The pedigree of telegraphy as the model of the communications process is clear. Indeed, for much of the first part of The Article, the telegraphy model is specifically in focus.

It might seem that this would limit the relevance and scope of Shannon's theory to the less interesting cases (like telegraphy), but Shannon develops the *discrete source* concept and generalizes it to include English text. It is here that he digresses into a study of the statistics of such sources (like English texts), and derives the philosophically interesting notion of the *entropy* of a source, which is essentially a measure of its predictability and, less obviously, its ability to undergo compression. Messages from a highly predictable source contain significant redundancy, which can be removed in order to skinny-down the message for transmission, and added back later at the receiving end. The example of q followed by u in English illustrates this point. The u does not carry any extra information, and we can drop these empty symbols from the transmission to save capacity. The receiver knows the rule to reinsert the u after every q.

Source compression is the first important practical lesson of Shannon's theory—the idea that the source can be studied and its messages compressed, based on carefully eliminating redundancy and exploiting statistical regularities to produce a more compact representation.

2.3.4 Channel Coding and Error Control

Returning to his main theme of how to communicate in the presence of noise in the channel, Shannon next develops the idea of channel coding, or error-correction coding, as a means of ensuring that the message can be received "with as small a frequency of errors … as desired" even when noise is present. It is here that he introduces more carefully the notion of channel capacity and relates it to the entropy of the source and the error criterion. His proof, as noted above, is an existence proof[8]—although he registers a certain defensiveness about this that may well be a trace of the influence of Hermann Weyl.[9] His comments on capacity are thus somewhat unsatisfying. But they lead to the second important practical implication of Shannon theory: the idea that errors can be controlled successfully as long as the

---

8. "The method of proving . . . this theorem is not by exhibiting a coding method having the desired properties, but by showing that such a code must exist in a certain group of codes" [2].

9. "The demonstration of theorem 11, while not a pure existence proof, has some of the deficiencies of such proofs" [2].

signaling rate stays under a certain threshold. The precise definition of the threshold, and the particular methods of encoding the signal to control errors, are not developed, but the concept is clear and important: The message can be "hardened" by extra coding to survive the effects of noise in the channel.

Note that the two processes described so far are antithetical to one another. The analysis of the entropy of a source leads to the conclusion that capacity can be gained by compressing the redundancy out of the transmitted message. On the other hand, the hardening of the signal to make it more error resistant requires adding back redundancy. The first process shrinks the message, whereas the second expands it again. Arguing that the underlying principles are different, Shannon makes the strong claim that the first process—which is generally called *source coding*—can and should be treated as completely independent of the second process—which is called *channel coding*. This assertion has strongly colored the entire field and shaped patterns of subsequent research down to the present day.

Nevertheless, Shannon has effectively approached the capacity problem from both ends. By developing the principles for compressing the source to the maximum, he sheds light on the question of how much capacity a given signal really uses. By considering the need for error control, he is able to define a relationship between channel capacity and the information rate.

2.3.5 Quantization of Analog Sources

The analysis of discrete sources might seem limited to text-like messages, like telegraph signals, but in the second part of The Article, Shannon announces that he plans to break through this apparent limitation and "consider the case where the signals or the messages or both are continuously variable." In other words, he intends to apply his ideas to signals such as voice telephony.

The way he gets to this result is deceptively simple: Just as the Greek geometers learned to approximate the area of a circle by studying polygons of successively larger numbers of sides and projecting the result to its asymptotic limit, Shannon proposes that a continuously variable signal can be approximated by using a discrete signal with fine resolution:

> To a considerable extent the continuous case can be obtained through a limiting process from the discrete case by dividing the continuum of messages and signals into a large but finite number of small regions and calculating the various parameters involved on a discrete basis. [2]

The effects of quantization are many and subtle, as we shall see. But Shannon plows through the matter in a direct line. He admits that, in fact, the theory does not apply to an analog source since "exact transmission is impossible"—but blithely argues that "practically, we are not interested in exact transmission when we have a continuous source, but only in transmission to within a certain tolerance." He develops the notion of a fidelity criterion to allow quantization to stand on its own feet, so to speak. He delves very briefly into the softer question of how the fidelity criterion is determined ("the structure of the ear and brain determine implicitly an evaluation"), and then concludes that "we are now in a position to define a rate of generating information for a continuous source."

We may question this assertion, but what he has accomplished is to bring at least digitized versions of analog sources under his theory. The mysteries of digitization may be glossed over, but at least we are no longer limited to speaking only about telegraph signaling. Shannon effectively shows that *if* a signal can be digitized, we can bring to bear on it the very powerful new conceptual apparatus that he has developed. Among other things, we can define the capacity of a channel for carrying *digitized* voice signals. We can apply the new ideas about compressing such signals, and we can use Shannon's general insights to guide us in finding ways to harden the signal to immunize it against noise and other forms of degradation in the channel.

All in all, Shannon provides us with a number of fruitful strategies for handling recalcitrant analog signals. Beyond that, he created a new way of looking at the communications problem, a new systematic vision. Before Shannon, communications engineering was basically a set of ad hoc rules and gimmicks for making certain kinds of circuits work more or less adequately. Some solutions could be engineered, and many could not, but everything that came after The Article was part of the new and dynamic discipline of communications theory, which prided itself on being able to bring mathematical power to the solution of the most intractable engineering challenges.

2.3.6 Power and Bandwidth

There is much more to Shannon's body of work than we have covered here, although I believe the essential practical points have been touched on: the importance of dealing with noise *within* the theory and the system design, the potential for source compression, and the possibility of controlling errors (and mitigating noise and interference) through channel coding. He had many other insights and suggestions, some of which we shall discuss later in the book. And there is one other key notion that is worth mentioning here: the idea of the *power/bandwidth tradeoff.*

Shannon observed that one effect of digitizing an analog signal is (often) to *expand* it. This expansion can be observed in either the time domain or the frequency domain, which are in some sense interchangeable perspectives on the physical signal. For example, the standard form of digitizing the voice signal—called *pulse code modulation* (PCM) (which was not Shannon's invention, but which he recognized as the perfect embodiment of his theory in many ways)—is created by sampling the analog waveform of the voice signal 8,000 times each second. The amplitude of each sample is encoded as an 8-bit number (i.e., as a string of eight 1's and 0's). If the sample amplitude were measured and sent "directly" (or with a very fine quantization scale of 256 individual steps) as a single amplitude figure, it would take only one pulse of the nearly precise amplitude to do the job. This would be very efficient. Such a signal would be very susceptible to being bumped by noise into an erroneous result. To ensure that it would be accurately received, we would have to make sure that the SNR was very, very good.

On the other hand, by sending the same value encoded not as a single pulse but as eight separate 1's and 0's, we find that the signal becomes more robust. Now the receiver needs only to detect each pulse as either a 1 or a 0 value, an "on" or an "off." The signal can withstand much higher levels of noise, or, to put it another way, the signal power can be drastically reduced.

The price paid is that the signal has expanded. If each pulse takes the same amount of time to send, whether it is a pulse measured on 256 different levels or a pulse measured on two levels, then the time taken to send the sample has expanded by a factor of 8. Alternatively, if we speed up the signaling rate and send the eight (two-level) pulses in the same time as we would have sent one (256-level) pulse, we find that the signal expands in the frequency domain—it occupies a wider bandwidth.[10]

Shannon's famous capacity equation describes a relationship between three variables: transmitter power (P), noise (N), and bandwidth (W).

$$C = W \log(1 + P/N)$$

If we take the noise as a given, the radio engineer has only two controllable variables: transmitter power and bandwidth, and the equation shows

---

10. The rationale for the time-frequency spreading phenomenon is discussed in Chapter 8.

that if one is increased the other may be decreased. This is the key to the robustness of PCM, which is able to function well at noise levels hundreds or thousands of times higher (that is, its signal-to-noise criterion is lower) than an analog voice circuit: "PCM requires more bandwidth and less power that is required with direct transmission of the signal itself.... We have, in a sense, exchanged bandwidth for power" [16].

The idea of purposefully spreading the signal, causing it to expand physically in time or frequency, to gain robustness, is one of the great counterintuitive insights of twentieth-century communications engineering. Counterintuitive because the straightforward way of communicating more accurately would seem to be to "speak up"—to increase the signal power to overcome the noise. The opposite approach, spreading the signal out to allow it to operate at much lower power levels, points to very different sorts of engineering solutions. Indeed, Shannon touches on the bandwidth expansion idea so lightly, almost in passing, that its significance was not fully appreciated for many years. It is only quite recently, with the development of true third-generation wireless technologies under the gun to generate maximum capacity, that the implications of this tradeoff are perhaps being fully understood and exploited.[11]

Shannon theory is really more than a sum of its individual ideas and insights and suggestions. It *is* a new philosophy, or at least the germ of one. It leads to new ways of thinking about how to communicate information in a real-world channel.

Oddly, it did not lead there directly. The initial impact of Shannon's ideas was felt by the intellectual community at the boundary of science and engineering. Shannon theory was an almost immediate sensation. It was brilliant, well written, simple, easy to follow, and apparently profound. Shannon's almost offhanded identification of information with the older concept of thermodynamic entropy was too intriguing to let pass.

---

11. CDMA advocates will argue that they knew this all along, and that the favorable power-bandwidth exchange is part of their philosophy of signal construction, embodied in the second-generation standard for CDMA (IS-95 and more). This is certainly true, and the system architects deserve full credit. But the implementation of CDMA in its second-generation form is arguably only a half-step in this direction. The bandwidth expansion is rather modest, and the most important manipulation of the Shannon equation comes through the reduction in signal power levels achieved through the use of a variable-rate voice coder. In other words, the most important impact of IS-95 is derived from a direct reduction of the P factor (signal power) rather than from the expansion of the W factor.

On the other hand, Shannon did not provide tutorials on system design. He set grand metaphysical limits, he defined "the possible" and "the necessary" for the new field, but did not provide the "how-tos" and the "wherefores." His existence proof for the ideal code was like a guarantee that a particular haystack does indeed contain the needle, but it offered little explicit guidance to coding designers (who nevertheless began working hard to find it). Moreover, the communications industry in 1948 and for several decades thereafter did not really need to solve the capacity equation. In a wireline world, capacity can always be added by simply stringing another wire. The coding schemes toward which Shannon was pointing required intensive computing capabilities that were simply not available at all in 1948, and not in a feasible package for most communications applications until the development of digital signal processors in the late 1970s.

Thus, for several decades, Shannon theory was something of an intellectual curiosity. Everyone knew about it, but very few engineers had any use for it. Meanwhile, Shannon himself had abandoned the field; it was not in him to be the Peter of this new church.

Then, beginning in the 1980s, with the first capacity crisis in cellular radio, the problems with which Shannon had wrestled, in theory, began to emerge in practice. Engineers were faced with a channel that was capacity challenged and extremely refractory. Shannon's ideas, which had lain on the shelf for so long were suddenly relevant again. By the 1990s, the second Shannon revolution was under way again. Coding, capacity, signal design—all began to show now the explicit signs of the application of Shannon's insights. Slowly, as with any theory that is finally exposed to the practicalities of real-world implementations, these insights began to undergo further development.

The attitude of the engineering community today is largely orthodox, when it comes to Shannon. He is invoked over and over as the patron saint or prophet of the field, and it is often claimed that everything we are now seeing is "merely" the straightforward working out of his original insights. Shannon theory is often presented as a complete, definitive solution to all problems of interest, if we are but allowed a few simplifying assumptions. The application of these ideas is treated as though it were nonproblematic, even obvious.

What can the phrase "post-Shannon" mean then? Well, in one way, we are all post-Shannon now. His ideas have at last acquired real relevance to serious practical engineering problems (which was not true in the 1950s, 1960s, or 1970s). We now approach these problems with his insights close at hand.

However, "post-Shannon" is also meant to imply something more. There are indeed ways in which the shortcomings, and even the errors of the 1948 formulations, are being exposed by recent engineering developments. The conceptual framework seems less impervious to questioning than it once did. To take one example, we shall see over and over again in the discussions in the following chapters how the addition of *noise* to the signal, or the purposeful creation of interference between signal elements, can have a beneficial effect. We are finding situations where the injection of noise can *improve* reception of the signal. It is hard to reconcile this with the rather stark model of the communications process presented by Shannon in the Article.

The pristine character of Shannon theory is, I believe, largely the result of its long state of suspended animation. A theory can only be seriously challenged by *facts*, and unless a theory can be applied it is difficult to develop new facts. For decades following Shannon's original articles, despite a great deal of philosophical interest and discussion, there were no new relevant *facts* to challenge his theory. That is now changing.

The capacity crisis in wireless is the impetus for this challenge. Why wireless? Because in every other field of communications engineering, it is possible—*in principle*—to avoid the challenge, intellectually if not practically. In every other field, an engineer may dream (at least) of perfect implementation, with Moore's Law as his guiding star (and its prospect of unlimited processing power) and the presumption of an affluent society ready to budget whatever is necessary to obtain perfect performance. Only in the world of wireless communications does an adverse reality intrude and begin to force a different set of conclusions. Let us see how.

REFERENCES

[1]   Bramwell, J. R., and M. Tomlinson, *An Introduction to Data Modulation and Error Correction Coding Techniques for Satellite Communications Channels*, BNSC Contract Report LPO/A/02936, March 8, 1991.

[2]   Shannon, C. E., "A Mathematical Theory of Communication," *Bell System Technical J.*, Vol. 27, July 1948, pp. 379–423.

[3]   "Obituary for Claude Shannon," *New York Times*, February 27, 2001.

[4]   Liversidge, A., "Profile of Claude Shannon," *Omni Magazine*, August 1987. Reprinted in Sloane, N. J. A., and A. D. Wyner, *Claude Elwood Shannon: Collected Papers*, Piscataway, NJ: IEEE Press, 1993, p. xx.

[5]   Shannon, C. E., "The Bandwagon," *IEEE Trans. on Information Theory*, Vol. 2, March 1956.

[6]   Boole, G., *An Investigation in the Laws of Thought, on Which Are Founded the Mathematical Theories of Logic and Probabilities*, London, England: Walton and Maberly, 1854.

[7]   Goldstine, H. H., *The Computer from Pascal to Von Neumann*, Princeton, NJ: Princeton University Press, 1993.

[8]   Sloane, N. J. A., and A.D. Wyner, *Claude Elwood Shannon: Collected Papers*, Piscataway, NJ: IEEE Press, 1993.

[9]   Shannon, C. E., "An Algebra for Theoretical Genetics," Ph.D. dissertation, Massachusetts Institute of Technology, 1940. Reprinted in Sloane, N. J. A., and A. D. Wyner, *Claude Elwood Shannon: Collected Papers*, Piscataway, NJ: IEEE Press, 1993, pp. 891–920.

[10] Kline, M., *Mathematics: The Loss of Certainty*, Oxford and New York: Oxford University Press, 1980.

[11] Weyl, H., *Philosophy of Mathematics and Natural Science*, Princeton, NJ: Princeton University Press, 1949, p. 234.

[12] Shannon, C. E., "The Bandwagon," *IEEE Trans. on Information Theory*, Vol. 2, March 1956. Reprinted in Sloane, N. J. A., and A. D. Wyner, *Claude Elwood Shannon: Collected Papers*, Piscataway, NJ: IEEE Press, 1993, p. 462.

[13] Proakis, J., *Digital Communications*, 2nd ed., New York: McGraw-Hill, 1989, p. 536.

[14] Nyquist, H., "Certain Topics in Telegraph Transmission Theory," *Trans. of the AIEE*, February 1928, pp. 617–644.

[15] Shannon, C., "A Mathematical Theory of Communication," reprinted in Sloan, N.J.A., and A. Wyner, *Claude Elwood Shannon: Collected Papers*, Piscataway, NJ: IEEE Press, 1993, p. 7.

[16] Oliver, R. M., J. R. Pierce, and C. E. Shannon, "The Philosophy of PCM," *Proc. IRE*, Vol. 36, 1948, pp. 1324–1331. Reprinted in Sloane, N. J. A., and A. D. Wyner, *Claude Elwood Shannon: Collected Papers,* Piscataway, NJ: IEEE Press, 1993, pp. 151–159.

3

Wireless Systems Design: Problems and Parameters

3.1 Three Unique Design Constraints

The architect of any wireless communications system must overcome three unique and fundamental obstacles that do not affect wireline systems:

1. *The physical channel is completely nonengineerable.* The wireless communication system (transmitters and receivers) must be designed around the natural or "given" characteristics of the radio channel.

2. *The channel is, so to speak, nonclonable.* In a wireline network, the growth in traffic between point A and point B can be addressed, at a cost, by adding new identical transmission facilities between those nodes: more fiber, more coax, more copper. In a wireless network, this option does not exist. The total available physical bandwidth between points A and B is finite.

3. *The wireless system architect must allow for the fact that the signal will experience significant, destructive interactions with other signals (including images of itself) during transmission.* These effects are produced, in part, by the physical channel itself. The system must

therefore be designed to perform in the presence of this persistent interference.

The first two constraints are obvious, although they are often overlooked or assumed without comment. The nonengineerability of the physical channel means that wireless systems, unlike other types of communications systems, must operate in a channel that *inevitably* produces a lot of transmission errors—even after aggressive countermeasures have been deployed; for example, "Residual transmission errors cannot be avoided with a mobile radio channel, even when FEC and ARQ are combined" [1]. The fact that the channel resource is finite means that the wireless designer must *inevitably* concern herself with capacity, efficiency, bandwidth, and tradeoffs in signal power and quality—in short, she enters immediately into the realm of Shannon's theory, while her colleagues in the wireline world can remain much longer in the pre-Shannon framework where transmitter power—and money—can solve almost any problem.

The third issue—the issue of interference, and how to handle it—is more subtle, and more controversial. As we shall see, the transition from pre-Shannon to post-Shannon solutions and techniques principally involves coming to grips with interference in a new and conceptually revolutionary manner.

Taken together, these constraints define the field of wireless technology and set it apart from the rest of the communications world. To understand the evolution of modern wireless technologies and architectures, we must understand these three problems and how they affect the designer's options for getting his message through the wireless channel from A to B.

3.1.1 The Basic Communications Link and the Nonengineerable Wireless Channel

Let us begin by outlining, on an intuitive basis, the conceptual framework within which the analysis of wireless architectures will be developed. To address this properly, we first need an adequate terminology.

The elemental "atoms" of any communications system are the transmitter, the receiver, the signal, the channel, and, of course, noise. Each "atom" presents itself intuitively as a "primitive term," although as with the atoms of the physical world, these communications "atoms" will eventually display a complex structure.

The *transmitter* and *receiver* are typically viewed as intelligent entities,[1] or their proxies, separated in space and time from one another.[2]

The *signal* is an information-bearing structure of some sort (such as a text message) that the transmitter intends to enable the receiver to replicate accurately. If this replication can be accomplished, communication has taken place.[3] (For the moment we can understand *information* and *structure* in intuitive terms as well.)

---

1. The theorists typically distill the issue of "meaning"—as something they cannot really handle—from the model of communication, and construe the transmitter, for example, as simply a *source* of symbols with a certain probability distribution. "In principle, the theory is simply an abstract mathematical theory of the representation of some undefined source symbols in terms of a fixed alphabet (usually the binary system) with the representation having various properties. In this abstract theory there is no transmission through the channel, no storage of information, and no 'noise is added to the signal.' These are merely colorful words used to motivate the theory" [2, p. 2].

2. Traditional terminology speaks of the source, which is coupled with the encoder, and likewise of the destination coupled with a decoder. These terms are linked to the original configurations studied by Shannon and his colleagues and reflect certain conceptual limitations that we need not accept today; for example, *encoding* (unless we reinterpret the term in extremely broad connotation) is only one of the technical processes that may be employed by the source to prepare the message for transmission.

3. The definition can be extended with some subtlety, to encompass varying degrees of purposefulness on the part of the transmitting and receiving entities. For example, Cover and Thomas write: "What do we mean when we say that A communicates with B? We mean that the physical acts of A have induced a desired physical state in B" [3, p. 183]. Or, shortly thereafter: "The communication is successful if the receiver B and the transmitter A agree on what was sent" [3, p. 183]. In these glosses (which admittedly are prefatory to their formal treatment of the subject), the authors introduce the complex notions of *desire* and *agreement* or intentionality and co-intentionality, into the definition of the act of communication. For a more thorough-going behavioral approach to the subject, see Ackoff [4], where communication is defined as a form of purposeful behavior of complex systems, and is measured or detected by the observation that the signal alters the preexisting "probability of choice" on the part of the recipient. I once asked Ackoff whether this definition might be extended to include the case of a ripening apple on a tree, signaling—by its change of color—to select seed-dispersers (those with the right visual equipment, i.e., color vision) that the fruit was now fully edible, thereby changing their "probabilities of choice." Ackoff pondered the question briefly but dismissed the possibility that the apple tree might be communicating with the orchard manager. I think the example illustrates the pitfalls of pursuing a fully teleological definition to its endpoint. In the "definition" offered in this text, I am defining communication simply as the ability of an observer to verify that the message input to the transmitter and the message output from the receiver are identical replicas (or nearly so—although in that "nearly" we place our foot on the slippery slope of functionalism once again).

The simplest conceptualization of the *channel* is nothing more than this space–time "gap" between transmitter and receiver.[4]

Noise, the serpent in this garden, is a kind of primitive attribute of the space–time continuum, the proxy of thermodynamic entropy itself,[5] and possesses a subtle nature[6] that will pose many fine questions later on, but its manifestation is straightforward: The tendency of the signal to become degraded as it passes through space–time, which in turn degrades the ability of the receiver to construct an accurate copy of the transmitted message.

The subtlety manifests itself promptly, however, in the fact that noise is not something we can define precisely. It is not a wholly deterministic process, and in most situations the best we can do is specify the statistical properties of the noise energy. The classical form of noise is based on the most random of all possible distributions of this energy across the frequency spectrum, and is known as *white* noise (by analogy with *white* light—similarly composed of all possible frequencies of light). White noise is favored in most theoretical discussions, because it is the most tractable for many statistical techniques. White noise is unfortunately a rarity; more commonly we encounter "colored noise" in which there are at least some elements of structure, or skewing of the statistics.

This in turn suggests a more abstract view of the channel, in terms of the statistical characteristics of the noise inherent in it. Discussions of different types of channels—such as the *fading channel*—are normally a shorthand for descriptions of different statistical models of the embedded noise [2, p. 139].

---

4. Shannon's original treatment glosses over the channel definition: "This is merely the medium used to transmit the signal from the transmitting to the receiving point. It may be a pair of wires, a coaxial cable, a band of radio frequencies, etc." [5].

5. According to *Encyclopedia Britannica,* 1998, "Entropy is analogous in most communication to audio or visual static—that is, to outside influences that diminish the integrity of the communication and, possibly, to distort the message for the receiver."

6. Noise is "random, unpredictable, and undesirable" in the standard theory. Shannon himself refers to noise a involving "statistical and unpredictable perturbations" [5]. Alternatively, another text speaks of "the uncontrollable ambient noise and imperfections of the physical signaling process itself" [3, p. 183]. Still another standard text refers to noise as "an erratic, random, unpredictable voltage waveform" and takes the metaphysical position that "here we need but to note that at the atomic level the universe is in a constant state of agitation, and ... this agitation is the source of a very great deal of this noise" [6, p. 1]. The reader can, I think, readily appreciate that there are deep and difficult issues embedded in this innocuous everyday language, from the definition of *randomness,* to the covert intentionality of words like *uncontrollable* and *unpredictable,* to the physics and metaphysics underlying the "constant agitation" of the universe.

The *process* of communication can take a number of forms. One way to pass the information over the channel is to make a physical copy, such as a letter, and to physically transport that copy across the "gap." But the communications methods of interest here are those we call *tele*communication; in these cases, the message is *encoded*—its information-bearing structure is mapped into some form of (electromagnetic) energy,[7] and the energy is transmitted in some fashion across the "gap" to the receiver. The physical characteristics of the form of energy employed to carry the signal determine additional relevant features of the channel, as well as the specialized tools employed by the transmitter and the receiver. Of course, typically the gap may be physically bridged with a specific medium, such as copper wire, connecting the transmitter and the receiver in a telegraph system. If so, some of the characteristics of this transmission medium also become relevant secondary features of the channel.

In Shannon's early work, the model is fundamentally this simple, and the problem is simply stated: How fast can the transmitter send information across this channel to the receiver? This problem developed out of the work of Nyquist and others on high-speed telegraphy, which was the root problem of electrical communications going back to Bell and Edison, and William Thomson (Lord Kelvin) in the nineteenth century: How fast could a telegraph channel (or something like it) be made to operate reliably? Or, as Shannon rephrased it: What is the *capacity* of the channel?

This is a very fruitful problem to study, as the evolution of a vast body of theory during the past 50 years has shown. It leads in particular to a consideration of the characteristics of the code used to carry the information, to map the information onto the transmitted energy signal. It also opens up the analysis of the relationship between the code required to carry the information properly over the channel, and the information required to accurately replicate the signal at the far end. Compression, error correction, source coding—all of these coding disciplines emerge from Shannon's simplified model of the basic link between a single transmitter, and a single receiver, across a discrete channel, with noise added.

This is a useful theoretical approach even for many situations where multiple transmitters are communicating with multiple receivers, for the simple reason that each transmitter–receiver pair can be allocated a separate and discrete channel. If there are 20 houses along a particular street, each

---

7. There are arguably forms of telecommunications that do not employ electromagnetic signaling; perhaps some forms of sonar can be considered examples. These are fringe phenomena.

house can be provided with its own copper-wire connection for access to the telephone network. The channel is not shared. If more houses are built, more wires can be strung. The only "capacity problem" is the *classical* Shannon capacity problem of the maximum rate of reliable transmission over a single channel. Since the true capacity of a wire circuit is far higher than the bandwidth required for a voice transmission, the issues that Shannon raised can be largely ignored for conventional telephony.

To restate this point, the normal framework for thinking about conventional wireline telephony networks is a pre-Shannon framework. Each link can be viewed as a point-to-point physical connection that is dedicated (at least for the length of the call) to a specific transmitter–receiver pair.[8] The requirements of any individual transmission (for a voice signal) are well within the physical capacity of the channel that the issues that Shannon wrestles with in theory simply do not arise in practice.

The design challenge in this framework is that of *engineering the link*. The signal leaving the transmitter must reach the receiver with an SNR that is good enough for the signal to be successfully detected and understood.

To achieve this objective, the wireline engineer can effectively lay his hands on all elements of the system: the transmitter, the receiver, *and the channel itself*.[9] The physical wireline channel—a copper-wire connection, typically—can be conditioned, replaced with a heavier gauge circuit, placed in a more secure environment, better protected against moisture or lightning, or shielded against sources of electromagnetic noise. If it is a high-

---

8. Originally, this was literally true. From end to end the telephone connection involved the linking of nonshared circuit elements, and the only economies were those of traffic engineering that allowed a smaller number of trunk circuits to support a larger population of local loop or access connections. Over time, of course, certain portions of the core network began to be subject to sharing among multiple simultaneous transmissions, with the emergence of T-carrier as a notable encroachment. Still, the coding for T-carrier was still so far below the physical channel limits, especially with the option of repeaters, that it is doubtful that Shannon's ideas came much into play.

9. From a formal perspective, the channel is sometimes regarded as unengineerable *by definition*. For example, consider this statement: "The channel is usually considered to be that part of the system which is beyond control of the designer" [7, p. 28]. This is a misleading perspective, however, for analyzing the practical issues involved in designing real-world communications systems, where the wireline engineer can almost always reengineer the channel (at least he has that option, in principle) if the transmission conditions of the channel are unacceptable. We need to distinguish between the part of the system that is left "as-is" in a particular implementation from what is truly "beyond the control" of the engineer.

speed link in a computer network, the physical connection can be upgraded to coaxial cable or fiber optics. Repeaters can be installed to reboost the signal at appropriate intervals. All of these options are available, at a cost, and indeed it is *only* cost (economics) that fundamentally constrains the choice of technologies. The engineer may choose a cheaper solution and attempt to get by with reduced margins of performance, but at the end of the day, a crucial assumption will stand: *The system can deliver essentially error-free transmission of information if we are willing to pay for it.*

The assumption of an engineerable link—meaning a link where it is possible to ensure essentially error-free transmission—is so pervasive that it has tended to distort the entire technology development process in the communications industry. This assumption accounts for the weaknesses of some of the basic protocols that have been developed for our most common communications needs, such as *asynchronous transfer mode* (ATM) and TCP/IP (the backbone protocol for the Internet), which are surprisingly fragile because their designers assumed that they would always be able to operate in an essentially error-free transmission environment.

In contrast, the wireless channel is a physical channel that cannot be engineered. It must be taken as it is found, in free space, in dense forests, in urban canyons, and deep in buildings where people live and work. The performance of a wireline channel—properly engineered—should be predictable and unvarying to within a few decibels of the SNR, and whatever changes may occur due to aging or temperature will be slow and should exhibit a graceful degradation. However, the SNR in a wireless channel, such as a typical mobile radio channel, will vary by a factor of 10,000 or more over very short time intervals. *Wireline* channels can be shielded and routed away from known sources of electrical noise. Wireless channels are subject to interfering radiation from countless sources, from other users, from computers and electronic equipment, from atmospheric disturbances such as rain or lightning, and from automotive ignitions. The noise affecting a wireless signal is not all, or even predominantly, "white." It tends to come in *bursts*, capable of knocking out significant blocks of the transmission, overwhelming simple error correction schemes, disrupting synchronization, and relentlessly taxing the system's margins for proper performance.

The net result is that it is impossible, practically speaking, to design a wireless transmission system to operate without a relatively high rate of (raw) channel-induced errors.[10] Indeed, the channel error rate in a good

---

10. The advertising of certain systems as the equivalent of "wireless fiber" should not mislead us here. There is no wireless channel that is stable enough to yield wireline

mobile radio link is typically millions or tens of millions of times higher than in a well-engineered fiber-optic channel.[11] Under actual operating conditions, the mobile channel must often tolerate much higher error rates—up to 10^{-2} or worse. Even with the most aggressive countermeasures of the sort discussed in the following chapter, it is often not possible to *guarantee* post-processing error rates better than 10^{-3} or 10^{-4} on a mobile channel.

This changes the system designer's entire perspective. All communications techniques and architectural alternatives must be selected and crafted to withstand this severe error-rate environment.[12] More than that, it forces the wireless engineer to deal with the physical environment of the transmission as an uncontrollable aspect of his overall technical model. In the world of *wireline* transmission technology, it is possible in effect to eliminate the physical environment from the model—to assume perfect transmission conditions, and to transfer the burden of the system design into the digital realm of software-controlled, logically definite outcomes. In a *wireless* system, we cannot do this; the system design remains firmly coupled to the real-world physical channel and its imperfections. The burden of ensuring the survivability of protocols, algorithms, and other logical devices deployed in the system cannot be engineered away.

The gap between the two styles of engineering is as great as the gulf between the hothouse gardener and the dry farmer of the high plains. One controls the environment and optimizes his technology under the assumption of near-perfect conditions; the other is forced to anticipate a wide range of contingencies, to select technologies that can withstand a great deal of punishment, and to build in margins. The implications are often not fully appreciated, as we will discuss in Chapter 5.[13] To take but one important

---

rates of raw bit errors. The issue of how strongly and intelligently the countermeasures of the sort discussed in later chapters in this book are applied is a valid one, and it may indeed be possible to sustain strong claims about the post-FEC, post-countermeasures error rate, but that is precisely the point of this book: That it is only by the aggressive employment of such countermeasures that one can overcome the fundamental penalty of the nonengineerable channel.

11. That is, a BER of 10^{-6} for a good-quality mobile channel versus 10^{-11} or 10^{-12} for a fiber link.

12. For example, in a recent article on *wireless LANs* (WLANSs), we find the following: "The IEEE 802.11 WLAN standard, commonly referred to as wireless Ethernet ... is similar to wired Ethernet in that both utilize a 'listen before talk' mechanism to control wireless access to a shared medium. However, the *wireless medium presents some unique challenges* not present in wired LANS.... *The wireless medium is subject to interference and is inherently less reliable* [emphasis added]" [8, p. 65].

example, the Internet-standard protocol, TCP/IP, is designed around an implicit assumption (born of "hothouse engineering") that any disruption in the flow of packets is due to network congestion, rather than to channel errors. TCP is equipped with embedded congestion management techniques that will thus respond to a lost packet by reinitiating the transmission process from scratch, reducing the throughput. This renders TCP/IP, at least in its current hothouse variety, as virtually unsuitable for certain kinds of wireless channels.[14] Similar, or even more serious weaknesses appear in other "hothouse protocols" from the wireline environment, which will make their adaptation to third-generation wireless networks difficult or even impossible (in their current form).[15]

3.1.2 The Nonclonable Wireless Channel and the Challenges of Multiple Access

What happens if we have more than one transmitter? More than one receiver? In the "normal" wireline network, as we have seen, the answer is simple: The channel is cloned, replicated. Build a new house, and the phone company will lay new wires to connect you (Figure 3.1).

This is not possible for the wireless engineer. His channel is finite, and cannot be replicated. This is a simple, obvious physical fact, and there is no

---

13. A bemused, and amusing, comment from the world of fiber optics: "The high performance of optical fiber systems is due in large part to the properties of the fiber itself. Remove the fiber, as in a wireless system, and the stable low-loss guided propagation path is no longer available. Conveying light between the terminal stations in a controlled, reliable manner then becomes a challenge . . . [hence] the transition from optical fiber to optical wireless is not straightforward, and new design solutions are needed" [9].

14. "A satellite channel may exhibit high bit-error rates due to factors including atmospheric conditions, RF interference, a weak signal, and so on. When, due to corruption, a packet is not successfully delivered to the destination and acknowledged, a TCP sender interprets this as network congestion and enters into a congestion avoidance state that can substantially reduce overall throughput. Unfortunately, *there is no way for TCP to know that corruption and not network congestion caused it to reduce its sending rate* [emphasis added]" [10, p. 77]. See also [11].

15. A good example is ATM, which probably cannot survive at all in its current form over wireless channels of even a higher quality than the typical conditions found in a mobile radio link [12]. The subject of protocol impacts of the wireless channel will be dealt with in greater detail in Volume 4 of this series, *Broadband Wireless Applications*.

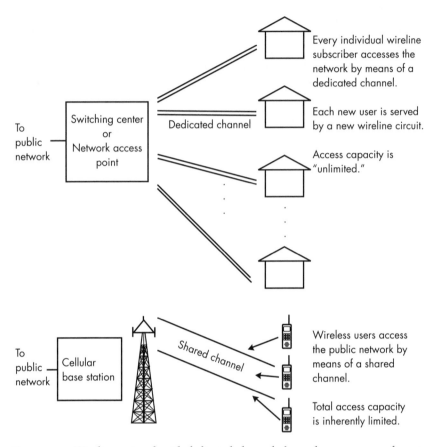

FIGURE 3.1 *Wireline network with dedicated channels for each user versus radio network with shared channels.*

way around it.[16] The engineer has to find some way to *share* the channel resource. He may choose initially to partition the total channel into subchannels and assign these to individual users on a permanent, dedicated

---

16. Subdividing channels on a frequency basis is not equivalent to manufacturing spectrum, as the analysis later in this chapter will make clear. Nor should the opportunity to expand into new frequency regions, which may be opened up by improvements in fundamental technology, be similarly confused with the replication of channels in the sense in which this is done routinely in the wireline network. We should not lose sight of the fundamental fact that spectrum, like land, is a finite physical resource; eventually we run into the limits of expansion, and we must address the issue of the efficiency with which the resource is being employed.

basis. This is how the TV broadcast industry operates, and it is how mobile radio systems used to operate (before trunking was introduced). Once again, however, a partitioned, dedicated channelization scheme pays a tremendous price in terms of inefficiency, at least for typical two-way point-to-point communications traffic. The capacity of nonshared mobile channels is typically less than 10% of the capacity of a shared, trunked system (depending on the number of circuits, the number of users, and the traffic patterns). Eventually, the FCC has decided that nontrunked systems shall be phased out and replaced with trunked systems in which the channel resources are shared by all users on an "as-needed" basis.

Assume then that there are at least two transmitters that desire to share, somehow, the same channel. Intuitively, it might seem that both transmitters cannot be active at the same instant in the same channel, or they will interfere with each other. (This intuition is not entirely accurate, as we shall see, but it is a starting point.) Moreover, a new problem arises: to coordinate the actions of the multiple transmitters so that they do not jam each other. We need some method for deciding how to allow access by any individual transmitter to the shared channel. In wireless architectures, this is called the *multiple access problem*.[17] In wireline networks in which there is a shared medium, such as LANs, the problem is referred to as *medium access control* (MAC).

Multiple access (or MAC) raises many new issues for the system designer. Intuitively, it is clear that the communications system now becomes more complex: There must be a higher layer in the communications protocols between the transmitters and the receivers to govern the sharing of the channel resource. In some systems, this access control can be accomplished by building the intelligence for self-regulation into the transmitters themselves. For example, the Ethernet protocol is an example of a multiple access system that is "uncoordinated," or self-regulating. In the pure contention concept, each transmitter is free to initiate a transmission over the shared channel at any time; if two transmitters happen to begin transmitting at the same time, they are able to detect this collision,

---

17. The multiple access channel presents a much more complex challenge to the Shannon theory: "A system with many senders and receivers contains many new elements in the communication problem: interference, cooperation, and feedback.... The general problem is easy to state. Given many senders and receivers and a channel transition matrix which describes the effects of the interference and the noise in the network, decide whether or not the sources can be transmitted over the channel... This general problem has not yet been solved" [3, p. 374].

whereupon they cease transmitting, "back off" for a certain waiting period (the length of the backoff is selected at random by each transmitter to minimize the likelihood that the same two transmitters will recollide), and then attempt to retransmit the original message. It is an elegant, simple, much modeled, and extremely popular system. It is also, in its pure form, extremely inefficient. More than 80% of the channel capacity is wasted. Simplistic, uncoordinated multiple access algorithms can also develop unstable behaviors under heavy traffic loading.

Most of the architectures we are interested in rely on a central controller to receive and arbitrate access requests to assign "channels" (really portions of the shared channel) to individual users on request, to manage traffic. The controller is typically a separate processor attached to the core network, to which are addressed the access requests, and from which emanate the various instructions involved in assigning and managing shared circuits for specific transmissions. Note that not only does this introduce a new level of control, but it also creates a new set of communications among the transmitters and receivers and the controller. These control and access messages are carried on separate channels (or segregated portions of the shared channel). There are now two kinds of channels, and at least two communications processes, taking place within the system. In fact, the whole architecture of the communications network changes; the "pure" point-to-point configuration is replaced by a double-hop network, in which both links are typically point-to-multipoint links.

The controller's job is to make sure as much of the total channel is utilized for user communications as possible. This problem is clearly located within the Shannon-theoretic framework. The central problem of multiple access design becomes capacity enhancement, the quintessential Shannon concern. Unfortunately, and ironically, the multiple access channel is not nearly as well modeled or understood within the theory. There is a great deal of seat-of-the-pants engineering in the development of multiple access systems, as most wireless engineers are all too aware. There is great debate over capacity claims of different multiple access architectures [13], and at the end of the day most capacity gains still come from engineering tricks and workarounds rather than the formal theory.

Note that the nonengineerable link and the nonclonable link tend to work against each other, narrowing the space within which the system designer can maneuver. If there were no capacity limit—if the channel were not finite and nonreplicable—the link engineer could afford to invest more in the countermeasures against channel errors. By the same token, if the channel were not so challenging, more optimized link protocols and more efficient coding schemes could be used that could alleviate to a greater degree the capacity constraints. Taken together, these constraints define the wireless challenge: to design a communications system that is *simultaneously robust and efficient*. How different it is for our wireline cousins, who do not have to build systems that are *either* especially robust or particularly efficient!

3.1.3 The Conundrums of Interference

In the real world—as opposed to the digital simulacrum constructed by software engineers—*everything radiates*. Every electrical or electronic device and every signal interacts with other signals in its vicinity. The compartmentalization promised by digital techniques is largely an illusion. Not long after Alexander Graham Bell's invention of the telephone, he was astonished to discover just how sensitive the device was. In the course of laying lines to several early customers, Bell found that if two telephone wires were bound together for a distance as short as 1 foot, the lines would induce audible cross-talk in each other even in the absence of a direct electrical connection. This led Bell to his second greatest invention, largely unsung: the twisted pair. By twisting the two wires around one another, Bell found that he could effectively cancel the effects of the mutual induction and eliminate the cross-talk phenomenon. Yet still he marveled that a signal of such a small magnitude—mere ghosts of ghosts, as he called them—could indeed carry information structures and interfere with the performance of his device. Of course, the problem did not end there. The bundling of large numbers of conductors into cables—a virtual necessity in economic terms as the telephone became popular, especially in urban areas—posed severe interference problems, cross-talk, and mutual induction. These problems were a continuing engineering challenge for decades, and were the subject of a series of engineering conferences held during the early years of the telephone industry, in the continuing search for cable designs that would work despite the intense interference environment they created.

Over time, and at great expense, the wireline world has gained a sense of mastery over the phenomenon. With the advent of T-carrier and its "miracle" of digital regeneration, and especially as fiber optics developed into the preferred trunking technology to replace all those bundled cables with hair-thin conduits of glass and EMI-proof optical signaling, a feeling has developed that interference has finally been vanquished. Bell's elusive ghosts are being dispelled, and as we go forward into the fiber age, wireline transmission can be assumed to be interference-free.

Leaving aside the question of how valid this view may be for wireline transmission,[18] it once again clearly demarcates the wireless world as an altogether different and wilder communications environment. In wireless transmissions, the signal is vulnerable to countless sources of uncontrollable interference in the channel and the environment itself. It is also subject to interference from other signals that are sharing the same communications channel (or portions of it, demarcated in time and frequency, for example). Most importantly, inescapably, *every wireless transmission interferes with itself*. The channel inevitably generates multiple paths of transmission between transmitter and receiver, resulting in multiple time and phase shifted images of the signal at the receiver. These images interact and can result in the near cancellation of the entire signal. We shall delve into this further in the next section.

The most interesting aspect of the technology revolution underlying what we have called the "second golden age of wireless" is *a changing attitude toward this problem of interference*. The traditional approach to interference, from Alexander Graham Bell on down, was to suppress it and to design communications systems to keep different transmissions well separated in space, frequency, and time—precisely in order to keep the level of interference under control. A huge penalty has been paid, in terms of the unused capacity ceded to create the physical margins or cushions between signals. As we shall see, a prominent objective of the third-generation systems now emerging is to recapture this lost capacity for the benefit of the users and system operators. To do this, it is necessary to rethink our approach to the issue of managing interference. The results of this rethinking are leading to some of the most intriguing new communications architectures, based on virtually turning the traditional interference management strategies on their heads.

Once again, before we delve more deeply into these intriguing new areas, it bears repeating that the wireless engineer has a different perspective on interference than his or her wireline cousins. He or she does not have the luxury of pretending to have succeeded in suppressing it. For the wireless engineer, interference—especially self-interference—is the cardinal

---

18. Clearly interference has only been vanquished if we restrict our interest to classical *electromagnetic interference* (EMI) from external sources. As we shall develop later in this chapter, the most important component of interference in modern systems is self-interference, in various forms. Consider the abiding problems of managing inter-symbol interference in digital transmission systems of all sorts, as one token of this abiding problem.

problem of the channel itself. The strategy he or she chooses for dealing with this interference is the most important design choice he or she will make. The major architectural alternatives—TDMA, CDMA, and so on—are defined by their choice of strategies for handling interference. The insatiable appetite for efficient utilization of the finite channel capacity is driving wireless architects to become ever bolder in their approach to the problem. It is not too strong to say that the more we can tame—not eliminate, but tame—interference in a wireless system, the more traffic-carrying capacity we can extract.

3.2 The Basic Parameters: Channel, Signal, and Noise

Thus, the charter for the wireless systems designer is to develop a communications system that is both robust in the presence of errors *and* efficient in its use of the finite spectrum channel, and which pays particular attention to the problem of managing the ever-present phenomena of interference. Optimizing his design under these three constraints is what sets him apart from his counterparts designing wireline systems, for whom none of these issues is paramount.

To organize the discussion of the various technological options, I want to recharacterize the signal/channel model of communications, in the very spirit of Shannon himself, as a geometrical model—a "space" defined by a number of relevant "dimensions." The selection of appropriate dimensions may illuminate important aspects of the issues at hand.

The idea of visualizing the communications process in geometrical terms was introduced by Shannon in his 1949 paper "Communication in the Presence of Noise" [14]. Essentially, it allowed the communications theorist to bring to bear a vast body of geometrical mathematics to probe different aspects of the communications process, with the added advantage of presentational clarity that geometry often affords. It has proven to be a very suggestive and fruitful metaphor for thinking about the topics of interest to Shannon, such as the design of coding schemes, signal detection, the treatment of errors, the efficiency of channel utilization, and so forth.

Shannon used geometry as a metaphor for talking about the structure of signals and messages. I want to use it to discuss the channel itself, within which those messages are transmitted, and within which they are exposed to such severe challenges. I would argue that another way of characterizing the charter of the wireless systems engineer is, simply, that he or she must *confront the physics, and the physical limits, of the wireless channel.* For Shannon, operating at a level of greater generality and abstraction, the channel was the least significant element in his model (and the smallest, unlabeled box in his famous diagram). However, if for wireless applications the channel is paramount, then it may be fruitful to start with the physical characteristics of the channel model.[19]

3.2.1 The Primary Signal/Channel Dimensions: Space, Time, and Frequency

A radio signal occupying an available wireless channel can be specified physically in terms of space, time, and frequency. These are the *primary dimensions* of the channel and the signal, and define a three-dimensional wireless communications space (Figure 3.2). We can locate and define the channel, and the signal within the channel, in terms of three-dimensional zones or regions or *subspaces*. Two different signals can often (but not always) be distinguished from one another by the fact that they occupy different subspaces.

This communications space is quite similar to the ordinary world of space–time that we perceive all around us. It is also related to Shannon's geometrical constructs of the *signal space* and the *message space*. Shannon presents his signal space, for example, as a product of the frequency bandwidth of the message W, and the time interval T, leading to the observation (deriving from a discussion of the sampling theorem) that "the $2TW$ evenly spaced samples of a signal can be thought of as co-ordinates of a point in a space of $2TW$ dimensions" [14, p. 32]. In creating his communications geometry, Shannon ignored the dimension of *physical* space because, in all likelihood, he was relying implicitly on wireline models, where the spatial dimension is much less meaningful (each wire is virtually its own universe, for communications purposes).[20]

It may be asked whether there are other relevant physical dimensions that belong in the primary set. One likely candidate is signal *polarity*. Systems have been described, and to some degree implemented, that rely on

---

19. Whether or not this approach is suitable for more general analysis of the communications process is a question beyond the scope of this volume, certainly, and quite likely beyond my scope as an author to fully assess. In 20 years of working with "advanced" radio systems architectures, I have become convinced that we can simplify and clarify a great deal if we begin thus with the physics of the channel.

20. As well, the use of "space" as both metaphor (the product of the selected dimensions) and as a real physical variable may tend to suppress its prominence in the abstract formulation of the theory.

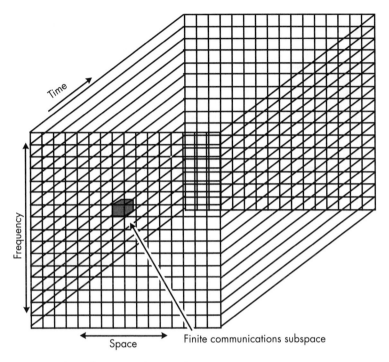

FIGURE 3.2 *A three-dimensional view of communications space.*

only a difference in polarization to distinguish between two signals that are in other respects (space, time, and frequency) identical. It has also been proposed as a dimension for the deployment of *diversity*—as a countermeasure against noise and interference—where it would take its place alongside the use of diversity in the other primary dimensions (see Chapter 5). Logically, polarization could be viewed as a primary physical dimension of the channel and the signal; however, I will leave it aside for now due to its relative novelty in current communications architectures.

The value of this space–time–frequency model is that it provides us with an easy visualization of the three-dimensional playing field for the wireless communications process. In any given application (e.g., cellular mobile radio), the available band is definable initially as a three-dimensional block of S, T, and F coordinates. Carving up this resource into individually assignable channels can be represented as the three-dimensional gridding out of the communications space into smaller, generally contiguous blocks, each defined in terms of S, T, and F coordinates.

Let us trace a single communications session (e.g., a cell phone call) through this communications space. It begins at a given time, in a given spatial zone (in this case, a given cell), on a particular assigned frequency. As time progresses, the call is traced through the communications space as shown in Figure 3.3: The S and F dimensions remain fixed, while T moves forward incrementally and continuously. At some point, a mobile call may reach the edge of a cell and be handed off to another subspace. There would be a jump from one S coordinate to another. It is likely that the handoff will also occasion a shift in the frequency (new F coordinates). We can also envision how a call in a simple TDMA cellular system would be mapped into this space: The T coordinates would not be continuous, but periodic.

This begins to give us a picture of what the system controller, which manages the multiple access process, is up to. The plotting of these circuits, as shifting patterns of S, T, and F, defines the available channel resources of the system to carry user traffic. Each call traces its path through this total communications space.

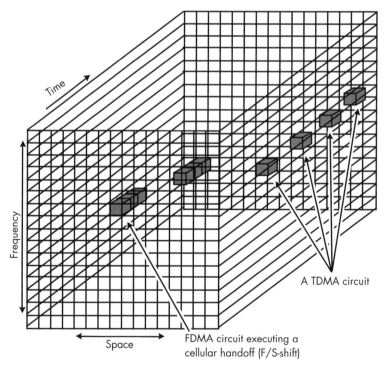

FIGURE 3.3 *Connection-oriented calls traced through the communications space.*

3.2.2 Noise and Interference

If we consider the total communications space, divided into STF subspaces within which unique signals can be inserted, we should ideally like to find each subspace completely clean, empty of any electromagnetic energy except for the signal that we wish to insert into it: an empty box to receive the payload.

Of course, this is never the case. There is *always* unwanted electromagnetic energy in the subspace. For one thing, there is always noise. The universe is apparently full of random, low-level electromagnetic energy, which forms a constant background to all communications processes. Additionally, we can observe that many separate processes interact in the sequence of producing and transmitting a message, but two sequential stages are often only "loosely coupled," which leads to a degree of uncertainty about the input/output relationships between two variables. In other words, the process itself is somehow error prone. This need not refer to electromagnetic disturbances; for example, the relationship between the typist's brain and his fingers is one such loosely coupled process, which can introduce errors into the transmitted message. In any case, by definition noise is unpredictable and it "produces a small region of uncertainty about each point" [14, p. 33] in the signal space, but it is *characterizable*, as a statistical distribution.

Most communications theorists do not devote much attention to the physical sources of this noise (although this can sometimes be relevant for explaining its special statistical properties). Noise is simply a statistical uncertainty or perturbation factor that is somehow added to the signal during its passage through the channel.[21] The particular statistical distribution of the additive noise is important because it determines the nature of the impact on the coding scheme employed. For example, if the noise statistics are independent from one sample of the signal to the next—that is, if the

---

21. For example, one classic information theory paper considers the following as a reducible case of signal and noise: "An atrociously bad telegrapher, despite his intention of sending a dot, dash, or pause, will actually transmit any one of an infinite variety of waveforms only a small number of which resemble intelligible signals. But we can account for this by saying that the 'channel' between the telegrapher's mind and hand is 'noisy,' and, what is more to the point, it is a simple matter to determine all the statistical properties that are relevant to the capacity of this 'channel.' The channel whose message ensemble consists of a finite number of 'intentions' of the telegrapher and whose received signal ensemble is an infinite set of waveforms resulting from the telegrapher's incompetence and noise in the wire is thus of a type [already] considered ..." [15, p. 92]. The modern reader may sense the rather thin ice underlying these speculations.

effects of noise on successive samples **are not** correlated—and they also share a certain distribution of amplitude values, then the noise assumes a particular pliable distribution known as *white Gaussian noise:* "A white thermal noise has the property that each sample is perturbed independently of all the others, and the distribution of each amplitude is Gaussian with standard deviation ..." [14, p. 36]. Most of the theory to date has assumed channels where the additive noise meets this criterion.

At least in passing, however, we should point out that the classical additive white Gaussian noise channel is rarely encountered in practice (except for certain satellite channels), and few would argue that typical terrestrial wireless channels are of this type.[22] Instead, the noise that is added to the land mobile channel by sundry partially specified processes is often highly correlated from sample to sample—that is, it is bursty, producing bursts or clusters of correlated errors.[23] This is highly relevant to the choice of countermeasures.

More importantly, unwanted energy is present that derives from other transmissions in other subspaces. In terms of our framework here, this energy has overflowed from signals occupying an adjacent communications subspace. Indeed, we must acknowledge that no signal is ever perfectly contained in its designated subspace. All signals will spill over to some degree (Figure 3.4). This spillover is the source of interference.

The importance of interference is that whereas noise can be assumed to be structureless, random, uncorrelated—even if this assumption later must be modified for non-Gaussian channels—the same is never true of interference. Interference always involves the interaction of the desired signals with another (undesired) signal, which has by definition a certain structure. This structure may have a much greater negative impact on the

---

22. Fading channels (that is, those affected by multipath self-interference) deviate from the white Gaussian model. The following comment is from one of the father figures of the field of coding theory: "It is worth noting that the world is getting less white and less Gaussian. Adjacent channel interference, accidental or deliberate interference, antenna misalignment all give conditions that are not intrinsically suited [to countermeasures which assume white Gaussian noise statistics in the channel]" [16, p. 49].

23. For example, Hamming states: "The model usually assumed for errors in a message [is]: (1) an equal probability p of an error in each position, and (2) an independence of errors in different positions. This is called "white noise" in a (poor) analogy with white light.... But in practice there are often reasons for errors to be more common in some positions in the message than in others, and it is often true that errors tend to occur in bursts and not be independent (a common power supply, for example, tends to produce a correlation among errors; so does a nearby lightning strike)" [2].

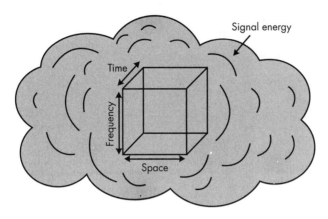

FIGURE 3.4 *Signal communications subspace energy spills beyond the finite.*

desired signal than its raw power level might indicate. For example, the phenomenon of crosstalk in improperly shielded wireline circuits can result in a very disruptive effect even at extremely "good" signal-to-interference power ratios. The same amount of interfering energy presented as random noise may be much more tolerable.

We may ask *why* the signals spill over into adjacent subspaces, and whether all signals are equally sloppy. There are indeed specific causes for many forms of signal spillover, and by understanding these causes we can design less sloppy signals; this is one broad and significant class of technological countermeasures (see Chapter 6).

The STF dimensions suggest a taxonomy of interference types, depending on which dimension is involved.

1. *F-interference:* spillover from one subspace to another along the frequency dimension; spillover from one frequency channel to an adjacent channel. This is normally referred to in wireless communications as *adjacent channel interference* (ACI).

2. *S-interference:* spillover from one spatial zone to another, such as when interference occurs between cells in a cellular system. This is normally referred to as *cochannel interference* (CCI).

3. *T-interference:* There are several manifestations of interference in the time dimension, including spillover between time slots in a TDM system, for example, or collisions between packets in a CSMA system; at the lowest level of the digital signal, an

intractable form of T-interference is *intersymbol interference* (ISI), which may be caused by a number of factors (see Chapter 7).

This taxonomy in turn suggests a preliminary classification of the traditional countermeasures. For example, F-interference (ACI) is traditionally addressed through the design of filters at both the transmitter and the receiver. T-interference (ISI) is addressed through equalization. S-interference has been handled, where the application permits, by the use of directional antennas. (We will address traditional, and nontraditional signal grooming techniques in Chapter 7.)

There are several reasons for emphasizing interference in the analysis of countermeasures:

1. Modern, high-capacity wireless systems are interference limited, not noise limited. That is, taken together, the effects of interference tend to be greater, more destructive to the desired signal, than conventional sources of additive noise.

2. Interference gets worse as we load the system with user traffic; noise, *qua* noise, does not.

3. Interference gets worse as we introduce new architectures to pack the signals more densely within the communications space; noise, *qua* noise, does not.

4. Many forms of interference, including all those described so far, are amenable to effective countermeasures; we can improve filtering to reduce F-interference, for example. By definition, there is much less that we can do about noise *in terms of eliminating its physical sources*.

In short, interference is a problem that grows in significance for third-generation high-capacity wireless architectures, as they pack signals and user traffic ever more densely in the available channel space. It is also a problem that we can do something about (again unlike most sources of noise).[24] A consideration of these possibilities will form the basis of a large set of interesting countermeasures (see Chapters 6 and 7).

---

24. Note that the terms *noise* and *interference* are sometimes used imprecisely and interchangeably. For example, wireless engineers commonly speak of the *noise floor* of a wireless base station facility, when they actually are referring to interference generated from other transmitters occupying the same facility. The unwanted energy is

3.2.3 Self-Interference

Managing the spillover of interfering signals from adjacent subspaces offers substantial gains in system performance, and such countermeasures play an important role in designing third-generation wireless systems. Unfortunately, there is an even more fundamental interference phenomenon that is not amenable to a direct frontal assault with straightforward solutions such as better filters or directional antennas: *In a wireless channel, the desired signal also interferes with itself.* This mysterious-sounding phenomenon is rooted, again, in the physics of the wireless channel, especially for mobile applications.

The transmitter radiates the signal in a wide beam, in many cases omnidirectionally. The signal is reflected by many features of the environment—buildings, vehicles, and hills. This creates a *multipath* geometry, in which the signal actually arrives at the receiver by many different pathways (Figure 3.5). These paths have different lengths, which means that the signal images arrive at slightly different times. The process of reflection may also change the phase of the signal, so that different images arrive in different phase relationships to each other. Thus, the received signal is actually a composite of multiple images of the transmitted signal, which are spread out in time (called the *delay spread*). The differing phases are summed at the receiver, which means that if, for example, we have two images—say, the image arriving on the direct path, which is in a given phase, and an image arriving by a reflected path that has been phase-shifted by 180°—the two images would mutually interfere and, in fact, would cancel each other.[25]

---

coming from known structured sources of interference. This suggests another difference: Interference sources can sometimes be turned off; noise sources cannot be.

25. I am assuming that the reader is generally familiar with the simple dynamics of interfering waveforms. "When two or more wave motions are present at the same place and time, the simplest assumption is that the resultant displacement is the algebraic sum of the individual displacements...." Regarding multipath in particular, according to *Encyclopedia Britannica,* "A particularly severe form of frequency-selective fading is caused by multipath interference, which occurs when parts of the radio wave travel along many different reflected propagation paths to the receiver. Each path delivers a signal with a slightly different time delay, creating 'ghosts' of the originally transmitted signal at the receiver. A 'deep fade' occurs when these ghosts have equal amplitudes but opposite phases—effectively canceling each other through destructive interference. When the geometry of the reflected propagation path varies rapidly, as for a mobile radio traveling in an urban area with many highly reflective buildings, a phenomenon called fast fading results." Here is a classical definition of

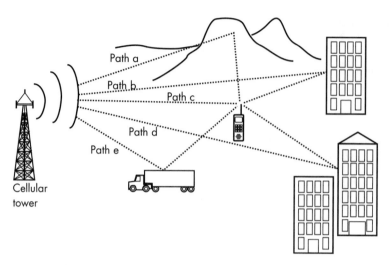

FIGURE 3.5 *Multipath geometry.*

There are other ways in which a signal may be said to interfere with itself (ISI could be considered a form of self-interference, for example). Multipath fading is by far the most important interference effect for many wireless applications, especially terrestrial mobile applications like cellular radio. Unlike other forms of interference, it cannot be simply suppressed at the source. In the case of multipath interference, the source is the channel itself, which, as we have stressed already, the wireless engineer must accept as she finds it. If we hold frequency constant, the fading phenomenon can be mapped out in the spatial domain as a series of *holes* or *fades*—points in physical space where the particular configuration of the physical environment (i.e., a particular wireless channel) interacts with the signal from a particular transmitter to produce a more or less static distribution of high

---

the phenomenon from one of the textbooks of the field: "At any point the received field is made up of a number of horizontally traveling plane waves with random amplitudes and angles of arrival for different locations. The phases of the waves are uniformly distributed from zero to 2. The amplitudes and phases are assumed to be statistically independent" [17]. Finally, the recognition that this is best viewed as self-interference is expressed succinctly by Viterbi, "Another serious [problem for mobile radio systems] is the self-interference caused by multipath. Whenever transmissions, propagating over two or more paths, arrive at the receiver separated in time by less than the inverse bandwidth of the receiver, there will be frequent instances of severe fading caused by phase cancellation between multipath components" [18].

and low signal amplitude summation points. The antenna of the receiver, if it is attached to a moving vehicle, passes through these fade points and produces a fluctuating signal level with a characteristic distribution of fades of varying depths. The effect is a highly varying SNR, variable bit error rates, and diminished signal quality (Figure 3.6). On the other hand, if the antenna is stationary—for example, if a person using a portable phone is standing still—the antenna may enter a deep fade and lose the signal altogether. (These fades may be as much as 40 dB down from the nominal signal level.)

There is much debate about when, and how severely, multipath interference affects the wireless channel, depending on frequency, bandwidth, and other parameters. We shall consider some of these positions in due course. It is fair to say that the approach chosen to mitigate self-interference becomes the cardinal attribute of any wireless architecture. External forms of interference can often be dealt with by means of measures external to the basic communications system—better shielding, additional filters, different antennas. The application of such measures is largely indifferent to the structure of the transmitted signal. However, self-interference, based on multipath propagation and manifest in delay spread, fading, phase distortion, is so inherent in the transmission and the channel itself that the countermeasures cannot simply be "bolted on," but must be "designed in." Self-interference is closely linked to the information structure of the transmitted signal itself. In this light, those who claim that multipath is not a problem for their particular system usually mean that they have already embedded what they take to be strong countermeasures in their design.

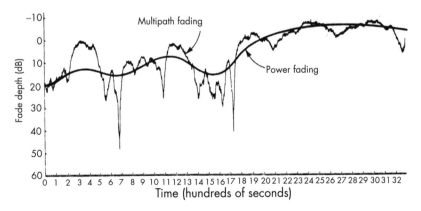

FIGURE 3.6 *Fluctuating signal amplitude due to fading. (From: [19].)*

3.2.4 Orthogonality

The traditional solution to managing external interference—that is, the external forms of STF spill over from adjacent signal subspaces—is to create buffers in all three dimensions. The objective is to keep all transmissions well separated from one another. Each subspace is intended to be occupied by only one signal, and the profile of this signal (measured along any one of the three basic dimensions) is reduced so that it is significantly less than the total volume of the subspace. In short, the signal is designed to fit within its targeted subspace with plenty of room to spare (Figure 3.7).

An architecture of this sort is often referred to, somewhat unhelpfully, as *orthogonal*. This simply means that there is no designed-in interference between different signals (leaving aside the issue of self-interference, which cannot be avoided by buffering). Each communications subspace (or channel) within the total communications space defined by the STF dimensions is uniquely assigned to only one active signal.

The use of orthogonal architectures is logical, straightforward, and has a long history. The early initiatives toward regulation of the wireless industry were focused on this issue—to make sure that users stayed in their own channels and did not cause destructive interference to other users. The FCC's posture toward wireless regulation almost down to the present day has been dominated by this concern, by the need to establish and police orthogonality. Interference avoidance is so strongly entrenched in the

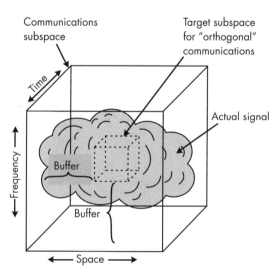

FIGURE 3.7 *"Orthogonal" signal within its communications subspace.*

wireless world that the idea of a nonorthogonal system seemed for a long time to be simply inconceivable, or at least strongly counterintuitive.

The problem is that orthogonality imposes a high price in lost capacity. Consider the standard example of leaving vacant TV channels between broadcast channels in use in a given city. That is, if TV channel 3 is assigned and in use in Philadelphia, channel 4 is taken out of the Philadelphia allocation permanently. It is left unused as a guardband for channel 3. This means that for every megahertz of occupied spectrum (in the broadcast TV application), there are two more megahertz (one on either side) that are unusable, because of the need for the guardband to protect against F-interference.

We will come back to this issue of the cost of orthogonality in Chapter 4, but it should be intuitively apparent that if we are looking for significant gains in capacity from a third-generation wireless system, one of the most promising places to look is at the unused buffers introduced by today's requirements for orthogonality. Indeed, although the framework that we are developing seems perhaps to enshrine the principle of orthogonality in the notion of dimensionality (STF), we will be using it eventually to set the stage for the discussion of nonorthogonal solutions.

3.2.5 Secondary Signal Dimensions: Amplitude and Angle

Once we have defined the communications subspace that defines where we expect to find the desired signal, we can look more closely at additional physical characteristics of that signal. For wireless signals, there are classically two such parameters of interest (Figure 3.8):

1. *Amplitude:* The power level of the signal, measured in watts; or, viewed in terms of the electromagnetic waveform, the height of the wave crest from the nominal baseline or zero point.

2. *Angle:* As reflected in either the instantaneous "slope" of the signal measured at a given point, stated in degrees of arc, which is called the phase of the signal, or, over a period of time (such as one second), in the number of cycles inscribed by the radio carrier, which is the frequency of the signal.[26]

---

26. Traditionally frequency and phase are regarded as separate parameters, and this may be appropriate especially where analog FM is being treated; however, many presentations now observe that digital implementations of both FM and *phase modulation* (PM) involve the modification of the angle of the signal on a very short-term basis to

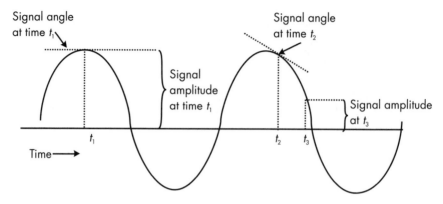

FIGURE 3.8 *Sine wave showing signal amplitude and angle (phase and frequency).*

These are normally the information-bearing dimensions of the signal. The message is encoded, or modulated, into the radio carrier wave by *varying* either the amplitude, the frequency, or the phase of the carrier. In analog modulation methods, this variation is continuous, in response to a continuous input (such as a voice signal). The two classical analog modulation methods are *amplitude modulation* (AM) and *frequency modulation* (FM). These are both old, established technologies that have been often described—and it is worth keeping in mind that even today when the interest and excitement is focused on newer technologies, many communications systems are still based on either AM or FM.

The trend, however, is decisively in the direction of *digital* transmission schemes, where the modulated information takes the form of discrete steps or shifts in either amplitude or angle, or both, representing specific digital symbols (such as a 1 or a 0) that are used to encode the information provided from the original source (which may be a voice signal, or some other kind of signal). In the realm of digital modulation, the preferred approaches to date have tended to be phase modulation schemes such as *binary phase shift keying* (BPSK) and *quadrature phase shift keying* (QPSK), or combined phase-and-amplitude modulation such as *quadrature amplitude*

carry the desired information, and FM methods and PM methods tend to blur together as they are employed in digital transmission systems. This has led some authors to prefer to group all of these techniques under the heading of "angle modulation," which is the basis of the presentation here.

modulation (QAM). In QPSK, for example, the signal information is encoded as a series of 2-bit symbols; there are four such 2-bit permutations—00, 01, 10, 11—and each of these permutations is mapped onto a phase value, such as 0°, 90°, 180°, or 270°.

The signal space for QPSK is represented in Figure 3.9. The nominal transmitted values are indicated in the middle of each quadrant. The receiver measures the received signal phase at the appropriate sample time, and locates it in the same signal space. Whichever quadrant it falls into—even if it has been knocked around a bit by the channel—determines the value that the receiver assigns to the received symbol. Note that in QPSK the signal amplitude—how far the detected signal is from the center of the space—is not a relevant information-bearing dimension. This means

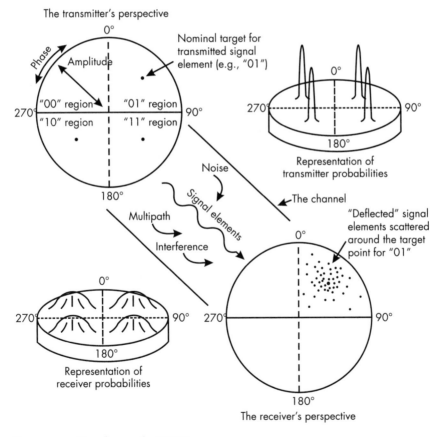

FIGURE 3.9 *Signal space for QPSK.*

that QPSK is largely immune to the random, unavoidable variations in amplitude introduced by multipath interference, as long as they are small.

QAM, in comparison, uses both phase and amplitude to define the relevant zones in its signal space (Figure 3.10). This can produce a more efficient transmission scheme—with more capacity, among other things—but it also means that QAM is potentially vulnerable to multipath-induced amplitude variations.

The choice of the method for imprinting the user information onto the signal has significant bearing on the robustness and capacity of the transmission. It also affects the amount of interference generated by the transmitted signal. Designing the secondary signal dimensions is thus one of the most important fields for technological innovation for high-capacity third-generation systems.

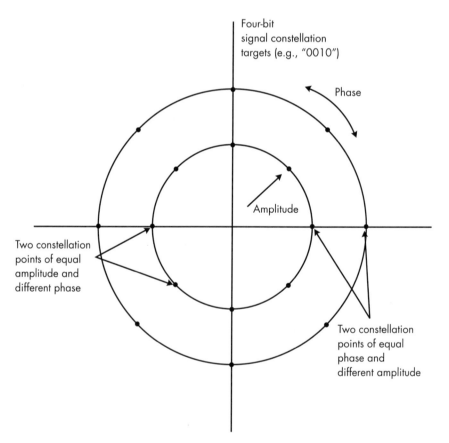

FIGURE 3.10 *Signal space for QAM.*

3.2.6 Tertiary Signal Dimensions: Signal Structure

The STF dimensions tell us where to look for the signal. The amplitude and angle parameters will allow us to read the raw information content of the signal—the digital bitstream.

The next relevant dimensions are those that organize the general structure of the transmission (viewed over a somewhat longer time window than the conceptually instantaneous sampling process that generates the raw information bits themselves). A single sample is meaningless unless its position in the larger signal structure is known. For example, a particular digital symbol detected by the receiver may be part of a framing sequence of bits; or it may be part of a pilot signal designed to provide the receiver with certain calibration information; or it may be part of an error correction block; or it may be a "guard symbol" that is not designed to carry any information per se, but merely to provide a T-buffer between meaningful symbols from two T-distinct transmissions that have the same S and F coordinates. It may also even be part of the payload, the meaningful component of the message that carries the semantically significant information. To understand how to handle and interpret the raw bits that are detected by the receiver sampling process, we have to know what sort of package to expect.

Consider the analogy of a letter sent and received through the mail. This letter arrives in a sealed envelope, and there are several different formatted aspects of the information package: the address of the recipient, which is itself divided into several fields (name, street, zip code); the return address of the sender; the stamp, which indicates proper payment for the delivery; the post office cancellation, which certifies the delivery on a particular date; inside, the letter itself contains other elements of message structure (headings, page numbers, salutations), to enable the reader to properly identify the semantic payload. An express delivery may include other structural elements, such as a signature exchange to verify delivery, and a bar code to allow tracking of the package through the transmission network.

A wireless signal typically has a significant amount of formatting overhead structure of the same sort: addresses, identifiers of sender and receiver, sequence markers, starting sequence and ending sequence markers, frame markers, and so forth (Figure 3.11). In addition, other structural elements may be associated with the embedded countermeasures against the problems encountered in the channel (such as noise and self-interference). These include:

- *Synchronization sequences:* to enable the receiver to acquire a synchronized clock reference, so that its sampling rate is aligned with the transmitter's symbol transmission rate;

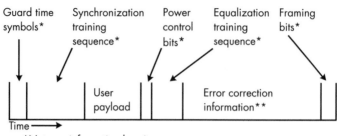

FIGURE 3.11   *Wireless signal format with various types of overhead.*

- *Error detection and correction:* additional information attached to the payload that is designed to allow recovery from a certain number of channel errors;

- *Pilot signals:* special sequences that are inserted to allow the receiver to maintain calibration of some important signal parameter, such as the amplitude of a fading signal in a modulation system (like QAM) that needs to track amplitude because it is an information-bearing parameter;

- *Power control:* an embedded control signal moving from the receiver back to the transmitter that causes the transmitter to adjust its output power;

- *Training sequences:* for example, the equalization algorithm (in the receiver) may require periodic calibration or training, involving the transmission of special bit sequences specifically for that purpose;

- *Hopping sequences:* in some systems, the transmitter and receiver shift from subspace to subspace, in either the frequency domain (frequency hopping) or the time domain (time hopping), and the transmitter and receiver need to exchange and synchronize information about the hopping sequence.

None of this is connected with the semantic payload per se. It is all "overhead", apparently reducing the traffic carrying capacity of the channel. Viewed from a somewhat broader perspective, each form of overhead is thus an investment—a certain amount of capacity is invested to support a particular countermeasure, and whether the investment is a good one or not depends on the often highly complex model of the payback in terms of improved performance. The designer faces numerous trade-offs. A system that can acquire synchronization more quickly, with fewer special synchronization symbols, will be capable of carrying more traffic than another system that takes longer and needs more overhead to achieve the same result, but the fast synch process may require more processing horsepower, and it may be less reliable under poor channel conditions.

These examples begin to suggest the complexity of the architecture of contemporary wireless systems. Indeed, probably the majority of new technological innovations in recent years have focused on building up the structure of the signal package, generally in order to gain greater control over channel errors and various forms of interference (especially self-interference).

3.2.7 Quaternary Signal Dimensions: The Structure of the Payload

After all of the structural countermeasures have been deployed, the ultimate package that determines whether the communication has been successful or not is the semantic payload. This is the user-originated message that contains intelligible, useful information of value to the receiver. In commercial systems, the payload is what the user expects to pay for. It is the *meaningful* portion of the signal.

A striking commonplace in many presentations of Shannon theory is a blunt denial that the theory has anything to do with meaning at all. In the opening paragraphs of his first and most famous paper, Shannon himself announces that "Frequently ... messages have meaning ... [but] these semantic aspects of communication are irrelevant to the engineering problem" [20].[27]

This overstated position is both true and not true. It is true that the focus of the theory is on the processing of messages, rather than the process of communication (which would, of course, address intelligibility, meaning, and many other semantic aspects). However, it is not true that all semantic

---

27. See also [2, p. 1]: "Although the text uses the colorful words 'information,' 'transmission,' and 'coding,' a close examination will reveal that all that is actually assumed is an information source of symbols s_1, s_2, ..., s_q. At first nothing is said about the symbols themselves, nor of their possible meanings. All that is assumed is that they can be uniquely recognized. We cannot define what we mean by a symbol. . . . Since we must use symbols to define what we mean by a symbol, we see the circularity of the process. Next we introduce the probabilities p_1, p_2, ..., p_q of these symbols occurring. How these p_i are determined is not part of the abstract theory. One way to get estimates of them is to examine past usage of the symbol system and hope that the future is not significantly different from the past."

aspects are "irrelevant to the engineering problem." In the follow-up paper, Shannon turns to the problem of transmitting an analog source—such as a voice telephone call—and admits that in fact "exact transmission [of such a source] is impossible" because it would require a channel of infinite capacity [21][28]. The solution is to introduce a *fidelity criterion*, which in effect specifies when a reproduction of an analog source is "good enough." By truncating the potentially infinite range of resolution associated with an analog source to a finite, quantized image managed as though it were a digital source, Shannon manages to bring all types of signals under the great tent of his theory. He convinces himself, and much of the field remains convinced, that once the fidelity criterion has been received (from somewhere outside the theory itself!), the rest is all mathematics [21, p. 26].

From where does the fidelity criterion derive? After discussing a number of purely formal candidates, such as the average error value of the received signal compared to the transmitted signal, Shannon finally lifts up the flap to let the camel in:

> The structure of the ear and brain determine implicitly an evaluation, or rather a number of evaluations, appropriate in the case of speech or music transmission. There is, for example, an "intelligibility" criterion in which $p(x,y)$ is equal to the relative frequency of incorrectly interpreted words when message $x(t)$ is received as $y(t)$. *Although we cannot give an explicit representation of $p(x,y)$* [emphasis added] in these cases, it could, in principle, be determined by sufficient experimentation. [21, p. 26]

In short, the criterion of "quality" or "intelligibility" is not after all a mathematical formula, but a mere "finding" of experimental psychology. In fact, it is worse: The quality criteria used in the field today are derived from loosely controlled subjective evaluations by small panels of human subjects, based on methodologies that are wide open to substantial criticism on grounds of lack of rigor, repeatability, and so on. The elaboration, defense, and critical analysis of various bases for different versions of the "fidelity criterion" are an emerging focus of research activity in a number of different applications.

---

28. At greater length: "A continuously varying quantity can assume an infinite number of values and requires, therefore, an infinite number of binary digits for exact specification. This means that to transmit the output of a continuous source with *exact recovery* [emphasis in the original] at the receiving point requires, in general, a channel of infinite capacity (in bits per second). Since, ordinarily, channels have a certain amount of noise, and therefore a finite capacity, exact transmission is impossible" [21].

I hope that the reader is not confounded prematurely by all this. We will return to this issue in a more tutorial fashion in Chapter 4. The point is that the actual payload of the transmitted signal may also be manipulated and engineered in order to mitigate some of the problems of the wireless channel. One very simple trade-off is *quality for capacity*; in general terms, the higher the quality needed by the application (especially analog sources such as voice, audio, video), the more channel capacity the signal will use, and the fewer users can be accommodated in the communications space. This is why aggressive source-coding strategies like very low bit-rate voice coders found their first real application in wireless systems.

3.2.8 The Layered Signal and the Primacy of the Physical Layer

In recent years the concept of a layered network architecture has become an occasionally useful generalization of the relationship between different aspects of the communications process. Essentially, a layered model such as the common seven-layer model (Figure 3.12), portrays an attempt to partition the engineering effort. In particular, the distinction between events, processes, and technologies operating at the *physical layer* and those

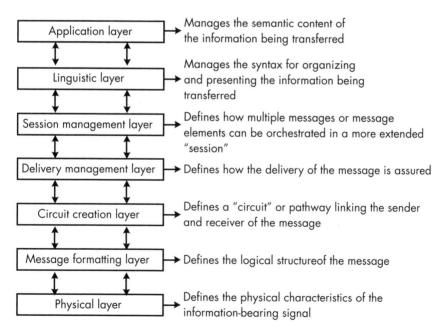

Application layer	Manages the semantic content of the information being transferred
Linguistic layer	Manages the syntax for organizing and presenting the information being transferred
Session management layer	Defines how multiple messages or message elements can be orchestrated in a more extended "session"
Delivery management layer	Defines how the delivery of the message is assured
Circuit creation layer	Defines a "circuit" or pathway linking the sender and receiver of the message
Message formatting layer	Defines the logical structureof the message
Physical layer	Defines the physical characteristics of the information-bearing signal

FIGURE 3.12 *Seven-layer communications model.*

operating at the high layers is often relevant to understand the problems that we face.

The physical layer is where the signal actually interacts with the channel, and where the issues of capacity, noise, interference, and self-interference originate. The physical layer is rooted in the physical reality of the signal and the channel. All the other "layers" reflect man-made (and thus arbitrary) distinctions drawn to help manage commercial interfaces. Whether seven, ten, or four in number, the layers are only useful if they provide a framework for classifying signal processing strategies.

The countermeasures that we shall catalog in Chapters 4 through 8 are thus said to be deployed at different layers. Some are physical layer implementations, such as the choice of modulation. Some are said to be implemented at layer 2—often called the data-link layer—like some forms of error correction. Some are implemented at the session or application layer, such as the use of discontinuous transmission for bursty sources like voice telephony. Such statements may sometimes be useful, but in the final analysis, the post-Shannon revolution has to do with a new understanding of the physical layer and how the information structures are laid down directly upon the physical substrate.

The model that I have outlined here is intended to clarify the presentation of the new spectrum of countermeasures that is revolutionizing the design of wireless systems. I find the notion of layering less useful than the idea of dimenionality of the signal. Once again:

- *The primary dimensions of the signal:* space, time, and frequency—tell us where to locate the signal in the total communications space;

- *The secondary dimensions of the signal:* amplitude, frequency, and phase—these are the information bearing parameters that will allow us to read the stream of bits being sent by the transmitter (in a digital transmission);

- *The tertiary dimensions of the signal:* those that have to do with the packaging of the signal, the structure that defines the relevant characterics of the signal to the receiver, and allows the receiver to reproduce with the best possible tools the original transmitted message; and

- *The quaternary dimensions of the signal:* those that are related to the semantic payload itself.

3.3 First- and Second-Generation Wireless Architectures

Before proceeding to the main topic—the third-generation wireless technologies, strategies, and countermeasures to deal with the unique challenges of the wireless channel—it is worthwhile to spend a moment justifying and clarifying the phrase *third generation* or 3G. What does it really mean? What were the earlier generations that we are now happily superseding?

3.3.1 First-Generation Systems: Power Versus Noise

The first radio systems were based on the principle of overcoming propagation losses, receiver insensitivities, and, if need be, interfering transmitters with sheer power. The first commercial wireless services were based on "spark" telegraph systems that created essentially small electromagnetic explosions to blast their Morse code pulses out in all directions as far as possible. Carrier-based modulation schemes—initially AM, later FM—combined with tunable transmitters and more sensitive receivers allowed the focusing of the transmitter power in the frequency domain and by the 1930s and 1940s the first modern mobile radio systems were developed [19]. For the next several decades, the basic architecture was refined within an essentially static technology framework.

For these systems, the main communications strategy was simply *more power*. Transmitters were placed in high towers and operated at high power. User terminals were large and "hot" 5-W, 10-W, and 20-W transmitters.[29] System engineering was one-dimensional. Too much noise? More power. Bad terrain? More power. Higher antennas projected the signal as far as possible....[30] Among other things, this also biased users against the higher frequencies, because of the declining return on investment from the "more power" scenarios.

The capacity penalty was very large, particularly in the S-domain. Transmitters high atop city skyscrapers were able to blast out 40 or 50 miles, saturating their frequency band over huge geographical areas with just one

---

29. Such systems are still in use widely by police and other industrial users. In late 1998, a policeman in New York City used a mobile radio to subdue a fleeing suspect—throwing the radio at the suspect (who was on a bicycle) and striking him in the head. (The suspect later died.)

30. This is not to deny that "equivalent distance at lower power" was not often preferred; the attractiveness of FM over AM was in part due to the lower power required to achieve a good SNR.

user signal. By the early 1980s, the congestion in mobile radio was so severe that there was capacity for only a few thousand users in a given area [19]. Even the introduction of trunking in the late 1970s did little to alleviate the situation. Mobile radio was essentially self-blocked from achieving commercial viability, by the inefficiencies of its transmission strategies.

3.3.2 The Cellular Revolution: Reorganizing the S-Dimension

The bottleneck was finally unblocked by the so-called "cellular" innovation, first deployed in the early 1980s in the United States. The basic concept had been proposed in papers from AT&T in the late 1940s (published internally at almost the same time as Shannon's first papers on communications theory). It rested on what was then a highly counterintuitive proposal: to improve performance by *reducing* the base station transmitter power in order to create multiple smaller coverage zones (cells) and to enable the reuse of the same frequencies in different cells in a metropolitan service area. This simple concept has become the basis of a huge industry.[31]

The less obvious and less felicitous implication of embracing the cellular concept was that effectively a new and problematic framework for designing, engineering, measuring, and expanding a mobile radio system was foisted on us. In the precellular systems, signals were said to be noise limited. This meant that in the space domain one could define coverage in terms of the SNR, which could be modeled statically as a function of the propagation losses due to the terrain. Engineering was simple and straightforward: More power meant better coverage. Higher towers meant better coverage. With the advent of the cellular concept, power and height were no longer beneficial, but harmful. The limit on coverage in a cell was no longer noise per se, but interference—S-interference, cochannel interference, from other cells. Engineering an interference-limited system posed a whole new set of problems. Indeed, designing a large cellular system has become a complex and dynamic problem. In fact, it is an NP-Hard problem, akin to the famous Traveling Salesman problem, where NP-Hard refers to the fact that although in principle a solution to an optimal design may exist, it is computationally unfeasible to find it.[32]

---

31. The cellular concept has been described in many places and will not be represented here. See [19].

32. According to *Encyclopedia Britannica*, "So-called easy, or tractable, problems can be solved by computer algorithms that run in polynomial time; i.e., for a problem of size n, the time or number of steps needed to find the solution is a polynomial function

Essentially, the cellular architecture is based on a technical strategy of gaining capacity by managing the S-dimension of the communications space more intelligently, at the cost of a *much more complex overhead* (cell planning, handoff implementation, backhaul networks). The gain in capacity is difficult to quantify, but is probably at least an order of magnitude greater than precellular mobile systems (allowing for the fact that cellular was staked to a relatively large new spectrum allocation at its launch). However, as cellular pushed hard on the S-dimension, the problems and challenges of interference loomed larger and came to dominate the engineering problem. In addition, the industry found that the full exploitation of the cell-splitting strategy was hampered by the statistical physical characteristics of the channel, the environment, and the phenomena of multipath, which combined to impose a sort of practical asymptotic limit on cell splitting. (The original forecasts of cells being split down to extremely small radii had to be tempered; also, the industry economists noticed that their infrastructure costs had a geometric growth path as cells shrank and multiplied. For both technical and economic reasons, the cellular approach of focusing strictly on the S-dimension did not carry us as far as we might have originally hoped.)

3.3.3 Second-Generation Systems: Digital Extensions—Reorganizing the T-Dimension

This set the stage for the emergence, beginning in the late 1980s and only gathering real momentum in the mid-1990s, of the so-called "second-generation" architectures. These were marked by the application of digital transmission techniques, which permitted further gains in capacity by allowing a restructuring of the *time domain* of the signal. Systems like GSM, developed in Europe and deployed eventually worldwide, took advantage of well-understood digital multiplexing procedures to *time share* the cellular channels among multiple active transmitters. Essentially, *time-division multiple access* represented a logical, straightforward extension to the wireless world of time-division techniques that had been in use in the wireline world since the 1960s (albeit with a few unique challenges in the transition to wireless). By the late 1990s, TDMA-based architectures had achieved

---

of *n*. Algorithms for solving hard, or intractable, problems, on the other hand, require times that are exponential functions of the problem size *n*." Grefenstette and colleagues have stated: "The TSP is NP-Hard, which probably means that any algorithm which computes an exact solution requires an amount of computation time which is exponential in N, the size of the problem" [22, p. 532].

preeminence as the direct successor to first-generation analog FM-based cellular systems.[33] However, the industry kept growing, and even the gains of three to eight times analog cellular capacity that could be garnered from TDMA soon appeared inadequate against the background of the wireless demand surge described in Chapter 1.

3.3.4 On the Threshold of 3G

The innovations of the 1980s and 1990s have nudged the wireless industry away from some of the old assumptions—in particular, reversing the assumption that more power is always better—toward interference-limited networks and the acceptance of digital source coding techniques. Great gains in spectrum efficiency have been realized by better management of the S- and T-dimensions of the communications space. Although it is difficult to quantify, I believe that the achievable capacity of a wireless system today is somewhere between 20 and 50 times higher than it would have been in 1980, measured in terms of erlangs per megahertz.[34]

These have been the easy gains, technologically speaking. Cellular concepts were middle aged by the time they were first applied, and digital transmission was almost three decades old before it came into use in wireless to any significant degree. The technological underpinnings of these strategies were well understood. The gains are not enough. The wireless industry is facing the practical requirement to increase capacity by another factor of 10 or more in the coming decade, just to keep pace with demand, and the potential impact of the Internet is a wildcard that can only point upward, traffic-wise.

So far, all the systems we have discussed strictly preserve the principle of orthogonality. They manage interference (other than self-interference) by buffering, suppressing, or avoiding it. They employ system designs that

---

33. TDMA is described in detail in both of my earlier books: *Digital Cellular Radio* [19] and *Wireless Access and the Local Telephone Network* [13].

34. Calculations of the gain from cellular reuse architectures are subject to complex sets of assumptions, especially regarding the minimum feasible cell size. Different assumptions can support different opinions. I base my estimate on the assumption that with 400 channel pairs (approximately the current cellular allocation at 800 MHz), a nonreuse system could support about 12,000 mobile telephone users (at a loading factor of 30) in a given metropolitan area. The implementation of reuse, plus a digital gain factor of, say, 3 from vanilla TDMA, brings me to my calculation of the gain factor. It may be somewhat low, although the loading factor may also be low (in practice, the ratio seems to be higher).

divide the communications space into subspaces that are uniquely assigned to one and only one signal (per cell, per frequency channel, per time slot). With regard to self-interference, they use fat margins to stay above the fading threshold as much as possible. They may throw in a little error correction. Otherwise, the countermeasures are traditional, and probably operating at the limits of their effectiveness.

As we move into the third generation, we shall see two important technological trends: First, the assumptions of strict orthogonality will be questioned. Greater levels of interference will be not only tolerated, but designed in. Indeed, some of the third generation architectures will move away from orthogonality altogether. Second, we will see an intensive effort to develop new types of countermeasures to enable systems to operate under conditions of much higher levels of interference.

References

[1] Girod, B., and Niko Färber, "Feedback-Based Error Control for Mobile Video Transmission," *Proc. IEEE*, Vol. 87, No. 10, October 1999, p. 1707.

[2] Hamming, R., *Coding and Information Theory,* Englewood Cliffs, NJ: Prentice-Hall, 1986.

[3] Cover, T., and J. Thomas, *Elements of Information Theory,* New York: Wiley, 1991.

[4] Ackoff, R. L., *On Purposeful Systems,* New York: Wiley, 1971.

[5] Shannon, C. E., "Communication in the Presence of Noise," *Proc. IRE*, Vol. 37, January 1949, pp. 10–21.

[6] Taub, H., and D. L. Schilling, *Principles of Communications Systems,* 2nd ed., New York: McGraw-Hill, 1986.

[7] Gray, R., *Source Coding Theory,* Boston, MA, and Dordrecht, Netherlands: Kluwer, 1990.

[8] Heegard, C., et al., "High Performance Wireless Ethernet," *IEEE Communications Magazine*, Vol. 39, No. 11, November 2001.

[9] Heatley, D., J., et al., "Optical Wireless: The Story So Far," *IEEE Communications Magazine*, December 1998.

[10] Metz, C., "TCP over Satellite ... The Final Frontier," *IEEE Internet Computing*, January/February 1999.

[11] Partridge, C., and T. Shepherd, "TCP/IP Performance over Satellite Links," *IEEE Network*, Vol. 11, No. 5, September/October 1997, pp. 44–49; see also Jacobson,

V., et al., "TCP Extensions for High Performance," 1992, available at http://info
.internet.isi.edu:80/in-note/rfc/files/rfc1323.txt.

[12] Pasternak, E., "Transmission of ATM and Frame Relay Traffic Using Line of
Sight Radio Links," White Paper, Netro Corporation, May 10, 1996, available at
http://www.netro-corp.com/Products/NTRAM.pdf.

[13] Calhoun, G., *Wireless Access and the Local Telephone Network,* Norwood, MA:
Artech House, 1992, Chapter 6.

[14] Shannon, C. E., "Communications in the Presence of Noise," *Proc. IRE*, Vol. 37,
January 1949, pp. 10–21. Reprinted in Slepian, D., (Ed.), *Key Papers in the Development of Information Theory*, Piscataway, NJ: IEEE Press, 1974, pp. 30–41.

[15] Feinstein, A., "A New Basic Theorem of Information Theory," *IRE Trans. on
Information Theory*, Vol. IT-4, September 1954, pp. 2–22. Reprinted in Slepian,
D., (Ed.), *Key Papers in the Development of Information Theory*, Piscataway, NJ:
IEEE Press, 1974, pp. 81–101.

[16] Berlekamp, E. P., et al., "The Application of Error Control to Communications,"
IEEE Communications Magazine, Vol. 25, No. 4, April 1987.

[17] Reudink, D. O., "Properties of Mobile Radio Propagation Above 400 MHz," *IEEE
Trans. on Vehicular Technology*, November 1974, pp. 143–159.

[18] Viterbi, A. J., "The Orthogonal-Random Waveform Dichotomy for Digital Mobile
Personal Communication," *IEEE Personal Communications*, Vol. 1, No. 1, 1992,
p. 19.

[19] Calhoun, G., *Digital Cellular Radio,* Norwood, MA: Artech House, 1988.

[20] Shannon, C. E., "A Mathematical Theory of Communication (Part 1)," *Bell System Technical Journal*, Vol. 27, July 1948, pp. 379–423. Reprinted in Slepian, D.,
(Ed.), *Key Papers in the Development of Information Theory*, Piscataway, NJ: IEEE
Press, 1974, pp. 5–18.

[21] Shannon, C. E., "A Mathematical Theory of Communication (Part 2)," *Bell System Technical J.*, Vol. 27, October 1948, pp. 623–656; Slepian, D., (Ed.), *Key
Papers in the Development of Information Theory*, Piscataway, NJ: IEEE Press,
1974, pp. 19–29.

[22] Grefenstette, J., et al., "Genetic Algorithms for the Traveling Salesman Problem,"
originally published in 1985. Reprinted in Fogel, D. B., (Ed.), *Evolutionary Computation: Selected Readings on the History of Evolutionary Algorithms*, Piscataway,
NJ: IEEE Press, 1998, pp. 532–540.

4

THIRD GENERATION SYSTEMS: PHYSICAL LAYER TECHNOLOGY STRATEGIES

4.1 From Interference Avoidance to Interference Management

Pre-Shannon wireless architectures are based on the principle of orthogonality, or interference avoidance, as discussed in the previous chapter [1]. Within the total space–time–frequency communications space, each subspace is uniquely assigned to carry only one signal. Any energy from other signals that leaks into a given subspace is classified as interference, to be suppressed.

This orthogonality is achieved in practice by designing the signal to be smaller (or narrower) in all three dimensions—space, time, and frequency—than the boundaries of the subspace to which it is assigned. That is, we create buffer zones between the signals: frequency buffers or guardbands in the F-domain, time buffers or guard times in the T-domain, and spatial buffers (e.g., geographical separation of transmitters) in the S-domain.

The problem—in terms of the Shannon framework, where *capacity* is the name of the game—is that these buffers represent a significant reduction in the percentage of the total communications space that is available to carry user traffic. Indeed, the amount of capacity lost (or "invested") to achieve orthogonality is often much larger than we may realize.

4.1.1 The Penalty for Orthogonality

Let us try to quantify the cost of achieving sufficient orthogonality[1] in each of the three primary dimensions of the signal.

T-Orthogonality

Achieving proper separation of temporally adjacent signals sharing the same frequency, in the same cell, is a standard problem for any wireless TDMA architecture. The need for the buffer between adjacent time slots is derived from at least three distinct requirements:

1. *Ramp-up and ramp-down of bursty signals:* In a TDMA channel (let us consider the uplink as the pure case), each mobile unit transmits in periodic bursts. These bursts do not physically begin and end instantaneously, but require a certain short but measurable period of time to ramp up and ramp off. Moreover, in some cases, the ramp-up is purposefully prolonged in order to allow the transmitter hardware designer a greater latitude on certain specifications (that may allow lower cost implementations, for example) or to mitigate potential interference deriving from sharp signal amplitude events.[2] This ramping process takes time.

2. *Different propagation delays:* Two mobile units assigned to adjacent time slots may actually be located at quite different distances from the cell site. The difference in propagation delay may be many tens of microseconds between a unit located very close to the base station receiver and one located at the fringe of the cell (each mile accounts for about 5 μs of propagation delay). Depending on the transmission rate in the channel, this may cause a potential shift equal to one or more data symbols. For example, in the original

---

1. Because in the real world there are no relevant absolute limits on interference in any of the three basic dimensions—all signals are physically *infinite* in space, time, and frequency from a theoretical standpoint—the definition of orthogonality subsumes a criterion of the amount of reduction of the interfering signal as a function of space, time, or frequency. For example, one may assert the achievement of practical orthogonality in the space domain if the interfering signal is reduced below an appropriate threshold—say, 18 dB or so in an analog FM mobile radio system at 800–900 MHz—relative to the strength of the desired signal.

2. In general, as discussed in Chapter 6 in more detail, sharp transitions in signal amplitude create more interference (in the F-domain) than smooth transitions, and are to be avoided.

U.S. TDMA standard (IS-54), the symbol time is a little more than 20 μs, which means that a difference between "near" and "far" users in adjacent time slots of more than 4 miles is sufficient to cause an overlap of these transmissions. In the *Global System for Mobile Communications* (GSM) standard, with a much higher transmission rate, the symbol time is less than 4 μs (Figure 4.1). Thus, if time slots were packed right up against each other, significant amounts of data could be lost. T-buffers must be provided.

3. *Synchronization:* Although synchronization is apparently concerned with receiver performance rather than buffering per se, the extra overhead bits needed by the receiver (and provided by the transmitter) to reacquire synchronization at the beginning of each transmitted segment cannot be used for carrying user information. It is a part of the price of achieving orthogonality in the time domain. In a *frequency division multiple access* (FDMA) system

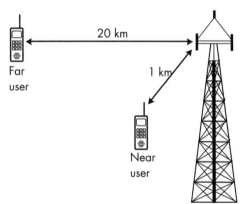

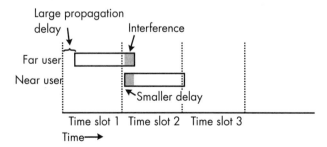

FIGURE 4.1 *Different propagation delay between signals originating from near mobiles and distant mobiles.*

with continuous transmission (for the duration of a session), synchronization is acquired once and is then maintained as a by-product of decoding the user data stream. That is, there is no additional overhead involved (after the first acquisition, which forms a negligible portion of the total transmission). In a TDMA system, the synchronization overhead must be reinvested every single time slot, and thus it can accumulate to a nonnegligible proportion of the total transmission.[3]

The structure of typical second-generation TDMA channels reflects the allocation of channel capacity to all three of these T-buffer requirements [4]. For the U.S. digital standard known as IS-54, the structure of the reverse link (mobile to base) results in an allocation of channel capacity as shown in Figure 4.2. According to this analysis, "the overhead part could be defined to be ... 16%" [5, p. 325]. The corresponding figure for the GSM channel is almost 30%.[4] In other words, the most popular second-generation TDMA systems operating today invest *between one-sixth and one-third of the total channel capacity* to buffer the signal in the time-domain (holding F and S constant).

F-Orthogonality

While the management of the time dimension of communications space is a relatively recent development in wireless systems, the apparent need to

---

3. During the process of choosing the standard for digital cellular in the United States, this extra TDMA overhead (compared to FDMA) was adduced by supporters of an FDMA standard as one their main advantages. See, for example, [2]. Later Lee even argued that this was one of the relative disadvantages of TDMA versus CDMA. Lee holds that one of the key advantages of CDMA is that there is "no guard time in CDMA: the guard time is required in TDMA between time slots. The guard time does occupy the time period for certain bits. Those waste bits could be used to improve quality performance in TDMA. In CDMA, the guard time does not exist" [3, p. 300]. Of course, the scope of the TDMA guard time and synch penalty is no worse than the investment in the pilot required for a CDMA channel—but that is another controversy altogether.

4. The GSM overhead figure includes training sequences for the GSM equalizer, which is necessitated in part by the intersymbol interference created by the propagation delay problem noted here (as well as the multipath delay spread). It could be argued that this is not really a part of the T-buffer; it could also be argued that it is, since it is necessary to the successful use of the TDMA channel and not as much channel capacity would need to be invested in this training sequence in an FDMA implementation, even at the same bit rate [5, p. 323].

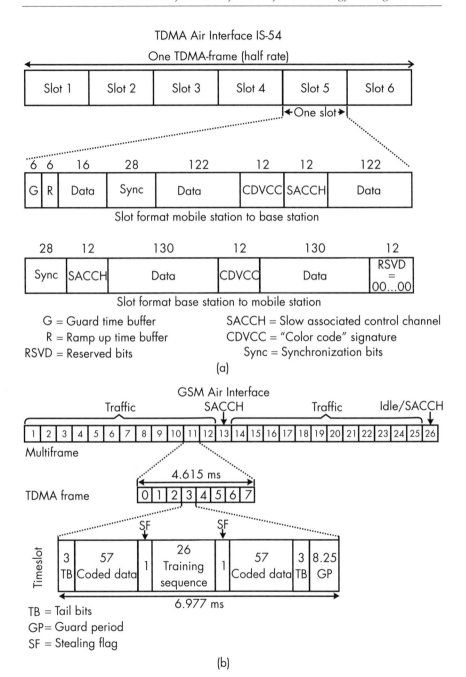

FIGURE 4.2 *Allocations of channel capacity to speech codec, forward error correction (FEC), and more for (a) IS-54 and (b) GSM. (Source: [5]. © 1986 IEEE. Reprinted with permission.)*

separate different transmissions in the frequency domain is as old as the commercial radio industry itself. We are all familiar with the concept of dividing the communications space into distinct *channels*—individual frequency bands uniquely assigned to different transmitted signals. This model has been employed for the regulation of spectrum used by TV and radio broadcasters, mobile radio operators, and probably most other commercial wireless systems (until very recently).

It is also generally accepted that *immediately adjacent* channels are often not usable in the same coverage area (S-dimension) at the same time (T-dimension). In other words, holding S and T constant, there is often a need for a buffer between adjacent F-subspaces (channels). Thus, we understand that if TV channel 3 is used for broadcasting in Philadelphia, we should not expect to find channel 2 or channel 4 in use. Similarly, until recently most, though not all, mobile radio systems were designed also to use nonadjacent channels in a given coverage area (e.g., a given cell), and to leave unused channels between as a buffer.[5]

This buffer is needed because all modulated carriers tend to spread out from the center carrier frequency. In addition, frequency-domain filters at the receiver are imperfect. The portion of the signal that spreads or spills over into adjacent frequency channels is the main source of adjacent channel interference in most wireless systems.

It may not be appreciated how large this required buffer can be. In the first-generation cellular architecture known as AMPS (based on FM signals), the nominal occupied bandwidth of the transmission is 30 kHz. However, the rolloff or power spectrum of the AMPS transmission is relaxed (to say the least), and the result is that to achieve the required ACI isolation of 64 dB it becomes necessary to impose a separation of *four unoccupied channels* between any two occupied AMPS channels in a given cell. This is not a function of cellular reuse (although it influences reuse planning); it is the result of the radio hardware design.[6]

---

5. One could say that nonadjacent channelization schemes were almost universal in mobile radio and related wireless applications. The only exceptions that I am familiar with until very recently were the *basic exchange telecommunications radio service* (BETRS), a wireless local loop service in the 454-MHz band (with initially 26 contiguous channels of 25 kHz); and the 900-MHz special mobile radio band (with bundles of five contiguous 25-kHz channels). I had the opportunity to confront the need for designing equipment to operate on an adjacent channel basis in both bands, for two different business applications.

6. See the analysis of ACI and frequency buffering requirements in [6]. This article dates from the analog era.

This is an enormous penalty. Instead of occupying 30 kHz, each AMPS transmission effectively occupies (in the sense of denying to other users) a full 150-kHz of frequency bandwidth.[7] This means that in the F-domain *nearly 80% of the channel capacity is wasted* on the buffering function to achieve F-orthogonality. Clearly, other systems exist that are less inefficient in this respect, although the need for some kind of F-buffer cannot be avoided in F-orthogonal systems.[8]

S-Orthogonality

The most celebrated innovation in the wireless world during the past 20 years has been the revolution in the management of buffering in the space domain—the advent of so-called "cellular" systems.

In the precellular era, spatial buffering was achieved by allowing the normal propagation losses (and, at higher frequencies, the curvature of the Earth) to attenuate the signal over relatively long distances, until at some point it was possible to reuse the original channel without mutual interference.

The cellular concept replaced this with a new spatial reuse concept based on small coverage zones—cells—and lower transmitter power. The idea has been described in many places, and we need not review the entire cellular idea in detail here. From this emerged a new methodology for designing the S-buffer, as reflected (in its orthodox version) in the writings of Bill Lee.[9] Following Lee, a set of simple calculations—starting from the

---

7. It is true that the inefficiency of this F-buffer is partially blended with, or masked by, the S-buffer in a cellular system—the reuse factor. It is also true that the use of such a relaxed F-buffer is possible in part because of the reuse scheme, and a much less relaxed RF design could have been mandated (at a higher cost). The interdependence of the F-buffering and S-buffering strategies in cellular architectures is clear, and it means that as systems designers attempt to squeeze capacity out of one of them, they will have to face the need to reengineer the other.

8. The most efficient F-buffer that I am aware of in a conventional, commercial wireless system was probably the BETRS system developed by IMM Corp. (later InterDigital Corp.), which had an occupied bandwidth of 20 kHz in a 25-kHz channel, such that the penalty is effectively about 20% of the total channel capacity. To achieve this, a highly linear system design was required with many unusual (and costly) features.

9. As CTO of Airtouch, Lee has been at the forefront of the standards wars in the wireless industry for many years. His many books and articles provide comprehensive, if somewhat idealized, treatment of the basic cellular concepts. A concise handling of the reuse-related principles is found in [6]. I would consider this version "orthodox" in the sense that it presents a view of the problem that is highly abstract, and in

stipulated signal-to-cochannel-interference ratio of 18 dB—leads first to a required spatial separation calculation (the so-called "D/R ratio"), and from there to the calculation of a cellular reuse pattern (N). This parameter, N, stands for the number of subsets into which the total set of available frequencies must be divided to ensure adequate S-buffering. It determines the framework for the frequency plan, and in principle dominates the entire configuration, including setting the upper limits on capacity and the basic system economics. In Lee's abstract, simplified model, using hexagonal cells, the reuse factor is set at 7.

Let us take these numbers at face value for a moment and consider the implications. Essentially, N becomes the denominator for the fraction of the available communications space (in the S-domain) that can actually be used for carrying traffic in any given cell. In other words, as a first approximation, if N = 7, then 1/7 of the communications space is being used to carry user traffic and the rest is invested in creating the S-buffer. This may sound like an overstatement (what about the capacity gained through reuse?), but it is not. The gains from reuse are still purchased at the price of leaving vast amounts of the communications space empty. Consider that if it were possible to reuse a given frequency in *every* cell (i.e., if N = 1), the number of channels available in a cellular system would increase by a factor of 7 over Lee's model. This holds true no matter how many cells there are in the system. Indeed, it is precisely this possibility—to achieve N = 1—that drove the huge capacity gains originally claimed for CDMA systems.[10]

The enormous cost of S-orthogonality is partially masked by the fact that through clever engineering the S-buffer can also be used as the F-buffer, or blended with it. Thus, because we must skip six channels out of every seven in any given cell, we can afford to use a very relaxed RF specification, and use the fallow channels also to provide the F-buffer against adjacent channel interference. The fact remains that even in the best orthogonality schemes in deployment today, huge amounts of spectrum that

---

some ways impractical (with assumptions of perfect hexagonal cells, smooth distributions of interferers, simplified geometries).

10. For example, Viterbi identifies the improved reuse factor as the dominant advantage of CDMA over FDMA and TDMA: "Most important is the gain in 'Reuse Factor,' a cellular parameter that is inversely proportional to the number of different frequency assignments necessary to guarantee that neighboring cells are assigned to disjoint frequency bands. Reuse factors of 1/19, 1/12, 1/7, 1/4, and 1/3 have been used or proposed for progressively more optimistic designs of FDMA and TDMA systems. CDMA systems employ ubiquitous frequency reuse and thus have a factor of 1" [7, p. 229].

could potentially be employed to carry user traffic are held out of service in order to provide the buffers for that orthogonality (and it makes little difference whether we attribute a particular unused channel to the buffer we need in the spatial domain or the buffer we need in the frequency domain).

I want to make two additional comments about the cellular concept and the penalty in the S-domain.

First, the classical hexagonal $N = 7$ geometry that has been portrayed so often[11] understates the interference problem based on average-case assumptions about interference, and not on worst-case scenarios that actually prevail in the field. In orthogonal systems, the *worst-case rules* the engineering equation.

This fact is widely recognized by field engineers. I am not sure how many systems today are using a reuse of 7, but I am aware of quite a few that use significantly larger values for the parameter N. In anecdotal discussions with a number of European operators, reuse factors of 12 to 16 were often cited—and this was for GSM systems, which use digital transmission and should be somewhat more resistant to cochannel interference. In any case, to the extent that nonideal conditions force cellular operators to adopt more conservative reuse patterns, the amount of idle spectrum in the communications space goes up as a percentage of the whole.

Second, the $N = 7$ calculation is based on Lee's *signal-to-interference requirement* (SIR) of 18 dB, which is "based on subjective tests and the criterion that 75 percent of the users say voice quality is 'good' or 'excellent' in 90 percent of the total covered area on a flat terrain" [6, p. 50]. Compare this "standard" against the voice quality standards of the wireline industry, and one can see that eventually the cellular operator may be forced by the market to improve this grade of service, and design for higher SIRs and poorer reuse factors. One may get by initially with a certain level of service, but this does not mean that the design can be sustained as the industry matures and becomes competitive.

Even the large price we appear to pay for spatial orthogonality in today's systems may be in some sense nominal. If we take into account implementation losses in the field, both natural and human-made, as well as the pressure to improve quality for the end user, it is likely that the S-buffer penalty is significantly understated by the textbook models described and referenced here.

So what is the total cost to achieve orthogonality? In today's first- and second-generation cellular systems, it would appear than the T-buffer takes

---

11. Including my own interpretations, in earlier books.

16% to 30% of the theoretically available capacity, and the F-buffer and S-buffers each would take up to 80% or more of the remainder, although there is some sharing of the buffer between the F-domain and the S-domain. Even if we assume that the S- and F-buffers are completely conjoint, the implication would be, as a sort of best case, that something like 85% of the communications space is wasted (or "invested") to achieve orthogonality for signals in the remaining 15%.

However, this number needs further adjustment, For, in addition to the obvious buffers, the signal itself is already partially invested in various countermeasures against cochannel and adjacent-channel interference. For example, the GSM signal uses a voice coding rate of 13 Kbps, to which is added error protection of 9.8 Kbps. This is a kind of built-in buffer, which effectively reduces the traffic carrying capacity of the channel even further in order to try to ensure orthogonality. In fact, of the total signal bit rate of 33.85 Kbps (per voice circuit), only 38% is real user payload information. The rest is devoted either to time-domain buffering or FEC (which is a form of embedded buffering). Combine this with an $N = 9$ deployment, for example, and GSM is actually making use of less than 5% of the available communications space to carry user traffic. The other 95% is being used to create the buffers for the orthogonality of the user signals.

It is not my intention here to get into a prolonged numbers game; I will concede that all of the specific factors are arguable. The drift of the calculations is clear: A very large proportion of the total available communications space is lost to implement the principle of orthogonality. It is also true that any system could gain capacity by, for example, reducing the bit rate of the codec—from 13 Kbps (which was more or less the benchmark for toll-quality voice 10 years ago) to perhaps half that today. However, we need gains of 5, 10, and 20 times the carrying capacity of the current systems and a mere doubling of capacity pales in comparison to the potential gains that might be had if we could find a way to reduce significantly the need for these massive buffers propping up the orthogonality of the individual signals.

4.1.2 Taming the Interference

What is required is a different approach to the problem of interference. Instead of fighting it at every turn (and mostly retreating from it, ceding the bulk of the communications space), we need to design wireless signals and systems to *work with* a much higher level of interference. More than that, we need to *design the interference itself*. This counterintuitive proposal is based on a certain logic. First of all, if we design something *in* to a signal, we can design it *out* later. That is, if we add to our desired signal a *known* interfering signal, we can use that knowledge to subtract the interference—perfectly in principle.

Fine, of course, but why add the interference in the first place? It turns out that adding the right kind of interfering signal can strengthen the transmission significantly, allowing it to withstand much higher levels of external interference than would otherwise be the case.

Second, interfering signals, which possess by their nature an inherent structure (whether designed in or encountered at large), can often be dealt with more effectively by *understanding* their structure and designing specifically targeted countermeasures. To take one important example, the self-interference created by multipath propagation has a definable structure—in theoretical parlance, it is non-Gaussian—which can be exploited against it. We know, for instance, that multipath-induced fades have a narrow frequency characteristic. If the transmitted signal can be *expanded* so that it occupies a larger frequency bandwidth than the bandwidth of the fade, the effects of multipath self-interference will be reduced.[12]

This approach is controversial. As we shall see later in this chapter, a disagreement in principle exists between those who view noise as the worst enemy of a communications system (essentially because it cannot be predicted) and those who view noise as "the most benign [form] of interference" [7, p. 229] or, to put it another way, between those who view the *structure* of non-Gaussian interference as the key to neutralizing it completely and those who would prefer to *remove* its structure and convert it into a structureless noise-like signal, and attack it with classical antinoise countermeasures. This debate has been largely resolved in favor of the latter camp (although there are still strong advocates of the minority view). We return to this debate in Section 4.5.

Increasing wireless capacity dramatically will require such conceptual boldness—to tame, and manage interference in its the various manifestations. Smart countermeasures have proliferated during the past 20 years or so into a cookbook of techniques employed opportunistically until now; with third-generation systems, we will see a more integrated design strategy.

These techniques can be divided into three broad categories:

1. *Signal hardening techniques:* These are countermeasures designed to increase the resistance of the transmitted signal to specific

---

12. This is, of course, one of the principles underlying *spread* spectrum architectures, as we shall discuss later in this chapter and in greater detail in Chapter 8.

forms of interference, allowing a reduction in the buffering required (Section 4.2 and Chapter 5).

2. *Signal shaping techniques:* These countermeasures are intended to reduce the profile of the transmitted signal, without compromising its information-bearing properties, so that all the signals in the system create less overall interference to one another (Section 4.3).

3. *Signal recovery techniques:* These techniques are applied typically by the receiver to clean up and restore a damaged signal, which again allows the transmission to absorb a higher level of interference and reduces the buffering requirements (Section 4.4 and Chapter 7).

All of these techniques point the way toward architectures that allow for signals to be packed much more densely in the STF communications space. Buffers are reduced, and the proportion of the space actually used for transmitting user traffic can increase. These techniques can also be applied, in many cases, without abandoning—*conceptually*—the goal of orthogonality.

However, at some point the whole frame of reference shifts, from orthogonality to something different. We are now seeing the emergence of the first truly unified approaches based on designing systems incorporating such countermeasures in a comprehensive, interlocking manner. These are the systems that we refer to as true post-Shannon architectures. Abstracting from the ad hoc successes of specific techniques, these architectures are based on the conscious understanding of a new set of design principles (Section 4.5 and Chapter 8).

4.2 Signal Hardening Techniques

Any signal has a certain threshold in terms of its SIR—its relative strength compared to an interfering signal—which it needs in order to provide acceptable quality at the receiver. Once such a criterion is defined, it dictates the amount of buffering that will be required to ensure orthogonality.

For example, Lee defines the SIR criterion for analog cellular (AMPS) as 18 dB.[13] From this "magic number," he derives a reuse pattern (the

---

13. In other words, the signal needs to be approximately 90 times stronger than the interference. This criterion is set, perhaps somewhat dubiously, on the basis that "normal cellular practice is to specify SIR to be 18 dB or higher based on subjective tests

S-buffer requirement), the adjacent-channel interference requirement, the need for channel skipping (the F-buffer), and many other features of the wireless network configuration.[14] Obviously, if the 18-dB threshold could be lowered—that is, if the signal can be processed by the transmitter so that it can tolerate a higher level of interference—the buffering requirements could be reduced and, hence, capacity gained. If the signal could be suitably hardened to withstand interference levels corresponding to an SIR criterion of, say, 10 dB to 12 dB instead of 18 dB, the reuse factor could be reduced for S-buffering from $N = 7$ to $N = 4$ or $N = 3$.[15]

There is an expanding portfolio of apparently disparate techniques that promise such gains. Chapter 5 offers a more detailed catalog; however, certain basic categories of signal hardening countermeasures stand out, including error correction, diversity techniques, and (partially overlapping both categories) convolutional techniques. Let us touch on these briefly.

4.2.1 Error Correction (Channel Coding)

A straightforward method of hardening the signal is to add redundant information to the transmission that allows the receiver to detect and correct channel-induced transmission errors. Many different coding techniques are available, but all have the effect of (1) expanding the size of the total information payload, typically by 50% or more for robust methods, and (2) improving the SNR and SIR performance (although the specific benefits vary depending on the type of scheme employed). Thus, the "size" of the transmitted signal is increased, but the size of the buffers needed is reduced, allowing signals to be packed together more densely.

The payoff from a well-designed error correction code can be quite significant. Some channel-coding systems are capable of achieving gains of up

---

and the criterion that 75 percent of the users say voice quality is 'good' or 'excellent' in 90 percent of the total covered area on a flat terrain" [6, p. 50]. Note also that, as mentioned earlier, the quality criterion from which this number is derived reflects a very relaxed concept of "acceptable service" that is probably only appropriate in the early stages of development of the cellular industry ([6] was published in 1986, just 2 years after the launch of cellular in the United States). A more demanding quality standard would push the required SIR up, perhaps significantly.

14. For example, in another article Lee [8] argues that a sectorized base station configuration can be deployed to gain approximately 2 dB of improvement—at the cost of tripling the RF infrastructure hardware, it may be noted.

15. These numbers are speculative, although they have been promulgated in the literature (e.g., [5], Table II, p. 329).

to 8 dB to 10 dB compared to uncoded transmissions; one of the most recent families of coding techniques, known as *turbo codes*, has been able to attain performance within a fraction of a decibel of the so-called "Shannon limit," achieving error rates of 10^{-5} at E_b/N_0 values of less than 1 dB [9]. These are truly impressive gains, which could be translated into drastically reduced buffering requirements in an orthogonal system.

4.2.2 Diversity Techniques

Unlike noise, which forms a constant background, the effects of interference are often concentrated in time, frequency, and space—severe at some locations, some frequencies, some periods, and not at others. To illustrate, let us consider the well-studied spatial distribution of multipath-induced fades.

For a given transmission channel (defined by a transmitter, a receiver, a frequency, and a propagation environment), multipath fades are distributed in space like radio "holes" where the amplitude of the received signal drops due to self-cancellation of multiple phase-shifted images of the same signal propagated along different paths. These holes have varying depths (measured in terms of the amount of signal cancellation) and are scattered throughout the physical environment. The relationship between the number of these holes and the depth of the holes is described by propagation engineers as meeting a particular statistical distribution pattern, known as a Rayleigh distribution (named for the nineteenth-century English physicist, John William Strutt, later the Baron of Rayleigh). Essentially, the deeper the fade, the less frequently it is encountered.[16]

We can view these holes as a semifixed feature of the channel. (The positions of the holes will change over time as reflectors change their positions, but since most reflectors are fixed, the distribution of the fades can be seen as spatially stable over short periods of time.) Thus, the time-varying phenomenon of "fast fading" is created by the movement of the receiver—and specifically, the receiver's antenna—through this fixed field of radio holes of differing depths. Given the mobile receiver's speed, the spatial distribution of the fades translates into a temporal rate of fading: "At a transmission frequency of 840 MHz, a mobile receiver traveling 30 mi/h through

---

16. "It is well-known that the envelope of the field strength seen by a mobile antenna obeys a Rayleigh distribution.... In such a distribution, fades of 10 dB or more below the [average] value of the envelope occur 10 percent of the time, 20 dB 1 percent of the time, 30 dB 0.1 percent of the time, and so on" [10].

[a Rayleigh-distributed fading environment] would expect 10 dB or greater fades 30 times per second, 20 dB fades 10 times per second, [30 dB fades once a second], and so on" [10, p. 81].

How large are these holes? This depends on the frequency and wavelength of the transmission. In general, however, if the antenna is located in a fade, and it then moves by a distance equal to half the wavelength, it will most likely have exited the fade and returned to a more normal signal level. In the 800- to 900-MHz region, the wavelength is about 1 foot (30 cm). Therefore, if a cellular antenna located in a fade is moved about 6 inches in any direction, it will probably emerge from the fade; or, to put it another way—if we have *two* antennas separated by 6 inches or so, the likelihood that they will *both* be in a fade of a given depth is much less than the chance that one antenna will be in such a fade. More specifically, "the probability of simultaneous fades of 20 dB or more from two such antennas is 0.01 percent, whereas it is 1 percent in each antenna individually" [10, p. 81].

In other words, we can improve the received signal quality significantly—by 10 dB in this simple case—if we have two appropriately separated antennas *and* the ability to select the better one at any given instant. Both antennas are experiencing the classical Rayleigh distribution of fades, but the statistics for each antenna are independent of and uncorrelated with the other. Thus, a selection scheme should realize a better average signal and a *much* lower frequency of really deep fades (Figure 4.3).

This concept of using more than one antenna has gained the name *space diversity* in the literature.[17] The use of two antennas is referred to as two-branch space diversity. The generalization of the concept speaks of *M*-branch diversity, and researchers have studied the improvement in signal quality resulting from 2, 3, 4, 6, and higher values of *M*. In practical terms, even two-branch diversity may be difficult to deploy (consumers tend to resist multiple antennas, although the idea is gaining greater acceptance).

We return to our discussion of diversity systems in detail in Chapter 5. The principle is fairly clear: By sampling the signal more than once, from appropriately separated points in space, and in some way selecting or combining the samples to maximize the signal quality, we can begin to bypass the Rayleigh statistics and return to a much more averaged signal. We can avoid the deep fades that wreak havoc with the transmission, and we can improve the overall SNR by many decibels.

---

17. This is admittedly an odd locution, but it has become entrenched and we shall use it. It also allows us to generalize the term diversity to characterize similarly motivated techniques in other dimensions (F and T).

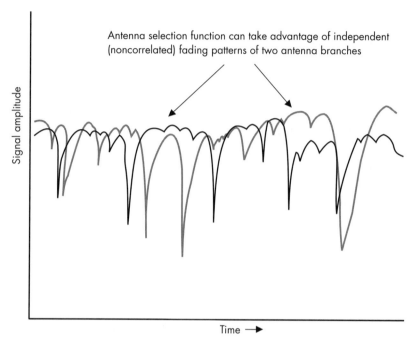

FIGURE 4.3 *Two-branch antenna diversity fading patterns.*

The distance that it is necessary to separate the two antennas in order to obtain uncorrelated multipath fading in the two samples of the same signal is defined, as noted, by the wavelength, and is sometimes called the *coherence distance*. If two S-samples of the signal are taken from points less than the coherence distance apart, they will show partially correlated fading statistics. If separated by more than the coherence distance, they will show uncorrelated fading statistics. To realize the benefits of diversity, the ability to achieve the required separation is paramount.

Space diversity seems to be an obvious countermeasure, easy to implement, and with no apparent penalty in capacity. It can be applied by the receiver without necessarily involving the transmitter. This might not seem to fit with the notion of signal hardening, which implies that some action is taken by the transmitter as it prepares the signal for transmission. For simple space diversity, this is true. However, the full exploitation of the potential for S-diversity will involve the transmitter as well as the receiver (Chapters 5 and 8).

Are there other forms of diversity?

The coherence distance describes the strength of the received signal if we hold F and T constant and plot the signal against changes in the S-dimension. What happens if we hold S and T constant, and evaluate the changes in received signal level as a function of frequency (F)? We find indeed that a similar phenomenon emerges. At a given point in space, the fading level is not the same for all frequencies. The signal may be in a deep fade at one frequency, but not at a different frequency. We can map the size of the multipath fade in the frequency domain, and we discover that such fades are typically rather narrow (Figure 4.4). At 805.0 MHz we may see a deep fade, but at 806.0 MHz there may be no fade at all (relative to the average received signal level). For this reason, multipath fading is sometimes referred to as being *frequency selective* [11].

Thus, we can also define a *coherence frequency bandwidth*, such that two samples of a signal from different frequencies separated by more than the coherence bandwidth will show uncorrelated fading statistics. Obtaining two or more such samples would constitute, by analogy, *frequency diversity*.

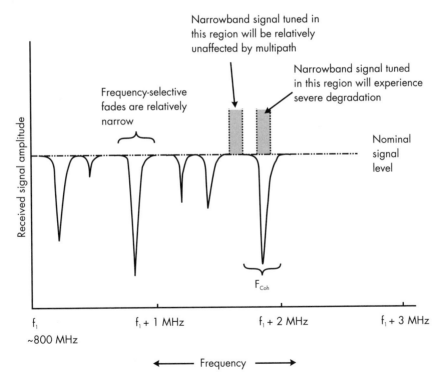

FIGURE 4.4 *Map of multipath fades in the frequency domain.*

It may be less apparent how to make use of this, because it would seem that transmitting the signal on more than one frequency would be wasteful of capacity and perhaps hard to implement in the radio system. However, as we shall see, several methods are available for doing just that and they are actually in use in commercially deployed systems. Perhaps the most straightforward approach, at least conceptually, is an approach called *fast frequency hopping* (Figure 4.5).

Let us imagine a system in which the information symbols to be transmitted are being generated at some reasonably modest rate, say, 800 per second, such that each symbol has a duration of 1.25 ms. Let us assume that instead of transmitting the symbol over just one frequency, the transmitter has the ability to divide each symbol into 10 "pieces" (which may be simply 10 copies of the same value) and to transmit each copy over a different frequency, hopping rapidly between them, such that the dwell time on each frequency is only 125 μs or so. The receiver possesses the same hopping sequence as the transmitter and, therefore, is able to collect all 10 copies

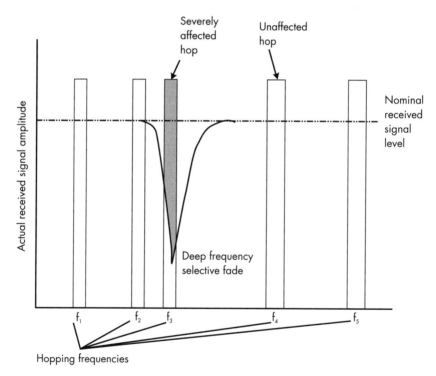

FIGURE 4.5 *Fast frequency hopping, as a form of frequency diversity over a wide bandwidth, relative to* f_{coh}.

from the various different channels. Now if the channels are selected such that the different frequencies are more widely separated than the coherence frequency bandwidth, then even if fading affects some of the transmitted "hops" it will almost certainly not affect all of them. If the receiver needs only, say, seven good samples to construct the received symbol, then we have created a stronger signal with significant frequency diversity.[18]

Figure 4.6 is drawn from an analysis of second-generation digital cellular systems such as GSM and IS-54. It projects a gain of 3 dB or more from antenna diversity and an additional 2 dB from a form of frequency diversity built into the GSM specification. These numbers are modest, and admittedly somewhat speculative, but they show that a signal that is constructed for diversity can withstand substantially more interference (up to 10 times as much) compared to a signal without diversity.

Finally, we note that the same phenomenon exists in the time domain (although characterizing it is more complex; we defer the presentation in detail until Chapter 5). There is indeed a *coherence time* for multipath signals, and systems that can derive multiple samples of the same signal from sampling instants separated by more than the coherence time can achieve the same sort of benefit from time diversity as we have seen from space diversity and frequency diversity.

Creating diversity in one or more of the basic signal dimensions is thus another way of hardening the signal against the effects of interference (especially multipath), which are definably concentrated in space, frequency, and time.

4.2.3 Convolutional Techniques

The third approach is based on the idea of strengthening the signal by "weaving" the individual symbol elements together into a new and stronger signal fabric. Traditionally, digital systems were designed to transmit, receive, and decode each symbol independently, one at a time. *Convolution*, from the Latin for "braided" or "twisted together," refers to a process of deliberately overlapping the transmitted information symbols to construct a new composite signal; just as a braided rope is stronger than the fibers from which it

---

18. Even from this example, it is not obvious how the capacity penalty of F-diversity is avoided or mitigated. We take up this issue in the next chapter. The reader may intuit, however, that if not all F-samples are needed, there is room for sharing of the total resource by multiple users.

Minimum required carrier-to-interference ratio in different 2nd generation
digital systems (under fading channel conditions)

	Without diversity	With diversity	Type of diversity
GSM	11 dB	9 dB	Frequency diversity, achieved through frequency hopping
IS-54	16 dB	12 dB	Space diversity, achieved through dual antennas
JDC	17 dB	13 dB	Space diversity, achieved through dual antennas

FIGURE 4.6 *Gains from space diversity and frequency diversity. (Source: [5]. © 1991 IEEE. Reprinted with permission.)*

is constructed, so a convolved signal can deliver superior interference resistance than a normal signal.

For example, during the past 30 years, the field of error correction has been dominated to a degree by a family of techniques referred to as *convolutional codes*. The basic idea is that a sequence of information symbols is passed through a convolutional process, which weaves them together to create a new sequence in which the information content of each original element has been blended through the entire sequence (or a significant portion of it). This involves expanding each information symbol out over several symbol time periods and overlapping these expanded symbols such that now the signal that is actually transmitted at a given moment in time (clock cycle) contains some data from a number of previous clock cycles (Figure 4.7). Each information symbol that is input to the convolutional process partially influences the value of a number of output symbols. Every output symbol is composed of elements of a number of input symbols blended together. The result is a very powerful code that is stronger bit for bit than nonconvolutional codes in many channels.[19]

The conceptual implications are complex. A common way to look at convolution is as a form of time diversity.[20] The signal is expanded in such a way that the effects of the noise are averaged. Other authors speak of convolutional techniques as a way to endow the signal with "memory."[21] That is, the probability of correctly decoding a given symbol is *linked* with the probability of having correctly decoded a certain number of previous symbols. We can in a sense leverage a generally clean sequence of decoded symbols to assist in decoding the occasional corrupted symbol. Only an "unlikely" *sequence* of errors can really lead to an incorrectly decoded symbol.[22]

Beyond this, however, there is the telling parallelism between convolutional processes, which strengthen a signal, and interference itself. Both involve a blending of two signals such that the resulting signal at the receiver is a composite of both of them. Some have even equated the two, terminologically at least:

> The samples are perturbed both by noise and by neighboring [samples]. The latter effect is called intersymbol interference. *Sometimes intersymbol interference is introduced deliberately*... [emphasis added] in so-called partial-response systems. [14, p. 270][23]

---

19. "Convolutional codes... have been applied over the past decade to increase the efficiency of numerous communication systems, where they invariably outperform block codes of the same order of complexity" [12, p. 751].

20. Referring to a particular convolutional encoding process with a constraint length $K = 7$, Berlekamp et al. comment that "a particular bit of user data affects the output for seven clock cycles; the encoder is adding time diversity" [13, p. 44].

21. Referring to continuous phase FSK, a "convolved" modulation technique, Forney writes that "the continuity of the phase introduces memory into the modulation process; i.e., it makes the signal actually transmitted in the kth interval dependent upon previous signals" [14, p. 270].

22. Of course, such a sequence of errors is "unlikely" only as long as the perturbations of the channel are noise-like—random, uncorrelated from symbol to symbol. If the perturbations are the result of structured interference, then bursts of errors in a continuous sequence become a real issue, and can cause havoc for some kinds of convolutional systems. See Chapter 5.

23. Blahut makes very similar comments, laced with somewhat the same sense of bemusement, in his book *Digital Transmission of Information*: "Rather than modulate one data symbol at a time into the channel waveform, it is possible to modulate the entire data stream as an interlocked unit into the channel waveform. The resulting waveform may include symbol interdependence that is similar to intersymbol interference but is created intentionally.... We can start out thinking of interdependence in a sequence as unintentional intersymbol interference, but once we have developed good methods for demodulating sequences, *we will be comfortable introducing intersymbol interference intentionally to improve performance* [emphasis added] [15, p. 139].

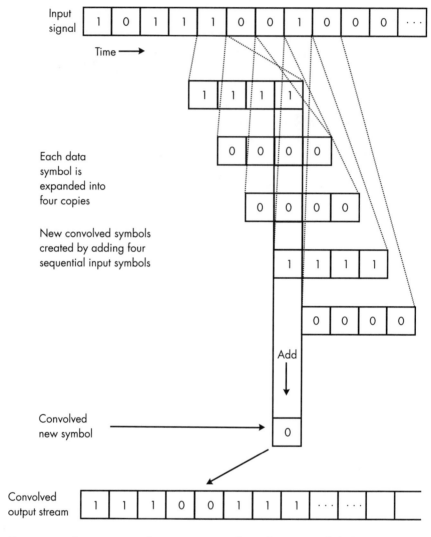

FIGURE 4.7 *A generic convolution process, with overlapping symbol elements.*

What a bizarre proposition! It points toward a deeper truth about the nature of communications, which only emerges openly once we leave the pre-Shannon world of strict orthogonality behind altogether. Managing interference and managing the signal itself are correlated tasks, and the best way to fight interference is often to *create* interference (or something very,

very similar). The best way to combat noise may be to *become* noise—but that is getting ahead of our story.

4.3 Signal Shaping Techniques

In military combat, or in the Darwinian struggle of the species, armor is only one of the strategies for surviving a hostile environment—and not always the best one. The knights of the late Middle Ages eventually discovered—like the tortoises of the Galápagos—that full-body armor ultimately creates its own vulnerabilities. Even the best error correction codes today can often be undone by specific, unanticipated error patterns.[24] Moreover, the cost (in terms of processing, hardware, and bandwidth) of achieving successive improvements in coding tends to become a game of diminishing returns. Even heavily coded systems often perform more poorly under certain channel conditions than do uncoded signals.[25]

Most graphs of "coded versus uncoded" bit error performance will show that the coded signals exhibit *worse* performance usually under conditions of very high channel error rates, which is often where the wireless channel is being required to operate. Like the knights in armor, heavy coding is favored in a relatively orderly environment (low channel errors); but when the mêlée became more chaotic, those slow-moving steel-plated tortoises were vulnerable to fleeter, less encumbered warriors, and so, by analogy, the armored-plated codes of today's elaborate signaling systems may be vulnerable precisely when they are most needed.

Another approach is that of *stealth*. Keeping a low profile, combined with precise targeting, can often be just as effective as adding extra layers of protection. In signaling terms, this means reducing the amount of signal energy that has to be transmitted in order to actually carry the information from point A to point B. The shaping and profile reduction of wireless signals has a long history. The very first wireless communications systems were

---

24. Convolutional codes, though powerful, are in some cases vulnerable to bursts of errors that "poison" the decoder, causing errors to propagate and even "kill" the decoder by forcing it into an unrecoverable state. For example: "A finite number of channel errors can cause an infinite number of decoding errors. This situation... is known as catastrophic error propagation...." [16, p. 59].

25. "At sufficiently low signal-to-noise ratios, one may observe that the coding gain actually becomes negative. This thresholding phenomenon is common to all coding schemes. There will always exist a signal-to-noise ratio at which the code loses its effectiveness and actually makes the situation worse" [17, p. 35].

based on *spark telegraphy*, which used a huge, essentially uncontrolled band-width, such that two transmitters attempting to operate in the same area would jam each other.[26] To solve this problem, it was necessary (over time) to move toward much more efficient technology based on carrier transmission, which uses a continuous radio wave that is fairly well contained (in the F-domain) to carry the information signal (modulated onto the carrier by means of either amplitude modulation or frequency modulation). This allowed the creation of channelized communication systems, to enable multiple users to begin to share the limited spectrum.

If we can find a way to reduce the amount of information that must be transmitted relative to a defined message, or to reduce the amount of RF energy that is needed to carry the information, and especially if we can reduce the amount of nonessential, wasted energy that is also transmitted along with the information, we can reduce the profile of the signal. If all signals in the system are slimmed down, less mutual interference occurs. There is more room in the communications space for additional signals to be packed in more tightly. We can create capacity.

A variety of strategies are available for signal shaping, described briefly here and in more detail in Chapter 6.

4.3.1 Compression: Source Coding

Shannon's invention of a mathematical theory of communication eventually divided into two formal disciplines, which, although twinned, have gone largely their separate ways. The study of the physical channel and of error management strategies has developed under the heading of *channel coding,* while the characterization of the statistical properties of the *message*—the sequences of information-bearing symbols to be transmitted—has developed under the heading of *source coding.*[27] Superficially, these disciplines have

---

26. The first commercial use of radio, at least in the United States, was the competition between Marconi and De Forest to provide real-time reports of the progress of the competitors in the America's Cup races being held off New York in 1901. The wireless telegraphers set up contracts with on-shore news services, and then sailed out in their own vessels to cover the race as it unfolded far from shore. Unfortunately, the two competitors discovered that they could not both reliably transmit reports simultaneously—perhaps the earliest commercial recognition of the need for orthogonality.

27. It is interesting—and sometimes confusing—that each field claims the family name *coding theory* for its own. Whenever one encounters a definitive statement about the nature of "coding" it is necessary to first ascertain which camp is holding forth in order to fully parse the proffered observations.

opposite aims. Channel coding would appear to be all about finding creative ways to *add redundancy* to the signal, to protect it against channel errors. Source coding, on the other hand, has to do with finding ways to *eliminate redundancy* from the signal without losing (essential) information.[28] Channel coding expands the signal. Source coding compresses the signal.

Today source coding encompasses a very diverse set of technologies, grouped according to the nature of the source (voice, video, still images, digital data, and English text). Many sources generate signals with very high levels of inherent redundancy; there is thus a lot of capacity to be gained from finding smarter ways to compress the redundancy out of the signal before passing it to the transmitter. In the case of voice coding, the bit rate required for a toll-quality voice transmission has fallen during the past 30 years from 64 Kbps to something around one-tenth of that rate. Other source compression techniques have shown similar progress. While it may be that most of the gains have already been realized, it is likely that in some areas there are still substantial gains to be expected in the next several years (as discussed in more detail in Chapter 6).

4.3.2 Baseband Shaping

Another field with quite a long history is the study of methods for shaping the actual pulses used to build the digital bit sequence (in digital systems). The *impulse response* of a channel—that is, the output of the channel in

---

28. One might regard the strict separation of these disciplines as counterintuitive; it would seem that two operations proceeding in opposite directions might benefit from a coordinated meta-strategy of the sort that was articulated (with excellent results) for blending the design of channel coding and modulation techniques that led to *trellis coding* (see Chapter 5). However, strict separation has been the rule, and the rule has been defended in principle (although the principle is not entirely clear) by many of the leading theorists. See [7, p. 34] where indeed this separation principle is formulated as one of the lessons referred to in the title: "Completely separate techniques for digital source compression from those for channel transmission, even though the first removes redundancy and the second inserts it." It is that "even though" that makes me wonder if we shall always regard this as true. I have the impression that what we tend to regard as "redundancy" in the source signal can, from another perspective, be regarded as the channel coding that was designed in by the source for the original transmission channel (for example, the redundancies in speech may be, in part, a form of channel coding for the acoustic channel within which speech originates). If so, I would expect that eventually it might be possible to adopt a higher level of abstraction and to address the possibility of coordinating the two strategies for even higher gains.

response to an input pulse—will tend to spread in time (and/or frequency) beyond the sampling instant. Shaping the input pulse so as to minimize the propagation of unnecessary and harmful reverberations in the channel was originally studied by Nyquist [18] in connection with telegraphy. The design of time–frequency pulses has become quite sophisticated. Substantial benefits are to be gained. Pulse shaping is the best place to control intersymbol interference, for example.

4.3.3 Spectrum Shaping

The same principle applies in the shaping of the radio transmission, especially in the frequency domain. Typically, the RF signal spreads into adjacent bands, causing interference in adjacent channels or communications subspaces. Design choices can mitigate this unwanted signal spreading. For example, different types of modulation schemes generate different amounts of adjacent-channel interference at the same power level (Figure 4.8).

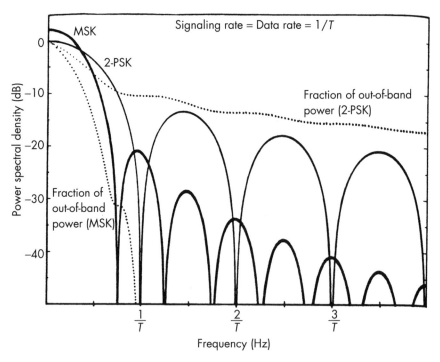

FIGURE 4.8 *Power spectrum of several modulation formats, illustrating different spectral spreading properties. (From: [19]. © 1982 John Wiley & Sons.)*

There are other ways to reduce the profile of the transmitted radio signal, some of them quite straightforward. For example, most first- and second-generation wireless systems employ some form of power control, implemented through a feedback principle of some sort, to enable the transmitter to reduce its power output to the level actually required by the receiver. The difference between maximum power and necessary power in many circumstances is to be counted in many tens of decibels, which suggests that a great deal of interfering signal energy can be suppressed. In Chapter 6 we will review a number of these strategies in more detail.

4.3.4 Beam Forming: Smart Antennas

One of the newest fields involved in the shaping of wireless signals to reduce the interference profile is based on the use of so-called "smart antenna" technologies that can focus the transmission (or the reception) along a narrow beam. In principle, huge gains could be realized, both in terms of reduced interference and dramatically better frequency reuse. Indeed, whereas source coding and the other techniques described here are all relatively mature, the spatial domain is only beginning to be exploited. There is reason to believe that some of the greatest gains in system capacity still to come in the wireless industry will come from the deployment of beamforming antenna systems.

4.4 Signal Recovery

Another family of innovative techniques focuses on different ways of repairing the damaged signal after it has reached the receiver. Error correction is an obvious example, which we have considered as a two-ended process: The correction information is encoded at the transmitter and decoded at the receiver, and as such it has a cost in transmission bandwidth. Other techniques are embodied mainly in receiver processing steps that do not impose a bandwidth penalty (or not very much of one).

In one way or another, these techniques are based on the information theoretic idea that if an interfering signal has structure (i.e., it is not noise-like), this structure can be discovered, and predicted, and the interfering signal can then be subtracted from the received waveform. This involves analyzing the received signal into its elements—the desired signal, plus the interfering signal(s)—and this analysis can be done offline at the receiver if we have enough processing power to do it quickly enough. Indeed, many of these techniques can be applied independently at the receiver, and can be implemented without requiring changes in the transmission format or the transmitter hardware and software. This makes them in some cases especially attractive for obtaining capacity improvements in existing networks. The generic term that is sometimes applied to these techniques is *interference cancellation*.

4.4.1 Cancellation of T-Interference (I): Equalization

Historically, the first interference cancellation technique to be deployed commercially addressed the pervasive phenomenon of ISI in digital signals. ISI occurs when individual symbol pulses tend to run together due to (1) the spreading inherent in the impulse response of the system and (2) additional distortions or attenuations introduced by the channel (e.g., multipath is one of the channel phenomena that can cause or exacerbate ISI).

The concept of equalization—the invention of which in the 1960s I regard as one of the most ingenious examples of a new design principle in the history of telecommunications—can now be stated with almost deceptive simplicity. The transmitter sends a known sequence of digital symbols to the receiver, and the receiver then compares the received waveform, which includes the distortion introduced by the channel, with what it knows the sequence to have been. From this comparison, the receiver can infer the adjustments necessary to restore the signal to its original form. By applying these same adjustments to the received signal containing the user data (which it does *not* know in advance), the receiver can cancel the distortion induced by the channel, and remove the intersymbol interference. Except for sending the training sequence, the transmitter does not participate in the equalization process,[29] nor does it use up channel capacity.[30]

4.4.2 Cancellation of T-Interference (II): Multipath Combining

A more radical concept of signal grooming in the time domain has recently emerged in certain spread spectrum architectures. Such systems employ superfast signaling rates (compared to second-generation digital TDMA systems). The basic signal element in a direct-sequence spread spectrum system is called a *chip*, and the chip rate is often many megabits per second.

---

29. Indeed, some schemes propose so-called "blind equalization," in which there is not even a training sequence. See Chapter 7.

30. If an equalization pilot is required, then some channel capacity is lost. See Chapter 7.

Each chip duration is less than 1 μs, which means that different multipath images of the chip, which are typically separated by at least several microseconds, appear not as blurring of a single symbol image, but as relatively distinct images of the same symbol (Figure 4.9).

This raises the novel possibility of combining the different images resulting from the different "rays" of the multipath signal. They are combined constructively ("coherently") to maximize the signal energy presented to the digital detection circuitry. Receivers have been developed to do just

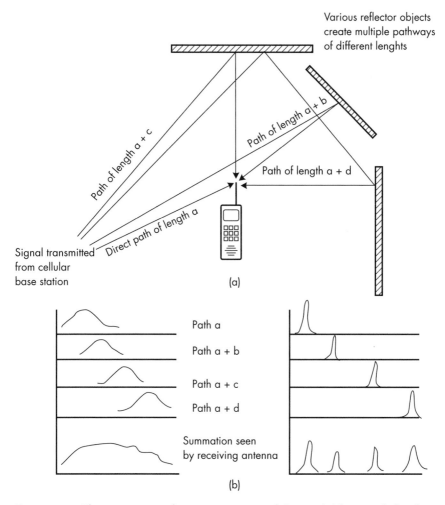

FIGURE 4.9 *Chip images in a direct-sequence signal format: (a) long symbol pulses, wich blur together; and (b) very short symbol pulses, which can be resolved.*

this, and the CDMA community now speaks of *multipath combining*, and even *multipath diversity*, as positive attributes of the CDMA architecture. The most destructive of all forms of wireless interference—multipath-induced self-interference—is neutralized, and even turned to advantage by this innovative receiver processing technology.[31]

4.4.3 Cancellation of S-Interference: Spatially Selective Receivers

Another approach to interference cancellation is to isolate the desired signal directionally with a steerable beam-forming receiving antenna (at the base station). Instead of accepting signal energy from all directions, the receiving antenna focuses its receiving beam narrowly and aims it at the transmitting mobile unit. A potentially interfering mobile unit transmitting from a different direction will be cancelled passively by the directional characteristics of the antenna.

Several different architectures are based on spatial processing technology—some for both transmitting and receiving base station antennas (addressing both the uplink and the downlink) and some for the uplink only. Some systems rely on steerable beams, while others use multiple fixed beams with handoff between beams.

4.5 Beyond Orthogonality: Convolved Wireless Architectures and Design Principles

The techniques described in this chapter have emerged piecemeal during the past 30 years and have been applied, here and there, sometimes as ad hoc solutions to particular problems, sometimes built into a more comprehensive transmission format. For example, GSM (a second-generation standard) incorporates technology elements such as equalization, frequency hopping (for diversity), error correction coding, power control, low-bit-rate voice coding, and low profile modulation [like *minimum shift keying* (MSK)] [5, 21]. However, the overall model of a wireless communications system has remained committed to the principle of orthogonal design. Even in sophisticated second-generation systems like GSM, the application of signal hardening and signal shaping techniques is still focused on buttressing the essential

---

31. For an early and balanced presentation of multipath combining, see [20]. For other references on multipath combining, see Chapter 8.

orthogonality of the system. Individual signals are intended to be kept apart, buffered, and to occupy unique subspaces in the STF framework.

The required conceptual leap beyond orthongonality develops from a perhaps unexpected direction: *electronic warfare*. In commercial applications of wireless, the concern for capacity that is the hallmark of post-Shannon thinking has developed relatively recently. However, in many military applications, an even more pressing problem has forced the designers of battlefield communications systems to embrace Shannon concepts much earlier and much more comprehensively: the problem of *jamming*.

Since World War II it has been recognized that modern military operations are vitally dependent on communications links, especially *wireless* communications. The modern doctrines of command and control assume a fully integrated communications capability. The smarter the weapons systems become, the greater the importance of the communications link. If we can jam the enemy's communications, we can gain an the advantage. By the same token, we must protect the integrity of our own communications links, in the face of aggressive attempts to disrupt them.

Electronic warfare (EW) has become a powerful technology driver. Jamming techniques have been matched by antijamming countermeasures, and systems have been designed that are capable of withstanding huge amounts of interference—not just accidental signal spillover, but hostile jamming signals that are intentionally designed to be as destructive as possible. At the same time, EW has encouraged the development of extremely stealthy communications technologies, capable of hiding the information signal deep in the noise background.

To achieve these challenging goals, EW systems designers had to take the principles of signal hardening, shaping, and so forth to a new level. They had to begin to think through the entire communications architecture from the post-Shannon perspective, and to design their countermeasures comprehensively, upfront (not as add-on enhancements). They had to discover how to mutually reinforce one set of techniques with another, so that the interference resistance became multiplicative (instead of merely additive).

This led to the abandonment of strict orthogonality. Keeping signals apart from the interference is no longer possible in the jamming scenario, and ordinary countermeasures are of little avail. Even a well-buttressed orthogonal system like GSM still presents an easy target for disruption. As jamming itself became a developed science, trying to maintain orthogonality was more and more like trying to swim without getting wet.

A new set of communications architectures are beginning to make their way into the commercial sphere. These nonorthogonal architectures begin from the idea that interference resistance should be *designed in* to the signal from the start. In Chapter 8, we will review some of these designs, but first let us try to formulate the "lessons learned" at the level of the counter-measures, which become the *design principles* at the systems level.

4.5.1 Signal Spreading

Multipath-induced self-interference (fast fading) is arguably the most perni-cious form of signal corruption introduced by the wireless channel (jamming aside). Unlike other forms of interference originating from other transmis-sions, which can, in principle, be controlled or buffered, self-interference arises from the physical propagation of the desired signal itself. However, multipath fades are narrow, in the dimensions of space and frequency. The parameters known as the *coherence distance* and the *coherence bandwidth* allow us to state, probabilistically, the boundaries of the typical fade in the S- and F-dimensions. Moreover, although fades are not viewed as changing rapidly in time, the passage of a *mobile* receiver through the received signal field introduces in effect a coherence time value as well (as a function of the average speed of the mobile).

These coherence values—S_{coh}, T_{coh}, and F_{coh}—define the size of the (typical) fade, in space, time, and frequency, respectively.[32] [Note that the coherence time is defined by the coherence distance and the vehicle speed for the case of a single-antenna receiver. In other words, for single-antenna receivers, we could use either S_{coh} or T_{coh} to define (with F_{coh}) a two-dimensional characterization of the fade. However, if multiple antennas are employed, then the S_{coh} parameter is relevant to determining the joint prob-ability of a fade for both branches, and T_{coh} is relevant for determining the length of time that either of the branches will remain in a fade. Thus, with-out promoting a specific formalism, the notion that there are indeed three relevant dimensions for the diversity analysis (and not just two, collapsing T and S) can be sustained for purposes of this discussion.]

If the fade thus has a definable physical extent in S, T, and F, it is pos-sible to conceive of a designing a signal that is *larger* than this in terms of the STF dimensions. If the signal is only somewhat larger, the fade will still destroy most of the signal. If we can find a way to make the signal signifi-cantly larger in any or all three of the primary signal dimensions than the coherence boundaries of the typical fade, we can see intuitively that more

---

32. I am following here the treatment of diversity and signal spreading introduced by Jung, Baier, and Steil [22].

and more of the signal will get through relative to the diminishing fixed percentage that has been lost in the fade (Figure 4.10).

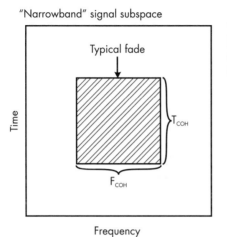

"Narrowband" signal subspace

Typical fade

T_{COH}

F_{COH}

Time

Frequency

The typical fade is bounded by the coherence time, T_{COH}, and the coherence frequency, F_{COH}

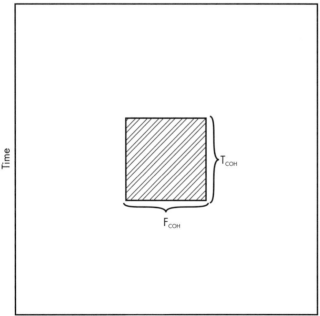

"Wideband" signal subspace

T_{COH}

F_{COH}

Time

Frequency

Same fade impacts a much smaller percentage of the signal

FIGURE 4.10 *Examples of two signals spread at different ratios to the average fade in the frequency domain.*

This is the first system-level design principle for a nonorthogonal architecture: to *spread* the signal so that it occupies a larger dimension of F than
F_{coh}, a larger dimension of T than T_{coh}, and/or a larger dimension of S than
S_{coh}. Actually, the transmitter naturally produces a signal larger than S_{coh}, so
the only real design issue is at the receiver. If we want to take advantage of
an S-spread signal, we have to employ an appropriate antenna diversity system. For both F-spreading and T-spreading, some specific steps must be
taken at the transmitter to prepare (spread) the signal, and other steps must
be taken by the receiver to collect or sample the spread signal over the larger
signal space.

This is clearly a generalization from the diversity techniques discussed
earlier in the chapter. The signal is sampled at several locations along one or
more of these primary signal dimensions; the sampling locations are separated by a dimensional distance greater than the coherence value of that
dimension. This architectural principle is being incorporated in much of the
thinking underlying third-generation wireless systems.

The notion of signal spreading suggests, of course, the concept of
spread spectrum, which is much in vogue. The technology of spread spectrum has a venerable history reaching back now more than 50 years, and it
has been the hallmark of military antijamming strategies for decades. The
introduction of spread spectrum into the commercial sphere of wireless
occasioned an earthquake in the industry in the early 1990s, because suddenly everyone was choosing sides in the CDMA versus TDMA debates. We
are still feeling the aftershocks, although by now it is clear that both systems
will survive and grow. However, in one respect CDMA has prevailed: Philosophically, it has redefined the goals of wireless communications away from
orthogonality, and toward more and more nonorthogonal systems.[33] Today
even the TDMA camp is active in developing and implementing techniques
that allow for greater use of nonorthogonal signals, or at least partially
orthogonal signals.[34]

---

33. As a TDMA partisan, it has taken me a long time to reach the point of acknowledg-
    ing this.

34. The scale between fully orthogonal signals and fully nonorthogonal signals corre-
    sponds (I would argue) to the gradual compression and eventually the elimination of
    the STF buffers. Certainly, the leap to true direct-sequence CDMA is a revolution
    insofar as it eliminates all three buffers at once and replaces them with a coded,
    convolutional signal space. Because TDMA systems allow for partial signal overlap,
    and/or drastically reduced STF buffering, it may be appropriate to refer to them as
    quasi-orthogonal systems (or, perhaps more appropriately, as quasi-nonorthogonal
    systems). At some point, of course, the issue is less about terminology than about

We will examine so-called spread spectrum architectures in Chapter 8, but here I want to underline the fact that signal spreading per se is a more comprehensive concept than spread spectrum (as it is often employed), and the principles of signal spreading to enable systematic and powerful diversity techniques can be used in service of system architectures that would not normally be classified as spread spectrum systems.

4.5.2 Interference Averaging

The second general post-Shannon design principle to emerge is based on a controversial proposition that the best way to manage interference is to find a way to average its effects, to remove as much of its inherent structure as possible, and to convert it into a noise-like signal.

This is closely connected with the diversity strategy, because one result of diversity is the randomization of the source of interference. For example, frequency hopping as a form of F-diversity will tend to expose a given signal to interference from one particular interferer i_1 on hop h_1 and to different interferer, i_2 on hop h_2. In a non-frequency-hopping system, two cochannel users operating at the same time will tend to jam each other completely unless buffered sufficiently in the S-dimension. Frequency hopping means that such continuous jamming cannot develop between pairs of cochannel users, and the statistics will work in favor of the system capacity by allowing smaller S-buffers and closer spacing of the reuse pattern.

However, the more radical step—full interference averaging—is implemented at a lower level of the signal architecture. Here, an effort is made to completely eliminate the structure of the interferer and turn the interference into noise (or something indistinguishable from noise). For example, assume that an interferer produces a burst of errors—a cluster of errors in a continuous sequence, in the time dimension (Figure 4.11). This type of error cluster is often fatal to some powerful types of coders (especially convolutional coders), However, with a T-diversity technique known as *interleaving* (see Chapter 5), we can spread those errors out over a longer time frame. After interleaving (at the transmitter) and deinterleaving (at the receiver), the errors appear to have a much more noise-like distribution. The structure of the burst has been removed, and now we can employ a convolutional decoder with soft decision processing.

---

the communications philosophy, and it is in this sense that we can say that CDMA has had a revolutionary impact on the whole industry, even those segments which are nominally, and politically, non-CDMA.

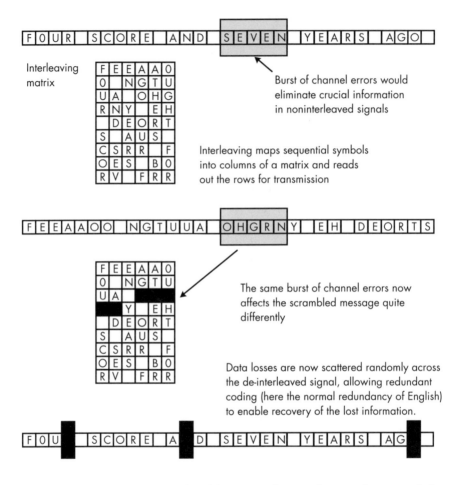

FIGURE 4.11 *Burst of errors produced by an interferer in the time domain and the effects of interleaving.*

Should we leave the structure in the interfering signal and use structure-oriented countermeasures to cancel it, or should we put it all in the blender, whip it to a fine consistency, and use noise-oriented counter-measures to clean out the errors? The question is controversial. Some of the most highly respected researchers have forthrightly denounced the interference-averaging strategy on the grounds of high theory. For example, Berlekamp et al. have stated the position so cogently that it is worth quoting at length:

A common argument is that codes can be interleaved so that un-interleaved received data [i.e., the data as it appears after interleaving at the transmitter and de-interleaving at the receiver] is more or less random and, therefore, more or less optimal for the convolutional decoder....

    This argument is appealing but fallacious. In a very precise information-theoretic sense, *the very worst type of noise is random noise* [emphasis added]. If the noise has structure, the structure can be exploited. In less high-brow terms, the effects of short interference bursts on a telephone line disturb reception less than a constant level of white noise. The reason for this is not profound. The receiver can locate when a burst occurs and make good use of this location information. In an interleaved code, the deinterleaver carefully tries to convert a less damaging type of noise (bursts) into a more damaging type of noise (random).... Interleaving is wrong on a philosophical level ... [13, p. 49]

Other authors writing from a relatively pure information theory perspective have forcefully supported this position. For example, it has been argued that it can be *proven* that interference is—theoretically speaking—utterly benign! This theoretical analysis of the "interference channel" is restated in a paper from 1980:

The interference channel has two senders and two receivers. Sender 1 wishes to send information to receiver 1. He does not care what receiver 2 receives or understands [and vice versa]....

    This channel has not been solved in general ... but remarkably, in the case of high interference, [it has been] shown that the solution to this channel is the same as if there were no interference whatsoever [i.e., there is no loss of capacity due to interference!] To achieve this, generate two codebooks ... each sender independently chooses a word from his book and sends it.... The first receiver perfectly understands the index of the second transmitter. He finds it by the usual technique of looking for the closest codeword to his received signal. Once he finds this signal, he subtracts it from his received waveform. Now there is a clean channel between him and the sender. [23, pp. 29–30]

In a sense, this is nothing more than an extension of a very reasonable, informal observation contained in one of Shannon's earliest papers:

Noise and distortion may be differentiated on the basis that distortion is a fixed operation applied to the signal, while noise involves statistical and unpredictable perturbations. Distortion can, in principle, be corrected by applying the inverse operation, while a perturbation due to noise cannot always be removed. [24, p. 31]

It is on the same grounds that Shannon later refers to white noise—the most completely structureless type of noise that we can define—as the "worst among all possible noises" [24, p. 40].

Yet Viterbi, one of the few contemporary scientific figures in this field who speaks with comparable authority to Shannon, has called white noise "the most benign of interference[s]" [7, p. 227]. Pursley, a leading authority on frequency-hopping architectures, agrees that "interference is generally much more disruptive than wideband noise" [25, p. 152]. Indeed, the practical consensus among engineers actually working on third-generation systems today undoubtedly favors this position over Shannon's. The strategy in practical CDMA systems being actually deployed today is to convert structured interference into unstructured noise[35] and to apply error correction and signal cleansing techniques that are powerfully optimized for white Gaussian noise. As Viterbi elaborates:

> In the presence of interference or jamming, intentional or otherwise, the communicator, through signal processing at both transmitter and receiver, can ensure that performance degradation due to interference will be no worse than that caused by Gaussian noise at equivalent power levels. [7, p. 228]

The immediate resolution of the contradiction is that there are opposite, unstated assumptions underlying the two positions. The Shannon camp assumes that "distortion [i.e., interference] can, *in principle*, be corrected by applying the inverse operation." The phrase "in principle" means that we can assume (unlimited) processing power, and perhaps time, to detect the structure of the interfering signal and then to "perform the inverse operation" to cancel it. Viterbi, approaching the problem from a practical perspective, assumes that what is needed is not a theoretically optimal solution, but a practically optimal solution. Average, known levels of disturbance are easier to deal with than unknown, highly variable disturbances, which is what we have if we have structured interference without the practical ability to analyze its structure. It is *much* easier to throw the interference into the processing blender and whip it into a noise-like state than it is to detect, analyze,

---

35. Or at least pseudonoise—that is, a deterministic signal that nevertheless has the statistical properties practically indistinguishable from those of random noise.

and invert the structure of a variety of contributing interference sources. From the practical perspective, trying to pick apart the messy, interference-ridden received signal into its individual components is like trying to unscramble a three-egg omelet into the contributions of three different chickens based on analyzing the DNA of each molecule—it may be theoretically possible, but who in the world would want to tackle it?

4.5.3 Signal Averaging: Noise-Like Signals

Shannon theorists and most practical spread spectrum architects hold that the same principle can be extended to averaging the *signal*, or hiding the *structure* of the desired signal itself in a noise-like envelope. (The ways in which this can be accomplished are discussed in Chapter 8.) Pseudonoise transmission is part of the credo of many third-generation developers involved in the development of direct-sequence CDMA architectures. They embrace the notion of making the signal *and* the interference both look as noise-like as possible. Again, Viterbi states:

> The jammer's optimal strategy is to produce Gaussian noise interference. [Note that this is actually congruent with Shannon's position that white noise is the "worst of all noises."] Against such interference, the communicator's best waveform should statistically appear as Gaussian noise. Thus, the "minimax" solution to the context is that signals and interference should all appear as noise which is as wideband as possible. This is a particularly satisfying solution when ... one user's signal is another user's interference [i.e., in multiple access spread spectrum systems like CDMA]. [7, p. 228]

As Shannon succinctly observes, "to approximate [the optimal] rate of transmission, the transmitted signals must approximate, in statistical properties, white noise" [26, p. 24]. This point of view has become widely held among CDMA advocates, and its roots in Shannon theory give it, at times, an almost religious coloration.

Yet it should be noted that this principle follows from and somewhat depends on the previous one: An averaged signal is optimum only in the presence of averaged (noise-like) interference, and is not necessarily the best if the interference is structured. The controversy is compounded, and as we shall discuss in Chapter 8, the proposition that noise-like signals are *always* superior may be questioned on both practical and theoretical grounds. Indeed, not all third-generation architects embrace this principle, by any means.

4.5.4 Interference Cancellation: Unscrambling the Omelet

In theory, we know (or can know) more about interference (which has discoverable structure) than about noise (which is random, *by definition*). If we had the power to act on this knowledge, we could use it to great advantage. As suggested above, one school of thought holds that indeed this is the proper goal—and in some sense the *ultimate* goal—of a wireless communications system.[36] Although most working engineers have dismissed this goal as unachievable, or at least highly impractical, there are some who see active interference cancellation as the real frontier of the technology, and the quarter from which the next quantum leap in capacity may be expected.

To illustrate by means of a conceit—let us recall Maxwell's demon.

James Clerk Maxwell, the great nineteenth-century British physicist, imagined his demon (from the Greek *daimon*, a "thing of divine nature") as an intelligent being situated as a kind of gatekeeper at the entrance to a small passageway between two vessels filled with an ordinary gas at room temperature. The channel between the two vessels is large enough to allow individual molecules of the gas to pass through from one vessel to the other. All the molecules are of course in constant random motion, and from time to time a molecule will happen to pass through the gateway from one vessel to the other. Some of the molecules are moving faster than others, as naturally happens in any gas (although the *average* motion may be stable—and is measured as the temperature of the gas).

Now, the demon has the ability to detect whether an approaching molecule is moving relatively fast or moving more slowly. By opening or closing the gate, he can (we assume) allow fast moving molecules to pass through in one direction, and slow moving molecules to pass in the other direction. Over time, the faster moving molecules (the warmer ones) will accumulate in one of the two vessels and the slower moving molecules (the cooler ones) will accumulate in the other. A temperature differential would be created between the two vessels, and this—Maxwell proposed—could be used to perform work. The system, as Maxwell conceived it, is a perpetual motion machine that violates the second law of thermodynamics (Figure 4.12).

Maxwell's demon was eventually acquitted of this charge, when it was realized that he would need to expend energy in order to observe and detect the motion of each molecule and to decide whether to open or close the gate, and this information processing overhead would consume more energy

---

36. See the elegiac opening paragraphs of Moshavi's article [27].

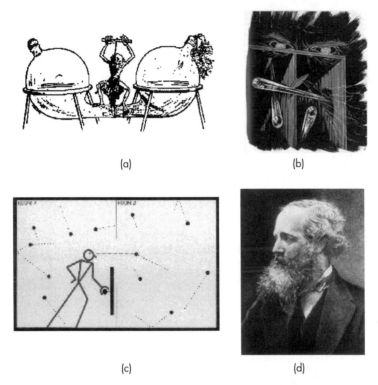

FIGURE 4.12   *Various depictions of Maxwell's demon: (a) From: [28] © 1955 American Journal of Physics; (b) From: [29]. © 1964 W.H. Freeman and Company; (c) From: [30]. © IEEE Reprinted with permission; and (d) Maxwell himself.*

than his actions would generate. It is among other things a nice illustration of the subtle connectedness of thermodynamics and information theory.

Let us now imagine our own interference-canceling demon, who is stationed at the front end of a radio receiver, where he is measuring the composite signal being received. The composite signal is made up of one desired signal, which was transmitted specifically to this receiver, and many, many interfering signals, including signals from other transmitters, and multipath images of the desired signal itself, all superimposed on each other. The interference demon has the ability to identify (never mind how) the sources and structures of all signals impinging on that receiver, and he has the ability as well to create a perfect *inverted* copy of each signal and feed it into the receiver. The effect, of course, is that each inverted copy cancels the component of the composite signal that comes from a particular undesired source, and after all the undesired sources have been cancelled, the receiver is left with ... a perfectly clean copy of the desired signal (Figure 4.13)!

It is not clear whether the interference demon is violating any physical laws. Indeed, there is a literature developing about him, in which he is called a *multiuser detector*.[37] Nevertheless, it is clear that the demon would be a very busy individual. We need to endow him with an enormous amount of computing power, and we need to allow him time to perform his tasks.[38] For this reason, even the supporters of trying to construct such a demon generally admit that "the optimal multi-user detector is much too complex" [31]. It is a practical question whether the enormous processing budget that this demon would require will offset in some way the gains in capacity that he could produce. Or, in the colorful language of one recent treatise: "The new proviso [is] real-time decodability without the consumption of a good fraction of the gross national product" [32, p. 21].

It is worth noting (as an antidote to such pessimism) that active cancellation techniques have been successfully developed and implemented in

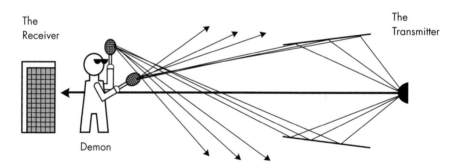

FIGURE 4.13   *The interference demon.*

---

37. See [27] for a summary of key concepts and a list of important references. See also the discussion of multiuser detection in Chapter 7.

38. "A precise, tractable model for the interference is necessary for an exact evaluation of the error probabilities for symbols in a corrupted segment, but for many practical applications such a model is not available. Even in [a] simple multiple-access example... the analysis is very tedious if the number of interference signals is larger than two or three, especially if the signals have unequal power" [24, p. 154].

39. In this field, the "noise" being canceled is in fact highly structured acoustic noise, such as the rotor noise of a helicopter or the siren sound of an ambulance, and is

a related field, known (confusingly) as *noise cancellation* (in the acoustic domain).[39] If the active interference cancellation strategy does develop into a practical body of techniques for wireless systems, it could reverse the current trend toward homogenizing everything, signal and interference alike, into a fine slurry of noise and pseudonoise. It could change the strategies for error correction, diversity implementation, and signal shaping. It will be an interesting area to watch.

4.5.5 Adaptive Signal Design

Another comprehensive design principle that has emerged in recent years is based on the idea that channel conditions (and other parameters of the transmission) are constantly changing. If we cannot track these changes, then we have no recourse but to define the statistics of the parameter in question, and to design for the worst case. In so doing, we almost always pay a huge price in the margin required to deal with extreme conditions, considering that (by definition) most of the time the system is *not* operating under extreme conditions, but under *average* conditions.

If we could somehow design the system to be optimized for the average case—with certain adaptation measures that would be called into action only when conditions diverged drastically toward the worst case—we could recapture most of this excess margin. Let us briefly touch on two examples.

First, we can consider *adaptive modulation*. The required SNR for a given modulation scheme tends to vary inversely with the spectrum efficiency of the modulation scheme. For example, if QPSK needs a SNR ratio of 10 dB to achieve a given error rate, then 16-ary PSK needs, say, 16 dB to achieve the same error rate. Translated into a coverage map for a wireless local loop application, the area covered by a 16-PSK system is much smaller than the area covered by a QPSK system (keeping all other parameters the same). Yet the capacity of a 16-PSK channel is, all things being equal, twice the capacity of a QPSK channel: 4 bps compared to 2 bps. If we are forced by worst-case engineering to design to the QPSK standard, we give up half the capacity of the system within the inner zone of potential 16-PSK coverage. If the population is denser toward the center of the wireless local loop coverage area (as it often is), the penalty in capacity is magnified.

Faced with this situation, the designer has another option. He can design the system to use adaptive modulation. On setting up the call, the signal quality is evaluated, and if a 16-PSK channel can be used, it will be. If a better SNR is needed, the system can automatically fall back to QPSK. In this way, the system can be designed for the typical or average case (at 16

PSK, for high capacity) with the fallback mode of QPSK for the worst-case link conditions.[40]

This example is similar to the familiar adaptation process used by Group 3 fax machines to negotiate and agree on a suitable bit rate for each transmission. If fax machines had to be engineered for a fixed bit rate, they would have to adopt the least common denominator approach and engineer for the worst-case line conditions. This would force the vast majority of fax users to put up with a slower-than-necessary transmission rate for the sake of the worst 2% or 5% of the lines in use.

A second example is the concept of *adaptive channel allocation*. In standard cellular architectures, each cell is assigned a set of usable frequencies and the assignment is fixed. If at some moment in time there is an overload of traffic in one cell, and there are unused channels in the next cell, there is no way to reassign the unused channels from the second cell to alleviate the temporary overload in the first cell. The problem is not only statistical (i.e., the occasional spike of traffic in one cell). It can also be systemic. For example, cells in the business district may be highly loaded during the daytime and lightly loaded in the early morning and evening. The cells covering major arteries into and out of the city center may have a complementary time distribution of traffic, heavy in the morning and evening rush hours, and lighter in the middle of the day. Once again, however, in a nonadaptive system, each cell must be designed for the worst-case load (or some compromise thereof), and the total system at any moment in time is bound to have a significant amount of unused, wasted, nonreallocatable capacity bound up to protect the worst-case scenario.

System designers have begun to look toward adaptive channel allocation, as described in the following excerpt:

> With adaptive channel allocation (ACA) there is no fixed frequency plan. Instead each base station is allowed to use any channel in the system. The ACA procedure is very attractive in a TDMA frame structure. The basic idea is to allocate channels depending on 1) the actual traffic situation and 2) the actual interference situation. Early work only took [traffic] into account with relatively poor capacity gain as a result. In

---

analogous to the structured interference in the wireless channel (rather than to "noise" as it is used by Shannon theorists).

40. The author was involved in the development of just such a system for wireless local loop service, using adaptive multilevel PSK modulation, in the mid-1980s, by IMM Corporation (now InterDigital Corporation) and its partners (including M/A-Com

more recent proposals it has been proposed to dynamically assign channels to every *call* instead of assigning channels to every *cell*....

Simulations of ACA have revealed large capacity gain. This is particularly true when the number of available channels is relatively few. In
a large system the capacity gain is 50% or more. In small systems capacity gains of 3–4 times have been found. [5, p. 331]

The principle of adaptive design is a general one. It can be applied to:

- *Vocoding:* coders operating at a variable bit rate depending on the
  information content of the source model at a moment in time[41];

- *Power control;*

- *Time alignment:* to adjust time of arrival in TDMA systems to
  reduce the guard-time requirements for mobiles in adjacent slots
  but at different distances from the cell site;

- *FEC:* assigning different levels of error correction as a function of
  channel conditions or other parameters[42];

- Even what might be called *adaptive signal triage:* determining which
  parts of a signal to jettison in the event of a network overload or poor
  channel conditions.[43]

Indeed, it is not conceptually difficult to apply the notion of adaptiveness to almost any aspect of a wireless system. The implementation, of
course, varies considerably. In all cases, the adaptation of the target parameter is tied in some way to the measurement of channel conditions. In some
cases, the relevant conditions of the channel can be inferred directly from
the transmission (e.g., from received signal, or BER) without additional
information or loss of capacity. In other cases, the feasibility of adaptive
solutions depends on the use of a pilot signal. Generically, a pilot is a special
signal that is inserted into the transmission to provide special information to

---

Linkabit, now Hughes Network Systems). That story is recounted in more detail in
[33, Chap. 10].

41. For example, the vocoder used in IS-95 (second-generation CDMA) is a variable-
rate coder, based principally on detecting voice activity and transmitting at lower
rates during periods of inactivity.

42. Adaptive FEC is being incorporated in many emerging 2.5G and 3G standards.

43. See Jayant's discussion of layered coding: "An important concept ... is layered coding. The output of the source encoder is divided into [ATM] cells of varying signifi-

the receiver about the channel. It does involve sending additional information, and it does use capacity, but it may be worth it.

One well-known example of a pilot is the power control pilot used in the forward link of the IS-95 CDMA standard. Actually this pilot is used not only for power control, but for other forms of adaptation such as the support of coherent demodulation. Indeed, the use of coherent demodulation on the CDMA forward link itself—which results in a gain of 2–3 dB—can be seen as an example of adaptive design. In a typical mobile channel, multipath phase shifts will make it difficult or impossible to use coherent PSK demodulation (in which the phase of each symbol is measured against a fixed, "coherent" reference). The worst-case design choice is therefore to use differential demodulation, where the information is encoded as the difference between the measured phases of successive symbols, rather than the absolute phase. By providing a coherent phase reference through the pilot, the IS-95 forward link enables the mobile receiver to use coherent demodulation and gain those valuable decibels of signal. In any case, the IS-95 pilot does use up about 20% of the capacity of the forward channel, but the 2- to 3-dB gain more than makes up for this loss.

4.5.6 Convolutional Signal Structures

Third-generation architectures will also make more systematic the principle of convolutional signal design, that is, signals that intentionally overlap one another physically. As discussed in earlier sections of this chapter, the judicious blending of multiple signal elements, and even multiple signals, can yield transmissions that are inherently more robust than nonconvolutional signals. This is true even though—in fact, precisely *because* of the fact that—the blending of these signals seems to resemble a kind of purposeful creation or introduction of interference.

Convolutional techniques are commonly used in the related fields of channel coding and modulation, where convolutional error coding and partial response signaling (also called correlative coding, although it is a modulation technique) have by now fairly long histories of use. Interleaving in the time domain can be viewed as a convolutional method in some applications.

The convolutional strategy is taken much further by DS-CDMA architectures, in which the, for example, base station blends together all the forward link signals destined for all users in a given cell into one composite transmission, which occupies the same frequency, time, and space for all its component signals.

References

[1] Viterbi, A. J., "The Orthogonal/Random Waveform Dichotomy for Digital Mobile Personal Communication," *IEEE Personal Communication*, Vol. 1, No. 1, 1994, pp. 18–24.

[2] Lee, W. C. Y., "Spectrum Efficiency in Cellular," *IEEE Trans. on Vehicular Technology*, Vol. 38, No. 2, May 1989, pp. 69–75.

[3] Lee, W. C. Y., "Overview of Cellular CDMA," *IEEE Trans. on Vehicular Technology*, Vol. 40, No. 2, May 1991, pp. 291–302.

[4] Goodman, D. J., "Trends in Cellular and Cordless Communications," *IEEE Communications Magazine*, June 1991, pp. 31–39.

[5] Raith, K., and J. Uddenfeldt, "Capacity of Digital Cellular TDMA Systems," *IEEE Trans. on Vehicular Technology*, Vol. 40, No. 2, May 1991, pp. 323–332.

[6] Lee, W. C. Y., "Elements of Cellular Mobile Radio Systems," *IEEE Trans. on Vehicular Technology*, Vol. 35, No. 2, May 1986, pp. 48–56.

[7] Viterbi, A. J., "Wireless Digital Communication: A View Based on Three Lessons Learned," *IEEE Communications Magazine*, Vol. 29, No. 9, September 1991, pp. 33–36. Reprinted in Abramson, N., (Ed.), *Multiple Access Communications*, Piscataway, NJ: IEEE Press, 1995, pp. 227–230.

[8] Lee, W. C. Y., "Smaller Cells for Greater Performance," *IEEE Communications Magazine*, November 1991, pp. 19–23.

[9] Berrou, C., A. Gavieux, and P. Thitimajshima, "Near Shannon Limit Error-Correcting Coding and Decoding: Turbo-Codes," *Proc. IEEE Int. Communications Conf.*, 1993, pp. 1064–1070. Reprinted in Rappaport, T., (Ed.), *Cellular Radio and Personal Communications, Volume 2: Advanced Selected Readings*, Piscataway, NJ: IEEE Press, 1996, pp. 387–393.

[10] Jakes, W., "A Comparison of Specific Space Diversity Techniques for Reduction of Fast Fading in UHF Mobile Radio Systems," *IEEE Trans. on Vehicular Technology*, Vol. 20, No. 4, November 1971, pp. 81–92. Reprinted in Rappaport, T., (Ed.), *Cellular Radio and Personal Communications, Volume 2: Advanced Selected Readings*, Piscataway, NJ: IEEE Press, 1996, pp. 301–311.

[11] Luise, M., et al., "Guest Editorial: Signal Synchronization in Digital Transmission Systems," *IEEE J. Selected Areas in Communications*, Vol. 19, No. 12, December 2001, pp. 2293–2295.

[12] Viterbi, A. J., "Convolutional Codes and Their Performance in Communication Systems," *IEEE Trans. on Communications Technology*, Vol. 19, No. 5, October 1971, pp. 751–772.

[13] Berlekamp, E. P., R. E. Peile, and S. P. Pope, "The Application of Error Control to Communications," *IEEE Communications Magazine*, Vol. 25, No. 4, April 1987, pp. 44–57.

[14] Forney, G. D., "The Viterbi Algorithm," *Proc. IEEE*, Vol. 61, No. 3, March 1973, pp. 268–278.

[15] Blahut, R. E., *Digital Transmission of Information*, Reading, MA: Addison-Wesley, 1990.

[16] Dholakia, A., *Introduction to Convolutional Codes with Applications*, Boston, MA: Kluwer, 1994.

[17] Clark, G. C., and J. B. Cain, *Error-Correction Coding for Digital Communications*, New York: Plenum, 1981.

[18] Nyquist, H., "Certain Topics in Telegraph Trasnmission Theory," *A.I.E.E. Trans.*, Vol. 47, April 1928, pp. 617–644.

[19] Bellamy, J., *Digital Telephony*, New York: Wiley, 1982.

[20] Pursley, M. B., "The Role of Spread Spectrum in Packet Radio Networks," *Proc. IEEE*, Vol. 75, No. 1, January 1987, pp. 116–134.

[21] Déchaux, C., and R. Scheller, "What Are GSM and DCS?" *Electrical Communication*, 1993. Reprinted in Rappaport, T., (Ed.), *Cellular Radio and Personal Communications, Volume 2: Advanced Selected Readings*, Piscataway, NJ: IEEE Press, 1996, pp. 485–494.

[22] Jung, P., W. Baier, and A. Steil, "Advantages of CDMA and Spread Spectrum Techniques over FDMA and TDMA in Cellular Mobile Radio Applications," *IEEE Trans. on Vehicular Technology*, Vol. 42, No. 3, August 1993, pp. 357–364.

[23] El Gamal, A., and T. M. Cover, "Multiple User Information Theory," *Proc. IEEE*, Vol. 68, No. 12, December 1980, pp. 1466–1483. Reprinted in Abramson, N., (Ed.), *Multiple Access Communications*, Piscataway, NJ: IEEE Press, 1995, pp. 26–43.

[24] Shannon, C. E., "Communications in the Presence of Noise," *Proc. IRE*, Vol. 37, January 1949, pp. 10–21. Reprinted in Slepian, D., (Ed.), *Key Papers in the Development of Information Theory*, Piscataway, NJ: IEEE Press, 1974, pp. 30–41.

[25] Pursley, M. B., "Reed–Solomon Codes in Frequency-Hop Communications," in Wicker, S. B., and V. Bhargava, (Eds.), *Reed–Solomon Codes and Their Applications*, Piscataway, NJ: IEEE Press, 1994, pp. 150–174.

[26] Shannon, C. E., "A Mathematical Theory of Communication, Part 2," *Bell System Technical J.*, Vol. 27, October 1948, pp. 623–656. Reprinted in Slepian, D., (Ed.), *Key Papers in the Development of Information Theory*, Piscataway, NJ: IEEE Press, 1974, pp. 19–29.

[27] Moshavi, S., "Multi-User Detection for DS-CDMA Communications," *IEEE Communications Magazine*, October 1996, pp. 124–136. Reprinted in Tantaratana, S., and K. M. Ahmed, (Eds.), *Wireless Applications of Spread Spectrum Systems: Selected Readings*, Piscataway, NJ: IEEE Press, 1998, pp. 137–149.

[28]   Darling, L., and E. O. Hulbert, "Maxwell's Demon," *American J. Physics*, Vol. 23, 1955, p. 470.

[29]   Hawkins, D., *The Language of Nature*, San Francisco, CA: W.H. Freeman and Company, 1964.

[30]   Duel-Hallen, A., J. Holtzman, and Z. Zvonar, "Multiuser Detection for CDMA Systems," *IEEE Personal Communications*, April 1995, pp. 46–58. Reprinted in Tantaratana, S., and K. M. Ahmed (Eds.), *Wireless Applications of Spread Spectrum Systems: Selected Readings*, Piscataway, NJ: IEEE Press, 1998, pp. 151–163.

[31]   Bennett, C. H., "Demons, Engines, and the Second Law," *Scientific American*, Vol. 257, No. 5, November 1987, pp. 108–116.

[32]   Reed, I. S., and G. Solomon, "Reed–Solomon: A Historical Overview," in S. B. Wicker and V. Bhargava, (Eds.), *Reed–Solomon Codes and Their Applications*, Piscataway, NJ: IEEE Press, 1994, pp. 17–24.

[33]   Calhoun, G., *Wireless Access and the Local Telephone Network*, Norwood, MA: Artech House, 1992.

[34]   Jayant, N., "Signal Compression: Technology Targets and Research Directions," *IEEE J. Selected Areas in Communications*, Vol. 10, No. 5, June 1992, pp. 796–818.

5

SIGNAL HARDENING TECHNIQUES

Golf balls need less protective packaging than eggs do, which means that more of them can be shipped in a given volume of transport space. By analogy, if we can find a way to *harden* communications signals, which are in some ways just as fragile as eggs, we should be able to pack them more efficiently into the available communications space.

In other words, if we can build resistance to noise and interference directly into the transmitted signal—if we can "armor-plate" the signal, so to speak—we can gain capacity. Hardened signals require smaller buffers in space, frequency, and time—less packaging overhead.

In this chapter we survey the growing catalog of signal-hardening techniques. It is not our goal to explain individual techniques in detail; references to the literature will be provided. The emphasis here is on the conceptual underpinning of the hardening strategies employed, grouped into three broad and partially overlapping categories:

1. Straightforward coding concepts, which involve adding redundancy to the transmitted signal, and are ultimately based on the principal of simplifying the receiver's discrimination problem in one way or another;

2. Frameworks for implementing signal diversity, which render the signal more resistant to highly structured forms of interference;

3. Convolutional techniques, for building additional structure into the signal, thus strengthening the correlation properties within the signal, relative to the presumed randomness of the noise sources.

5.1 Coding: A Vast Philosophy

Our starting point is an *analog* signal such as human speech. The signal varies *continuously* and *infinitesimally* in amplitude and frequency. Such a signal is exquisitely vulnerable. It begins to degrade as soon as it is created. Even within the transmitter hardware itself, noise is already accreting to the signal as it passes from stage to stage: From the transmitter, through the combining network, through the duplexer, out the cable through various imperfect connectors to the antenna, over the physical radio channel to the base station receiver and through the similar chain of receiver hardware elements, and thence on through switches and landline transmission channels to its ultimate destination—noise constantly accumulates. Moreover, as Alexander Graham Bell had observed, two circuits laid close together—even for a distance of a foot or two—generate destructive crosstalk in each other. Running many circuits through cables and conduits in office buildings and under city streets, although unavoidable, was a recipe for the propagation of intense interference effects. The proximity of high-power electrical transmission systems and devices, which were growing in popularity right alongside the growth of the telephone network created still more electromagnetic threats to the integrity of the fragile analog signal.

Of course, once the noise blends with the signal, the two cannot be easily unblended.[1] This sticky composite of signal-and-noise is passed on through the network, getting dirtier and dirtier as it goes along. The relative strength of the desired signal steadily diminishes, and the level of noise steadily increases.[2] Entropy prevails, relentlessly.

---

1. Of course, our interference demon—described in Chapter 4—could perhaps unscramble even an analog signal from the noise and interference (or at least from the interference, depending on how we define and endow the demon in information theoretic terms), but we are talking in this chapter about real-world systems.

2. For example, from an early Shannon paper: "In most transmission systems, the noise and distortion from the individual links cumulate. For a given quality of overall transmission, the longer the system, the more severe are the requirements on each link. For example, if 100 links are to be used in tandem, the noise power added per link can only be one-hundredth as great as would be permissible in a single link" [1, p. 154].

Historically speaking, this was a very serious constraint on the construction and operation of a telephone network. Fifty years ago, when the
entire network was analog, a long-distance call would have proceeded over
just such a cascading series of noise-adding, signal-corrupting segments connected in series. First, acoustic noise would have been added even at the
input of the telephone handset. The imperfections of the handset microphone and the in-home wiring would have contributed more noise. The local
loop connection from the caller's home to the central switch would have
degraded the signal further. The switching equipment in the central office
would likely have been quite noisy itself (electrically speaking), especially in
older switches. Then the call would have been routed over a series of long-
distance segments, of varying quality, hop by hop across the country to the
destination switch (more noise), out over the local loop at the far end (more
noise), to the receiver's handset (more distortion and noise depending on the
condition of the equipment). The SNR of the entire wireline link is the composite of the SNRs of each and every individual segment. Like all entropy
processes, the arrow only points one way—toward progressive deterioration.
The only real recourse was to try at the very outset to add as much power to
the transmission as possible, and to try to maintain the network segments in
good condition to keep the noise down as much as possible. In fact, for many
decades, it was not possible to transmit an analog voice signal reliably much
further than about 1,500 miles over wire.[3]

The solution was revolutionary—*digital* transmission.[4] It began with a
handful of deceptively simple ideas—quantization, threshold detection,

---

3. The first transcontinental long-distance lines in the United States were not constructed until 1915—almost 40 years after the invention of the telephone—and they
   had to use wires as thick as pencils, carried on more than a hundred thousand poles
   across the interior of the country. Even so, these were in many ways showcase solutions—the cost of a 3-minute coast-to-coast call was $20 (in 1915 dollars—equiva-
   lent to several hundred dollars today) for a voice quality that must have been barely
   intelligible (the measured bandwidth was only 900 Hz). The routine availability of
   transcontinental telephony had to wait quite a bit longer, until the development of
   microwave radio and, later, digital T-carrier transmission technology. (See "The
   Electrical Century," *Proc. IEEE*, Vol. 84, No. 4, April 1999, pp. 691–694.) "In trying
   to extend communications distances, amplifiers were needed to compensate for
   transmission-line attenuation.... Distances of a few hundred miles were routinely
   achievable, and, with great care, perhaps 1000–2000 miles ... but the quality was
   poor. After a tremendous amount of work, a crude transcontinental telephone service was inaugurated in 1915, with a 68-year-old Alexander Graham Bell making the
   first call to his former assistant, Thomas Watson, but this feat was more of a stunt
   than a practical achievement" [2, p. 387].

regeneration—embedded in a technology called *pulse code modulation*, or PCM, the ur-technology of all of today's digital communications [1]. Suddenly, surprisingly, PCM offered a way to escape from the noise-entropy trap. Among many other things, it made routine (and cheap) long-distance telephony a reality. Although it took many years for PCM to penetrate the commercial network, it is fair to say that this technology, more than any other, is the demarcation line between modern telecommunications and what went before. PCM was as important in this sense as the original invention of telephony itself.

But as with many an engineering *tour de force,* these solutions were somewhat opaque, theoretically speaking. Shannon and his collaborators deserve great credit for deriving certain profound insights right at the beginning, but the full penetration of the "mysteries" of digital transmission has been a continuing process. Indeed, the slow unraveling of the implications of PCM has been, and still is, the dynamic basis for the whole Shannon framework. PCM is a protean solution—a philosophy more than a technology, to recall the title of another famous Shannon paper—which assumes a different significance in different contexts. The early papers are full of remarks and asides that have gained full meaning only in light of subsequent developments. When Shannon comments that with PCM "we have, in a sense, exchanged bandwidth for power," we hear the faint pre-echo of today's thunderous CDMA polemics [1, p. 154]. (We will return to this in Chapter 8.) When he muses that "it is interesting to note that the noise in the recovered [PCM] message is actually produced by a kind of general quantizing at the transmitter and is not produced by the noise in the channel" [4, p. 27], we can now see the prefiguring of source-coding issues that are still very much alive in today's controversies over voice and image compression techniques. The rather Delphic pronouncement that "it is evident that [any digital system] must be highly nonlinear in character" [5, p. 36] hardly hints at the signal processing challenges that such transmission

---

4. The qualifiers that must be added are, first, that digital techniques had been of course widely deployed for the transmission of data, as telegraphy, and indeed the technical and conceptual roots of PCM and other post-Shannon digital techniques are to be found in the study of telegraph systems during the first several decades of this century. Note also that many of the earliest attempts to develop a voice transmission system were quasi-digital in character, derived as they were from telegraphy. It was revolutionary, in fact, for Alexander Graham Bell to recognize that analog voice transmission was a feasible and, in fact, superior alternative for voice communications. See the historical discussion of Bell's analog revolution in Chapter 5 of [3].

techniques have given rise to in practical applications (including, notably, spectrum-efficient digital radio).

The "analog versus digital" construct—with the moral overtones of "digital good, analog bad"—has since become a mythic staple of the communications business, repeated in countless textbooks over the years, and there is no need to provide a detailed résumé of the basic ideas here. I want to emphasize here the aspects of the PCM revolution that are related to the concept of signal hardening, or *ruggedness*, as Shannon denotes it in his exposition of the PCM philosophy:

> One important characteristic of a transmission system is its susceptibility to interference. We have seen that noise in a PCM circuit produces no effect unless the peak amplitude is greater than half the separation between pulse levels. In a binary (on–off) system, this is half the pulse height.... The presence of interference thus increases the threshold required for satisfactory operation. But, if an adequate margin over the threshold is provided, comparatively large amounts of interference can be present without affecting the performance of the system at all. A PCM system ... is therefore quite "rugged." [1, p. 155]

If we can imaginatively reenter the mind-set of the practical telecommunications engineering community at the time this article was published in 1948, which was struggling with the noise-entropy problem, we may be able to appreciate quite how astonishing the claim that PCM could withstand "large amounts of interference ... without affecting system performance at all" must have seemed back then. The community of engineers to whom this manifesto was addressed had grown up with the problem of managing extremely fragile signals in a hostile environment, and here was a philosopher promising them that they could, if they chose, process the signal in such a way as to render it practically impervious. This was not just an assertion; it was a mathematically demonstrable fact.

On the standard life-principle that nothing is ever really free, we may wonder what Faustian price must be paid for this deliverance from noise and interference. The answer(s) are still not entirely clear. Arguably, at least part of the answer is mathematically obvious from Shannon's original formulations, but the conceptual and practical implications have proven to be complex and interesting. In the next few sections of this chapter, we will touch on several of the key concepts. The goal of this presentation is to support the survey of ruggedizing techniques that then follows. A full treatment of the fundamental trade-offs of coding technology—to reintroduce the highly ambiguous but unavoidable term that is normally applied to name this field—is a book-to-be-written in itself.[5]

5.1.1 Coding: The Standard View

What, then, does it mean to digitally encode a signal?

Let us first recapitulate the standard answer as it is conventionally given. The basic sequence of operations in a digital circuit comprises the following steps (Figure 5.1; using now the terminology that was introduced by the PCM revolution):

- *Sampling:* Normally the first step is to divide the time-continuous analog signal into a series of discrete samples. In standard PCM, the sampling rate is 8,000 samples per second.

- *Quantization:* The second step is to round off the continuous amplitude of the signals (which is still present in the discrete time samples) to the nearest of a set of predefined amplitude levels or thresholds. In 8-bit PCM, where each sample is quantized with an 8-bit number, there are 256 predefined amplitude levels, and each sample is rounded to the nearest of these. (How the levels are defined in the first place is an issue that we shall defer for now.)

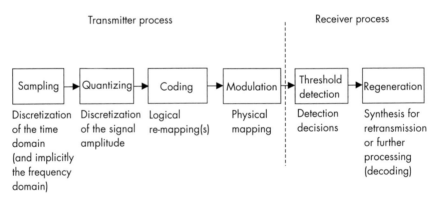

FIGURE 5.1 *Sequence of digital stages in PCM: sampling, quantization, coding, modulation, threshold detection, and regeneration.*

5. I presume to say "to be written" because although there are hundreds of books devoted to coding theory in its many forms, there is still no single well-grounded and integrated treatment of the philosophical problems and issues involved. I use the word *philosophical* in the sense of Shannon's article on "The Philosophy of PCM"—sort of a combination of answers to "Why?" and "At what price?"

- *Coding:* Each quantized sample is next coded into a sequence of eight 1-bit numbers. That is, a string of 1's and 0's is produced. This is the stage at which we speak of the signal as having been fully digitized.

- *Modulation:* In the simplest systems, the 1's and 0's created by the sampling and quantization process are finally translated into electrical pulses (there are many schemes for doing this) representing, again, only two logical states. The modulated signal is transmitted across the channel to the receiver. In radio systems, the 1's and 0's might be represented, for example, by two different frequencies. A tone pulse at frequency A would represent a 1, and a pulse at frequency B would represent a 0.

- *Threshold detection:* The receiver has the task of detecting each incoming signal event, which has been somewhat altered or corrupted by noise or interference during transmission, and deciding which symbol or set of symbols was actually transmitted. In the simplest systems, the receiver is attempting to discriminate between 1's and 0's.

- *Regeneration:* And now comes the *payoff* (Shannon's exuberant term) of the whole process. As long as the noise in the channel is not too large relative to the amplitude of the signal pulse itself, the receiver can accurately determine whether a 1 or a 0 was sent. The receiver can then feed to the end user, or to another transmitter, *not* the original, noise-degraded signal as it was received, but the receiver's *re-creation* of the original signal—a clean copy that is indistinguishable from the original. The receiver *regenerates* the original 1's and 0's. All the noise that was accumulated on that particular transmission link is wiped out (under reasonable assumptions), and the regenerated signal can be retransmitted over the next link. The process, of course, can be repeated indefinitely, and if reasonable care is taken, the regenerated signal at the far end will still be a perfect copy.

That is the story in its conventional nutshell: Regeneration is the magic of PCM and digitization generally. As Shannon exults, "Practically, then, the transmission requirements for a PCM link are almost independent of the total length of [the circuit]. The importance of this fact can hardly be overstated" [1, p. 154].

5.1.2 A Deeper Look

Anyone who has read any of the hundreds of textbooks on digital communications has come across some version of the nice, compact presentation we have just presented. It seems simple, straightforward, and eminently comprehensible. So what is wrong with this picture? What is left out of the standard explanation? What is really going on here?

Well, we can start with a crucial problem posed by the subject matter of this book: In mobile radio systems, serial signal regeneration as described by Shannon is impossible. This is the corollary of the nonengineerable channel we discussed in Chapter 3. If regeneration were the main payoff, then it would seem that radio engineers were not invited to the party.

There is something else important going on here that has nothing per se to do with regeneration. From our perspective in this chapter, it is the *coding* step that produces the other payoff—by significantly *hardening* the signal for transmission over any given link. Just as amazing as the conquest of distance (produced by regeneration) is the conquest of noise within any individual link. Shannon, for example, estimated that a quantized signal like PCM could achieve good telephone voice quality at a signal-to-noise level of about 20 dB, whereas to realize the same output quality with an analog signal would require "the 60- to 70-odd dB ... for high-quality straight AM transmission of speech" [1, p. 153]. In other words, by his estimate a quantized signal could survive with noise levels *10,000 to 100,000 times higher* than the level that an analog signal could tolerate.

Let us pause a moment to consider this point. If Shannon's conquest of distance by regeneration is impressive, it should be at least as astonishing that it is possible by means of what looks like a very simple procedure to turn an egg as hard as a golf ball, so to speak. We are so used to digital techniques that they have lost the magic they probably ought to have for us. Quantization—is it more than just rounding? Followed by a simple translation to the binary system? Indeed, if rounding and base 2 translation can have such a transformative effect, it should perhaps alert us to the likelihood that what is happening may not be quite so simple. There is real alchemy at work.

Benchmarks of signal quality, and analog/digital comparisons in terms of equivalent decibels of SNR, do vary considerably. Bellamy cites an equivalence of 45–46 dB in analog wireline transmission to 15 dB or so for digital systems [6].[6] In mobile radio systems, where the analog standard is FM (not AM), a more robust technique, the signal-to-noise level for good

---

6. A similar number for wireline AM telephony is cited by Kucar [7].

quality has been variously defined between 18 dB and 25–30 dB, and corresponding digital systems are said to need 10–12 dB to achieve the same quality.[7,8] Other studies tend to push the digital numbers somewhat higher.[9]

Such quality comparisons are difficult in part because the perceptual nature of noise in an analog system is quite different from that produced by digital systems (as we shall touch on further later). Yet the fact remains that even if the gap between analog and digital is reduced to only a few decibels, this still translates into series of enormous advantages in the practical deployment of *wireless* systems. A few decibels may mean reducing the number of cell sites by half, or doubling the capacity of a wireless network. Those familiar with the recent history of the wireless business know that huge debates have been waged, and billions of investment dollars wagered, over claims of a few decibels of relative advantage between technological alternatives.

Thus, the first important step toward signal hardening is the quantization process itself. Simply by using digitized signals we can achieve, in principle, a significant gain in capacity and performance, because the signals are more resistant to noise and interference. What is it about quantization (and coding) that really produces this gain? It may seem obvious, but it is really quite complex, and one of the driving forces behind the post-Shannon revolution has been the gradual unraveling of the mysteries of this simple, straightforward procedure. From this inquiry have flowed many of the most important technologies vital for third-generation wireless systems that—to achieve high capacity—must be based on hardened signals.

5.1.3 Quantization: Many-to-One Mapping

As writers have struggled to find a high-level characterization of the processes underlying the creation of a *digital* signal, one of the most general

---

7. For example, see [8] where a figure of 18 dB is said to be equivalent to "a subjective test with a criterion that 75 percent of the listeners evaluate the voice quality as good [or] excellent while driving in a mobile radio fading environment with various vehicle speeds" [p. 71].

8. Later Lee makes this equivalent to 10–12 dB SNR for an unspecified digital system. Unfortunately, this kind of loose formulation is typical of mobile radio transmission quality benchmarks.

9. "In summary, the speech quality perceived in analog with 18 dB C/I is about the same as for digital at 18 dB C/I. Defining coverage with the same quality, both systems need about 17 dB C/I" [9, p. 326]. This sort of result, which runs counter to the basic argument in this chapter, requires a fuller treatment in connection with channel coding concepts.

and useful models has been to view digitization as a process of *mapping*.[10] Mapping is itself a suggestive and somewhat ambiguous term, and one way to view it is as the construction of a table of equivalences or substitutions between the elements of two different sets. The first set here is the set of possible messages produced by the source, and the second set is the set of signals that can be created by the transmitter. The mapping of one set to the other tells us which signal to transmit based on which input message has been observed (Figure 5.2).

For the first mapping in a digital communications system (there will be others), the second set—the target set—is typically smaller than the first set. That is, multiple different messages (or message elements) are mapped onto,

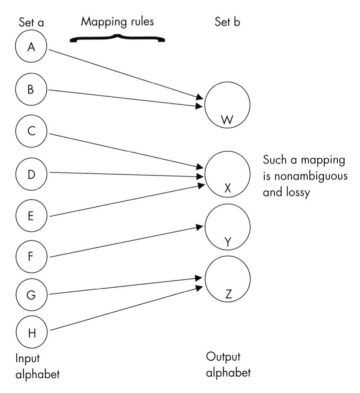

FIGURE 5.2 *Mapping from a larger set a (messages) to a possibly smaller set b (signals to be transmitted).*

10. As with so much else, Shannon was probably the first to introduce this metaphor [5, Sec. IV].

and can be substituted by, just one signal element. In fact, the range of variability in the message set may be practically infinite. This corresponds to an analog source. This infinite (or at least very large) set of message elements is thus *compressed* substantially as it is mapped onto a much smaller set of signals elements.

This compression has several effects. First, the signal set is now finite, and the target set is normally quite small. It is commonly referred to as an *alphabet,* and this is an appropriate metaphor. Our common alphabet is in fact a digital signal set that converts the immense variability of a spoken language with millions of speakers and different pronunciations into a standardized written form.

Second, as the size of the signal set is reduced relative to the source message set, the receiver's task of determining the correct message is made easier and the likelihood of making an error is reduced. It is a classification problem, sorting a large number of input messages into a relatively small number of bins or categories. Intuitively, the fewer the number of categories, the fewer the errors. If we are trying to classify a set of animals by their species only (zebra, shrimp, and so forth), we will probably make fewer errors than if we also have to classify them by sex or age.

Third, the mapping should have a certain logic to it. If we are sorting animals, we probably want to sort male and female horses into one category, and male and female bears into another, or we might sort female animals into one category, and males into another. It would somehow make less sense to create one category for female horses and male bears, and another category for male horses and female bears. In other words, there is an underlying *semantic* criterion that is developed in order to construct the categories. Indeed, we can go one step further and acknowledge that the *design* of the categories is likely to be one of the most important characteristics of a particular coding scheme. This is worth emphasizing, because it is so often ignored in formalistic treatments of communications systems.

Fourth, the mapping itself can be subtly optimized to further reduce the likelihood of making an error. The design of a digital system involves designing a whole series of these mappings (Figure 5.3) The source messages are mapped onto a set of transmittable signals, which are mapped onto modulation symbols (e.g., phase zones in the case of a phase-shift-keyed system, or frequencies in an FSK system). The received signals (which are now corrupted by noise) are remapped onto a set of possible received messages (which is normally equivalent to the set of transmittable signals at the transmitter).[11] The concept of mapping thus extends to a whole series of substitutions of signal elements—symbols, sequences of

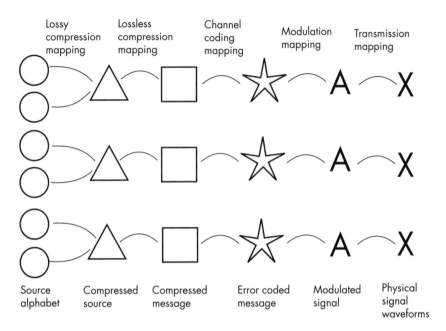

FIGURE 5.3 *Series of mappings involved in constricting a digital signal: some compressing and some expanding. Each mapping involves a set of rules for translating from one alphabet to another. The first stage here is a many-to-one mapping (involving information loss). The other stages are all one-to-one mappings (no loss of information).*

symbols, alphabets—as the original message is groomed for transmission and/or cleaned up by the receiver.

Even for the same number of target categories—that is, for the same size alphabet or codebook[12]—different mapping strategies can result in higher or lower error rates for a given channel. In particular, if the mapping

---

11. Jayant and Noll [10] illustrate a typical use of this term, studded with formalisms: "A digital system consists of at least three mappings. At the transmitter, the real-valued amplitude x is mapped onto an L-ary number k. The channel map k into k deterministically, if the transmission is error free; if not, it maps k into another L-ary number k', probabilistically. At the receiver the L-ary number k or k' is mapped into representation level y_k or $y_{k'}$." [p. 116]. I cannot pass on without observing how, in a book that is really quite excellent in so many ways, the formalistic style can be so dysfunctional.

12. When the signal set is defined in terms of sequences of shorter symbols, these "vectors" are sometimes referred to as codewords, and by analogy, the dictionary of these codewords is referred to as a *codebook*.

is designed with an understanding of typical channel errors in mind, better results can be achieved. A very basic idea of this sort is incorporated in what is known as *Gray encoding*, in which adjacent modulation phase categories (in a PSK system, for example) are mapped from signal elements that differ by only one binary digit. Because phase errors in a received signal are most likely to cause the detector to shift to one of the adjacent phase regions (rather than to a more distant phase region), this will limit the impact of the most likely errors to a single bit[13] (Figure 5.4). The idea of manipulating the mapping strategy is also behind one of the more recent revolutions in coding theory: so-called "trellis coding," which is based on a sophisticated handling of the mapping between the coded information bit sequences and the signal points in a modulation constellation.[14]

To recapitulate, the practical importance of coding, as many-to-one mapping, is that it begins the fundamental process of ruggedizing the signal by reducing the significance of variations in the input signal. Many variations will be mapped into a few categories—the alphabet—and the receiver will have far fewer and easier discriminations to make. In the most radical of all mappings, the infinitely variable analog signal is mapped into a series of on–off pulses, 1's and 0's in the binary alphabet. The receiver's job has been simplified to the extreme: All it has to do is correctly decide whether the signal is "on" at a given moment or not. The infinite variety of the analog signal has been reduced to a series of basic binary choices. Indeed, we are all by now quite familiar with the notion of digital representation of input signals possessing great variety and subtlety. With enough ones and zeros (and the right display engine), we can paint the Sistine Chapel. Because it is encoded in binary form, such a digitized painting can be reproduced perfectly, transmitted perfectly, over and over.

---

13. "The mapping or assignment of k information bits to the $M = 2^k$ possible phases may be done in a number of ways. The preferred assignment is one in which adjacent phases differ by one binary digit ... This mapping is called *Gray encoding*. It is important in the demodulation of the signal because the most likely errors caused by noise involve the erroneous selection of an adjacent phase to the transmitted signal phase. In such a case, only a single bit error occurs in the k-bit sequence" [11, p. 259]. See also [12, pp. 97–100].

14. See Proakis [11] on trellis coding: "The key to this integrated modulation and coding approach is to devise an effective method of mapping the coded bits into signal points such that the minimum Euclidean distance is maximized. Such a method was developed by Ungerboeck, based on the principle of *mapping by set partitioning* [emphasis in original]" [p. 489].

Decimal number to be encoded	"Natural" binary code	Note that a change from "3" to "4" changes all three digits	Decimal number to be encoded	Gray code	With Gray code, each code value differs by just one digit from adjacent values.
0	000		0	010	
1	001		1	011	
2	010		2	001	
3	**011**		3	000	
4	**100**		4	100	
5	101		5	101	
6	110		6	111	
7	111		7	110	

(a)

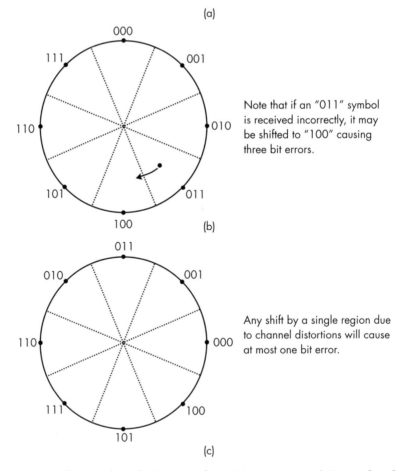

Note that if an "011" symbol is received incorrectly, it may be shifted to "100" causing three bit errors.

(b)

Any shift by a single region due to channel distortions will cause at most one bit error.

(c)

FIGURE 5.4 *Phase circle with Gray encoding: (a) comparison of Gray code values, and phase circle for 8-PSK modulation with (b) "natural" encoding and (c) Gray encoding.*

But something even more profound has taken place, which is implicit in the last statement. The possibility of *error-free transmission* has been established. Shannon showed, or admitted, that the true analog signal could not be transmitted with perfect accuracy at all. Errors are *inevitable*. As Shannon deduces, almost whimsically:

> A continuously variable quantity [i.e., an analog source] can assume an infinite number of values and requires, therefore, an infinite number of binary digits for exact specification. This means that to transmit the output of a binary source with exact recovery at the receiving point requires, in general, a channel of infinite capacity (in bits per second). Since, ordinarily, channels have a certain amount of noise, and therefore a finite capacity, exact transmission is impossible. [4, p. 26]

Once the signal has been mapped onto a finite number of categories—a finite alphabet—the theoretical possibility of error-free transmission is created, *ex nihilo* it would seem. The mapping process has transformed a very difficult discrimination problem with an unavoidable probability of error into a (large) number of easy discrimination problems. The probability of error has not just been averaged out or somehow apportioned among this larger number of little decisions. It has been (under certain reasonable assumptions) eliminated almost entirely. Even in the realm of practice, as opposed to theory, this seemingly simple mapping substitution produces striking results.

5.1.4 Nonlinearity and Threshold Effects

This miracle has been received somewhat uncritically by the communications industry, and perhaps that is as it should be. The deliverance from noise, which had been beyond reach, is now something that can be designed, achieved, and managed as a cost function in the network. Especially in wireless networks, where serial regeneration is not normally feasible, the quantized signal is significantly hardened compared to its analog counterpart, which should allow smaller buffers and denser packing of the communications space with user traffic. Unfortunately, the price paid for this is also complex, and to some degree the gains are given back (unless further steps are taken, as we shall see).

In an analog system, the relationship between channel quality (noise levels) and signal quality is (broadly speaking) what the engineers call *linear*. A small amount of noise will degrade the signal by a small amount, and a somewhat larger amount of noise will degrade the signal by a somewhat larger amount, and so on. The degradation is gradual, cumulative, and is even said sometimes to be graceful.

With a quantized, digital signal, the effects of noise are very different. The relationship is nonlinear: A *threshold* has now been created by the mapping process.

> As we change the message a small amount, the corresponding signal will change a small amount, until some critical value is reached. At this point the signal will undergo a considerable change. In topology it is shown that it is not possible to map a region of higher dimension into a region of lower dimension [i.e., to map a larger set into a smaller set] *continuously*. It is the necessary discontinuity which produces the threshold effects ... [5, p. 165]

It can be stated today even more forcefully. Essentially, up to a certain threshold value the noise has no effect whatsoever; at the threshold, the effect is catastrophic. The system degrades in a precipitous, nonlinear fashion (Figure 5.5). At the level of an individual symbol or bit of information, an error means that the receiver has assigned the received signal event to the wrong category. More interestingly, a similar phenomenon occurs at the system level. If the average level of noise (relative to the signal strength) is below a certain threshold, the whole system will work smoothly. If the noise level crosses the threshold, the system breaks down. In a sense, the whole system now has two states: an "on" state in which there is essentially error-free transmission, and an "off" state in which there are so many errors that reliable transmission is not possible.

Nevertheless, the digital strategy pushes us to embrace nonlinearity. As Shannon further observes, "it is evident that any system, either to compress TW [time and bandwidth] or to expand it and make full use of the additional volume, must be highly nonlinear in character ..." [5, pp. 165–166]. In other words, to maximize efficiency and capacity utilization, the thresholds and nonlinearities should be made fewer, hence sharper.

The introduction of nonlinearities, or thresholds, creates or poses a series of new issues. For one thing, it alters the perceptual nature of noise to the end user. The system tends to become an all-or-nothing proposition. In an analog voice communications application, noise is experienced as a fading, intermittent phenomenon, and the acoustic processing systems in our brains are well equipped apparently to handle this kind of interference (for speech signals). We can tolerate a lot of analog noise and still understand what is being said. In a digital system, the all-or-nothing phenomenon means

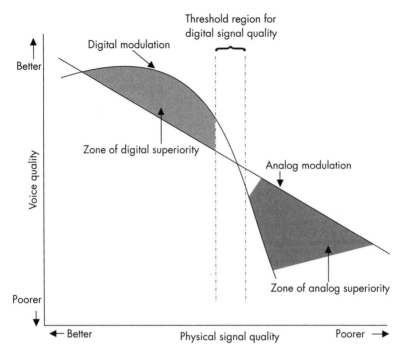

FIGURE 5.5 *Analog versus digital degradation—the effects of nonlinearity.*

that the speech signal will tend to break up instead of merely fading or going through short periods of noisiness. Around the threshold value, the structure of the signal energy is quite disrupted and the speech goes from being okay, to being unintelligible, often quite abruptly.

It also creates a new kind of signal degradation. As we have seen, the process of quantization involves rounding off the analog samples to the nearest threshold value in the alphabet of signal elements. In this rounding process, some of the information is lost, irretrievably. An analog signal sample that lies *between* threshold A and threshold B, or category A and category B, *must* be assigned to one or the other. Once it has been assigned (quantized), all information that would distinguish it from all other samples assigned to the same category is gone (Figure 5.6). A deviation from the original signal has been introduced into the quantized representation of that signal.

Is this a new form of noise, or a new type of error—it has been called both[15]—or is it something else? It has often been modeled (to my mind,

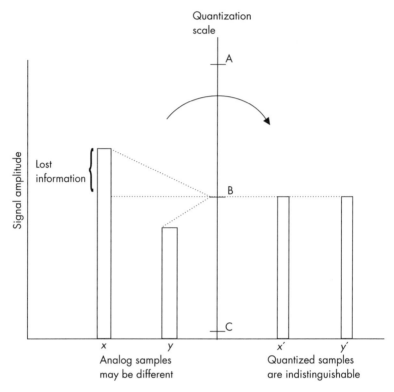

FIGURE 5.6 *Information lost during quantization.*

naively) as an additive white noise source [11] (Figure 5.7).[16] It is better seen as an artifact of the quantization process, something that we create. It is an aspect of the signal itself, which has a definite, deterministic structure like

---

15. In "The Philosophy of PCM," Shannon mixes his metaphors in a way that became characteristic of the entire field: "Representing the signal by certain discrete allowed levels only is called quantizing. It inherently introduces an initial error in the amplitude of the samples, giving rise to *quantization noise*" [1, p. 151].

16. Gray [13] is more specific about the limits on this assumption: "The white noise approximation for quantization noise is originally due to Bennett and it has become the most common approximation in the analysis of quantization systems. Note that the … approximation really only makes sense if: 1. The quantization is uniform. 2. The quantizer does not overload. 3. The rate is high. 4. The maximum size of the quantizer cells is small. 5. The input joint density is smooth. Unfortunately, the assumption is often made even when these conditions do not hold" [13, pp. 134–135].

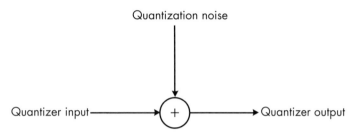

FIGURE 5.7 *The naïve view: additive quantization noise. (From [14]. © 1996 Morgan Kaufmann.)*

the signal, rather than the property of randomness we associate with noise. As one recent author remarks:

> The difference between the quantizer input x and output y ... besides being referred to the quantization error is also called the *quantizer distortion* or *quantization noise*. But the word *noise* is somewhat of a misnomer. Generally, when we talk about noise we mean a process external to the source process [that is, the signal itself]. Because of the manner in which the quantization error is generated, it is dependent on the source [signal], and therefore cannot be regarded as external to the [signal]. [14, p. 173][17]

It bears emphasizing: Quantization noise (so-called) is an artifact, a result of the purposeful modification of the signal. "Unlike channel noise or thermal noise, quantization noise is deliberately introduced ..." [10, p. 118]. The creation of this quantization artifact opens a new Pandora's box for the communications engineer. First of all, it does degrade the signal, and if we are willing to ignore its structure, we can (and must) accept additional decibels of degradation (in addition to noise, interference, and other forms of signal degradation). We must engineer for it. Next we discover that different quantization procedures—that is, different ways of dividing up the input signal range into threshold values—produce different amounts of quantization noise. Even for basic PCM, the difference between a dumb quantizer (based simply on uniformly spaced amplitude thresholds) and a moderately smart quantizer that takes into account some of the properties of human speech

---

17. See also Gray [13]: "[Quantization noise] is not really noise in the usual sense: it is a deterministic function of the input and hence cannot be statistically independent of the input" [13, p. 135].

and hearing can mean as much as a 24-dB reduction in quantization-induced signal degradation, which is quite a lot [13, p. 91].

Second, this so-called "quantization noise" is actually correlated with the signal itself. It has structure, and this structure is accentuated by the use of structured (nonuniform) quantizers. The existence of this structure means that the noise is not at all white but "colored and input-correlated" [10, p. 164]. The structure can assume the form of perceptually significant interference, which can be far more undesirable than the level of the noise value alone would indicate. To break up this structure, one common solution is to add even more noise. A technique called *dithering* is used to randomize these quantization artifacts more thoroughly, to make them more truly noise-like and thus less offensive to human eyes and ears (for audio and image data).[18] Indeed, designing smart quantization schemes has become a well-elaborated and active field of communications research. The sometimes odd and unlooked-for effects of quantization-induced nonlinearities, and the methods of managing them, have assumed central importance in many subdisciplines of digital communications.[19]

Even just skimming the surface of this subject, it should be clear by now that quantization (coding), far from being a simple, intuitive signal conversion procedure (as it is often portrayed), is in fact a step through the looking glass. It is indeed a new philosophy—and perhaps more than that. We have done more than simply map a complex reality into a set of categories. We have embedded strange nonlinear boundaries in our previously smooth phenomenal fabric, and these nonlinearities will henceforward become a source of continuing engineering challenges. Philosophically speaking, we should recognize that we are actually injecting noise (or something like it) into our signal, on purpose. The hope is that by so doing we can inoculate our signal against the various wild forms of noise that prey on it. Thus, like medical researchers designing vaccines against infectious diseases, we find that we must take great care to design this artificial noise. We must apply to

---

18. "When quantization is coarse ... [there is] a colored and input-correlated error sequence. ... The ... autocorrelation and cross-correlation properties are reflected by perceptually undesirable signal-dependent patterns in the error sequence. By adding appropriate high-frequency signals to a waveform prior to quantization, it is possible to break up these undesirable patterns. The high-frequency perturbation signal can be a pseudorandom signal; it can also be a carefully designed deterministic signal; in either case, it is referred to as a *dither* signal ..." [10, p. 164].

19. See, for example, Chapter 6 in [15] for a discussion of the many types of quantization artifacts that must be dealt with in designing a digital filter.

the design an increasingly detailed knowledge of the interaction of our coding processes with the signal, the channel, and the receiver. We find as well that we must apply the cure in complex regimens, in layers, to ensure that each remedy does not create a new illness.

As the belated, though by now overwhelming, success of digital techniques would suggest, this strategy is highly productive. We can produce immunized signals that are much more resistant to higher levels of interference. With this ability to design hardened signals, we are launched on the path toward true high-capacity networks, which was the basis of many of Shannon's insights, but it is a complex path, with many forks and choices. The design of hardened signals involves new types of trade-offs and conceptual challenges that often seem open ended, even 50 years after the Shannon revolution. The immunology of communications signals is as complex a subject as we can find in engineering.

5.1.5 Coding as Redundancy Construction: One-to-Many Mapping

The quantized signal is tough, but brittle. It can withstand a high level of noise compared to its analog precursor, but when the noise level reaches a certain threshold, the signal deteriorates abruptly and catastrophically. Where analog bends, digital snaps. In an engineerable channel (such as a wireline T1 link with repeaters), the noise level can be managed such that it will never reach this threshold and the digital signals are transmitted with virtually perfect accuracy. In the nonengineerable wireless channel, where the noise and interference fluctuates violently and rapidly, the basic digital signal is paradoxically vulnerable.[20]

The structure of the coded signal also creates vulnerability. For example, in 8-bit PCM, the *most significant bit* (MSB) accounts for half of the entire amplitude range, while the *least significant bit* (LSB) accounts for only 1/256th of the range. An error in the LSB may pass unnoticed, but an error in the MSB will completely alter the signal (Figure 5.8).

As a result, basic quantized signals show a sharp degradation in error performance at some crucial noise level (Figure 5.9) [16]. This "knee in the curve" defines the overall suitability of the coding scheme for a particular channel. Again, for clean (engineerable) channels like those in most wireline networks, where we can be sure that the noise level will stay below the

---

20. Unprotected PCM is unusable in mobile systems. There is no reasonable way to guarantee sufficient margins to make sure that the catastrophic threshold will not be breached.

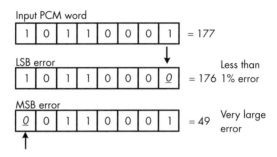

FIGURE 5.8 *LSB and MSB error impacts.*

threshold value, the use of unprotected PCM and similar schemes is unproblematic. In a wireless channel where extremely high bit error rates are likely from time to time, we need additional protection. Fortunately, once the signal is in code, (digital) form, it is highly amenable to additional processing, which can further harden the signal, and very substantially so.

The basic concept underlying the most straightforward protection schemes is to add *redundancy* [17]. The very simplest solution is to repeat the coded signal more than once. If an odd number of repetitions are transmitted, the receiver can use a simple majority decision to compare the various received copies of the underlying information bits (Figure 5.10). Because the channel noise is often assumed to be a random process that affects each bit independently, the chances of getting a correct decision on, say, a three-out-of-five decision are considerably better even for a long string of such expanded information bits than for unprotected bits decided on the basis of a single transmission each.

Generically, this type of process is called *channel coding*—it is additional coding of the signal that is designed to add redundancy and thereby mitigate the effects of channel noise. It is also often referred to as forward error correction coding, or FEC, where *forward* refers to the fact that the transmitter *forwards* to the receiver the additional information needed to detect and correct some errors. Indeed, it is also often called simply *coding* (ignoring the other uses of the term).

One way to look at this channel coding process is as a second mapping stage.[21] However, whereas the quantization mapping was a many-to-one

---

21. "With linear block codes the redundancy is added by taking an input sequence of k information symbols and mapping it into a transmitted sequence of n symbols (nk)

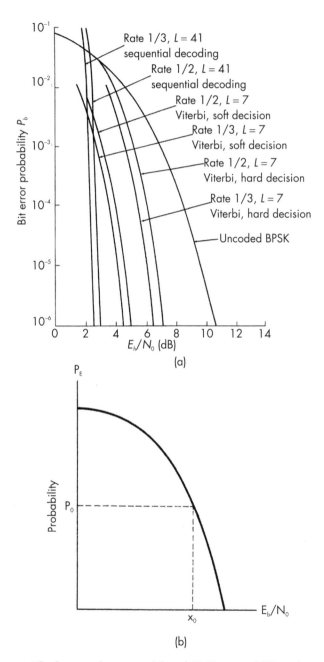

FIGURE 5.9 *The knee in the curve of digital SNR versus BER performance:* (a) *performance curves of various convolutional codes (From: [11]. © 1989 McGraw-Hill Inc.); and* (b) *general shape of the probability of error versus E_b/N_0 curve (From: [16]. © 1982 IEEE. Reprinted with permission.)*

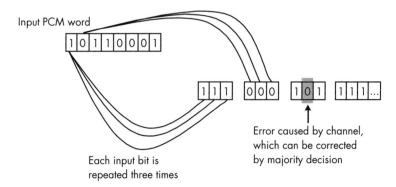

Input PCM word

Each input bit is
repeated three times

Error caused by channel,
which can be corrected
by majority decision

FIGURE 5.10 *Simple repetition coding.*

process, whereby the analog signal was in some sense *compressed* and certain information rounded off and thrown away, the second-stage mapping involved in channel coding is a one-to-many mapping, which results in an *expansion* of the signal and the addition (if not per se the creation) of new information. As we shall see, the mappings are usually quite complex (the repetition code described above is far too simple to provide enough protection for most purposes). In some cases, the mappings produce distinct extra bits that contain error correction information about the payload. In other cases, the mapping commingles the information payload and the error correction redundancy information such that the output of the channel coding process consists of bits or symbols that blend both functions. Indeed, such mappings are increasingly generated by means of advanced mathematical transformations, such that the redundancy is not readily discernible as a separate part of the transmitted code, other than by the obvious expansion of the size of the signal.

In a sense, this expansion is the obscure key to the whole matter. Conceptually, and practically, the one-to-many mapping *expands* the signal. It also causes an expansion of what we may call the entire *message space*. Both expansions (signal space and message space) have to do with accentuating the distinction between the *structure of the signal* and the (presumed) *structurelessness of the channel noise*. The core idea is that the signal has structure,

_____

which are linear combinations of the k information symbols. The mapping defines the particular code.... Redundancy is added in a similar fashion for convolutional codes ..." [18, p. 46].

whereas the noise does not (or should not). The effect of expanding the signal is to accentuate the randomness of the noise. The effect of expanding the message space is to accentuate the uniqueness of the signal relative to other signals. Let us look at each of these processes in turn.

5.1.6 Signal Expansion: Channel Coding as Noise Averaging

The repetition code described above introduces the idea of spreading the signal out in time, which, intuitively, ought to average down the effects of the channel noise. Assume that we are transmitting an 8-bit PCM word as a sequence of eight 1's and 0's. Assume that the chance of receiving any given pulse incorrectly due to channel noise is 12.5% (to make the math easy). For each 8-bit PCM word, therefore, we should expect to see an average of 1 bit error.

If we do nothing more than send the same PCM word three times and let the receiver take the majority decision on each bit, the probability of making an incorrect two-of-three decision on each bit goes down to 4.3%, and we should expect to see an average of about 0.34 bit errors per decoded 8-bit PCM word—or about one uncorrected bit error every three 8-bit words.

If we expand to a best three-of-five, the probability of an incorrect decision on a bit drops to 1.6%, and we will see an uncorrected bit error about once every eight 8-bit PCM words.

With seven repetitions and a best four-of-seven decision rule, the probability of an uncorrected bit is about 0.6% and uncorrected errors will sneak in only about once in every 20 PCM words (Figure 5.11). These gains are achieved without any heavy math or ingenious algorithms. There is no real code space or distance function being created here. Just by brute force repetition of the same transmission a number of times, and allowing a majority decision by the receiver, we can drive the error probability into the ground (as long as the noise is truly a random process).[22]

Coding theory has sometimes referred to this as *noise averaging* [18] and it underlies, explicitly or implicitly, many of the powerful error correction techniques that we shall survey below. For example, another way of

---

22. This is very close to the concept underlying Shannon's famous and infamous "proof" of the possibility of error-free transmission in the presence of noise. "It is a rather surprising result, since one would expect that reducing the frequency of errors would require reducing the rate of transmission, and that the rate must approach zero as the error frequency does. Actually, we can send at the rate C [the capacity of the channel] but reduce errors by using more involved encoding and longer delays at the transmitter and receiver. The transmitter will take long se-

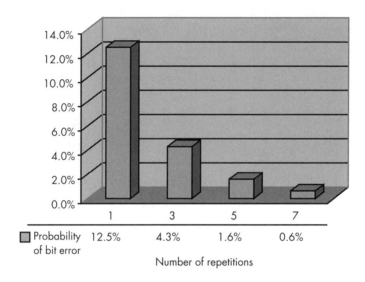

Number of repetitions	1	3	5	7
Probability of bit error	12.5%	4.3%	1.6%	0.6%

FIGURE 5.11 *Noise averaging: declining probability of error as a function of the number of repetitions for an 8-bit word with majority decision on each bit.*

looking at this is to view the decoding as proceeding not on a bit-by-bit basis, but in blocks of symbols that are probabilistically decoded (e.g., three of five). Clark and Cain essentially reformulate the results of our repetition example in terms of the block-size (Figure 5.12):

> As the block size increases ...not only does the fraction of symbols that are in error in a [given] block approach the average channel rate, but more importantly the fraction of blocks that contain a number of errors that differ substantially from the average becomes very small.... If we are willing to process symbols in blocks rather than one at a time, it might be possible to reduce the overall error rate.... The potential for

quences of binary digits and represent this entire sequence by a particular signal function of long duration. The delay is required because the transmitter must wait for the full sequence before the signal is determined. Similarly, the receiver must wait for the full signal function before decoding into binary digits" [5, p. 166]. See also Hamming's [12] comments: "Shannon's main theorem, which used very long block codes, shows that we can signal at a rate arbitrarily close to the maximum rate and with arbitrarily few errors. The proof is based on random encodings which use very long blocks, and in any one of the long words there will probably be many errors to be corrected at the receiver, which means that to be efficient we must work at high error-correction rates" [12, p. 209].

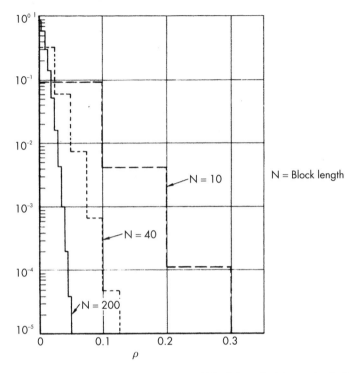

FIGURE 5.12 *Probability of error as a function of block length. (From: [18]. © 1981 Plenum Press.)*

performance improvement that is obtained through noise averaging ... increases with block length. [18, p. 2]

This line of reasoning leads, on the one hand, to sophisticated block-encoding schemes such as vector quantization (see Chapter 6). On the other hand, it suggests that better decisions can be made by allowing the receiver to proceed probabilistically, making so-called "soft decisions" based on *sequences* of symbols (instead of hard symbol-by-symbol decisions). Indeed, much of the emphasis in recent coding theory has been on finding ways to soften the decisions at all levels of signal processing, and allow the noise to average itself down. When the noise plays along (by being sufficiently random), the strategy works very well.

Credit for the insight belongs once again to Shannon himself, and he has a marvelous, easygoing way of expressing it:

Actually, two signals can be reliably distinguished if they differ by only a small amount, provided this difference is sustained over a long period of time. Each sample of the received signal then gives a small amount of statistical information concerning the transmitted signal; in combination, these statistical indications result in near certainty. [5, p. 166]

There, now we all must have known that!

5.1.7 Message Space Expansion: Channel Coding as Signal Geometry

The penalty for this improved signal hardening is the wanton waste of channel bandwidth (or so it would seem). By expanding the signal by a factor of 3 (in the repetition example), we reduce the number of error-containing 8-bit words by a little *less* than the inverse of that factor. The five-times repeating code gives us a moderately better return on investment, and by the time we reach a seven-times repeating code we are getting back a respectable multiplier in terms of error reduction.

However, we must admit that the very parameter we set out to optimize, capacity, has suffered grievously. To achieve a modest additional hardening of the signal, we have suffered huge reductions in capacity. Although the hardened signals can be packed more densely (in space, time, and frequency), it is not intuitively obvious that we have gained much ground. Indeed, for many years it was a (false) cliché in the telephone industry that digital signals always required more bandwidth than analog signals, and although we are past it now, I can remember in the early and mid-1980s having to overcome this objection to the introduction of digital techniques for cellular radio.[23]

The solution, in part, is to begin to design the expansion more intelligently. For example, in the three-repetition code discussed earlier we were sending 24 bit symbols of information for each original 8-bit PCM word.

---

23. "Analog communication of speech with commercial quality will require ... a signal-to-channel-noise ratio of at least 30 dB. Digital speech coders, on the other hand, are very robust ... and they require signal-to-channel-noise ratios of only 10.8 dB [for BPSK] and 18.7 dB [for 16-PSK].... The saving in terms of required channel quality is, of course, accompanied by greater demands on bandwidth. In place of the 4 kHz bandwidth required for analog speech communications, digital speech systems will require ... bandwidth expansions ... over the analog system.... Digital communication is based on an exchange of bandwidth for (channel) signal-to-noise ratio" [10, pp. 19–20]. The specific numbers cited here are not relevant to the general point of view.

That is, we were sending 3 bits to represent each original bit. Yet we know that an error in the MSB has a much worse effect than an error in the LSB. So assume that we can rob the 2 extra bits from the expansion of the LSB and use them to expand the MSB to 5 bits. The LSB is now "naked" and its error probability (in this example) is back to 12.5%, but the MSB enjoys the extra armor and a lower error probability of only 1.6%. In this skewed three-times repetition code, the overall probability of a bit error has increased slightly, but if we weight the errors by the significance of the damage they do, the skewed code is about 50% harder than the flat code.

This is merely suggestive of the power of applying more intelligence to the design of the channel coding scheme. Let us now make a great leap, conceptually, to the other end of the process altogether. We have been looking at all this from the perspective of the signal. Let us look at it now from the perspective of the code. The real issue is not how to tweak this bit or that, but how to design the entire code space or signal space in a comprehensive way that maximizes the *uniqueness* of each transmission. We want to expand the whole signal space in a way that provides an efficient and effective buffer between individual transmitted coded signals, making it easy to discriminate them from one another. We want to spread the whole code out.

The best way to capture the perspective of the code is by means of another of Shannon's powerful conceptual innovations: to view the code in *geometrical terms.*

The full presentation of the geometrical metaphor is contained in Shannon's paper "Communications in the Presence of Noise" [5] and it bears reading, and rereading, as one of the most brilliant and enduring conceptual revolutions in our century. In many ways, I regard the geometrical metaphor as a more important constellation of insights than some of the other, more popular Shannon concepts (such as the much-abused definitions of entropy and average information and even channel capacity). Whereas information as entropy is one of those double-back-flip sort of ideas that never fails to astonish the beginning student, I am not sure I have ever given it much thought since or heard it applied in discussion by others. On the other hand, the geometric metaphor is the common language of communications engineers when they are talking at a conceptual level. It has even recently penetrated (as popular engineering terms will do) into the general business jargon, and in the late 1990s it was common to hear businesspeople and even stockbrokers talking about this or that space (product space, applications space), in ways that I often find neither inappropriate nor unenlightening.

Nevertheless, I will not attempt a close derivation of Shannon's specific ideas here. We can begin (for purposes of this chapter) with a much simpler presentation of the concept of the code space, drawn from one of the better textbooks [19, p. 235].

Assume that there is a source that is generating a string of letters of the English alphabet, uppercase only. The digital encoder maps each letter onto a different sequence of five 1's and 0's. There are 32 such sequences, which is enough to accommodate all 26 letters plus a few punctuation marks (Figure 5.13). The channel coding in this case involves replacing the letters with 5-bit blocks, or words, which are then transmitted to the receiver. The entire field of available 5-bit words is utilized; every possible permutation has meaning. The code thus appears to be very efficient.

At the other end, the receiver detects and decode these sequences and recreates the letters sent by the source. Now, what happens if there is an error in transmission? Let us say that the word 01101 (which is supposed to represent the letter *N*) suffers a change of its last bit to become 01100 (which corresponds to the letter *M*). The receiver has no way of knowing that an error has occurred. Because all of the permutations are valid, *any* transmission error will cause an incorrect decoding decision. The maximally efficient code is a very vulnerable code.

Now let us assume that each letter is mapped onto an 8-bit word (Figure 5.14). Two things are obviously different: (1) the total code space has been greatly expanded—there are now 256 different permutations or codewords—and (2) not all of these codewords are valid; in fact most are not valid. Only about 10% of the possible codewords are valid.

Suddenly, we can detect errors (at least some errors). If the receiver decodes an invalid codeword, then it knows for certain that there was an error in transmission. *Expanding the code space* to become larger than the number of valid codewords is the crucial step in channel coding:

> It is obviously not possible to detect an error if every possible symbol, or set of symbols, that can be received is a legitimate message. *It is possible to catch errors only if there are some restrictions on what is a proper message* [emphasis in original]. [12, p. 21]

The follow-up question is clear: If we are going to designate only some of the sequences in the code space as valid codewords, which ones should we select? We could decide to locate all the valid codewords in the center of the code space (Figure 5.15). This design would allow detection of some errors, but it would be rather ineffective in reducing the vulnerability of the code. Because most codewords would have only one error, the clustering of valid codewords together would still lead to a lot of undetected errors getting

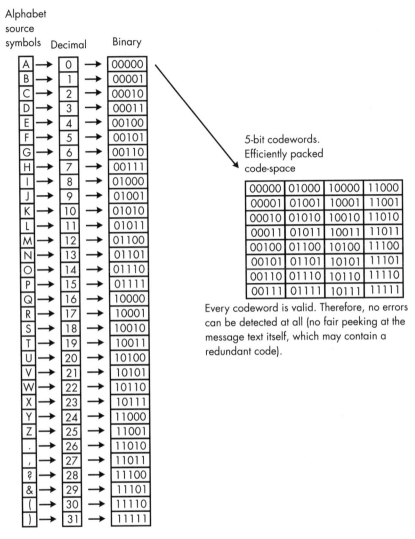

FIGURE 5.13 *An efficient code.*

through the system. It should be intuitively apparent that the best way to select the valid codewords is to spread them throughout the entire code space, and to keep every valid codeword as far apart from all other valid codewords as possible (Figure 5.16). In this way, many errors can not only be detected, but also *corrected*. For example, if there is only one bit in error

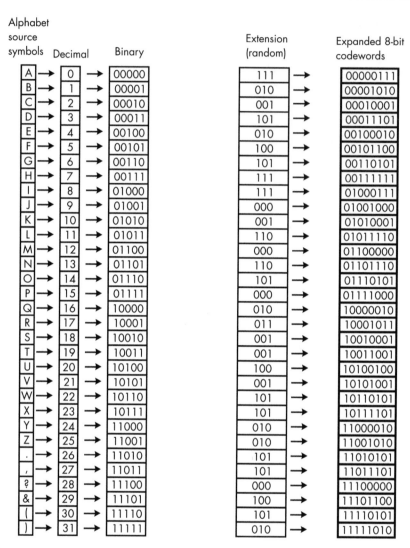

Alphabet source symbols		Decimal		Binary		Extension (random)		Expanded 8-bit codewords
A	→	0	→	00000		111	→	00000111
B	→	1	→	00001		010	→	00001010
C	→	2	→	00010		001	→	00010001
D	→	3	→	00011		101	→	00011101
E	→	4	→	00100		010	→	00100010
F	→	5	→	00101		100	→	00101100
G	→	6	→	00110		101	→	00110101
H	→	7	→	00111		111	→	00111111
I	→	8	→	01000		111	→	01000111
J	→	9	→	01001		000	→	01001000
K	→	10	→	01010		001	→	01010001
L	→	11	→	01011		110	→	01011110
M	→	12	→	01100		000	→	01100000
N	→	13	→	01101		110	→	01101110
O	→	14	→	01110		101	→	01110101
P	→	15	→	01111		000	→	01111000
Q	→	16	→	10000		010	→	10000010
R	→	17	→	10001		011	→	10001011
S	→	18	→	10010		001	→	10010001
T	→	19	→	10011		001	→	10011001
U	→	20	→	10100		100	→	10100100
V	→	21	→	10101		001	→	10101001
W	→	22	→	10110		101	→	10110101
X	→	23	→	10111		101	→	10111101
Y	→	24	→	11000		010	→	11000010
Z	→	25	→	11001		010	→	11001010
.	→	26	→	11010		101	→	11010101
,	→	27	→	11011		101	→	11011101
?	→	28	→	11100		000	→	11100000
&	→	29	→	11101		100	→	11101100
(	→	30	→	11110		101	→	11110101
)	→	31	→	11111		010	→	11111010

FIGURE 5.14 *Expanding the code.*

in a received sequence, it will be closer to the correct codeword in the code space than to any of the other codewords.

> Some reflection ...will reveal that if one is going to attempt to correct errors in a message represented by a sequence of n binary symbols, then it is absolutely essential not to allow the use of all 2^n possible sequences

0	0	0	0	0	0	0	0
0	0	0	0	0	0	0	0
0	0	0	0	0	0	0	0
0	0	0	0	0	0	0	0
0	0	0	0	0	0	0	0
0	0	0	0	0	0	0	0
0	0	0	0	0	0	0	0
0	0	0	0	0	0	0	0
0	0	0	0	0	0	0	0
0	0	0	0	0	0	0	0
0	0	0	0	0	0	0	0
0	0	0	0	0	0	0	0
0	0	Valid	Valid	Valid	Valid	0	0
0	0	Valid	Valid	Valid	Valid	0	0
0	0	Valid	Valid	Valid	Valid	0	0
0	0	Valid	Valid	Valid	Valid	0	0
0	0	Valid	Valid	Valid	Valid	0	0
0	0	Valid	Valid	Valid	Valid	0	0
0	0	Valid	Valid	Valid	Valid	0	0
0	0	Valid	Valid	Valid	Valid	0	0
0	0	0	0	0	0	0	0
0	0	0	0	0	0	0	0
0	0	0	0	0	0	0	0
0	0	0	0	0	0	0	0
0	0	0	0	0	0	0	0
0	0	0	0	0	0	0	0
0	0	0	0	0	0	0	0
0	0	0	0	0	0	0	0
0	0	0	0	0	0	0	0
0	0	0	0	0	0	0	0
0	0	0	0	0	0	0	0
0	0	0	0	0	0	0	0

Each cell represents a potential codeword. Here, valid codewords are in another valid word (and hence goes undetected).

FIGURE 5.15 *Codewords in the center of the code space—an ineffective design.*

as being legitimate messages.... Carrying this reasoning a little further, it becomes clear that if one wishes to correct all patterns of t or fewer errors, it is both necessary and sufficient for every legitimate message sequence to differ from every other legitimate sequence in at least $2t + 1$ positions. For example, if one wished to correct all single and double errors, it is necessary that all pairs of message sequences differ in at least five symbols. Any received sequence which contains two errors and, therefore, differs from the correct sequence in exactly two places will always differ from all other message sequences in at least three places. We refer to the number of positions in which any two sequences differ from each other as the *Hamming distance*, d, between the two sequences. The smallest value of d for all pairs of code sequences is called the *minimum distance* of the code ... [18, pp. 3–4]

The geometrical metaphor allows us to speak thus not only of the code space, but also of the *distance* between codewords, and to easily visualize the receiver's decision process. Figure 5.17, drawn from Blahut [19], gives us an

Valid	0	Valid	0	0	Valid	0	Valid
0	0	0	Valid	Valid	Valid	0	0
0	Valid	0	0	0	0	0	0
0	0	0	0	Valid	0	0	0
0	0	0	0	0	0	0	0
0	Valid	0	0	0	0	0	0
0	0	Valid	0	0	0	0	0
0	0	0	0	0	0	0	0
0	0	0	0	0	0	0	0
0	Valid	Valid	0	0	0	0	0
0	0	0	0	0	0	Valid	0
0	0	0	Valid	Valid	0	0	0
Valid	0	0	0	0	0	0	Valid
0	0	0	0	0	0	0	0
0	Valid	0	0	0	0	0	0
0	0	0	0	0	0	0	0
0	0	0	0	0	0	0	0
0	0	0	0	0	0	0	0
0	Valid	Valid	0	0	Valid	0	0
Valid	0	0	0	0	0	0	0
0	0	0	0	0	Valid	0	0
0	0	0	0	0	0	0	Valid
0	0	0	0	0	Valid	0	0
0	0	0	0	0	0	0	0
0	0	0	0	Valid	0	Valid	0
0	0	0	0	0	0	Valid	0
0	0	0	0	0	0	0	Valid
0	0	0	0	0	0	0	0
0	0	0	Valid	0	Valid	0	0
Valid	0	0	0	0	0	Valid	Valid
0	0	0	0	Valid	0	0	0
Valid	Valid	0	0	0	Valid	0	0

Here the valid codewords are scattered out through the total codespace. Therefore, many errors will convert a valid codeword into an invalid one, allowing the error to be detected.

FIGURE 5.16 *Codewords well spread throughout the code space—a more effective design.*

idea of a code space from the receiver's point of view. Each valid codeword is like a bull's-eye at the center of a small circular target. The transmitter aims each transmitted signal element at one of these bull's-eyes, but it may be deflected by the noise in the channel. If it hits the bull's-eye or comes fairly close (within the circle), the receiver can decode the signal correctly. If the noise is large enough to deflect it into the circle of another valid codeword, the error will not be detected. If the decoded signal lies in the no-man's land between the circles, the receiver knows that an error has occurred, but does not have enough information to correct it [19, p. 235].

In a fundamental sense, that is what channel coding is all about: spreading the code space out, and locating the codepoints appropriately. The vast complexity of encoding and decoding procedures is, conceptually, at the

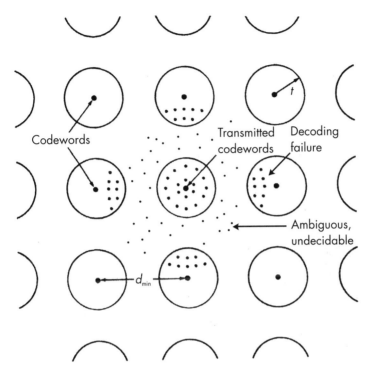

FIGURE 5.17 *Code space diagram. (From: [19]. © 1990 Addison Wesley Inc.)*

next level of detail. Signal hardening is achieved by building in the error detection and error correction capabilities, which is a result of the remapping of the signal into the expanded code space.

To summarize, coding is a strategy for signal hardening that is based on a two-stage mapping, or transformation, of the signal into a specialized (coded) form. Broadly speaking, the first stage involves a compression through quantization of the analog signal. This embeds logical (and physical) nonlinearities in the signal, which allow the receiver to employ threshold detection strategies that can operate in the presence of much higher levels of noise (compared to the original analog signal). In the literature, the first stage is referred to as *source coding*, and we shall deal with this subject at greater length in Chapter 6. The second stage involves reexpanding the quantized signal by mapping it into a larger code space, such that the receiver can detect and in some cases correct errors caused by channel noise. It is generally referred to as *channel coding*.

The subtleties we have touched upon include the following:

1. The use of one kind of noise to defeat another—quantization artifacts that are intentionally introduced into the signal, to counteract and control the effects of naturally occurring noise; yet these artifacts themselves must be managed carefully, which sometimes requires the addition of even more intentional noise to break up structured interference patterns.

2. The change in the nature of the degradation of the signal from a generally linear and often psychologically more tolerable deterioration (with analog signals) to a highly nonlinear, catastrophic failure mode characteristic of digital systems; to some extent, this phenomenon takes back what we gained from digitization in the first place.

3. The strange trade-off of bandwidth for signal hardness (resistance to noise).[24]

In the following sections, we will survey some of the popular channel coding techniques that have been developed from these concepts during the past 50 years.

---

24. Shannon presented this trade-off in (I think) somewhat simplistic terms: "We have seen that PCM requires more bandwidth and less power than is required with direct transmission of the signal itself.... We have, in a sense, exchanged bandwidth for power" [1, p. 154]. This observation, couched as it is in empirical terms—he is calling attention to a correlative relationship, which he then suggests "in a sense" may be causally connected—has become a mantra among spread spectrum theorists, who will often start (and try to finish) any argument over wireless architectures by citing Shannon's "proof" that wideband signals are inherently superior, and so on. My own sense is that it is the *relative* power that matters, or, to put it another way, the amount of power required by the receiver to discriminate signal from noise. A hardened signal, built up out of strong nonlinearities that tend to "contain" the noise in defined compartments, can tolerate higher levels of ambient energy than a nonhardened transmission where the noise permeates the entire signal. I also think it is not clear that digitization, properly exploited, necessarily expands the bandwidth (prior to channel coding). PCM is particularly wasteful of bandwidth, and does not produce a very hardened signal. Modern digital radio coding schemes are approaching the equivalent frequency bandwidth of analog systems (six digital channels in a 25-kHz channel is close to the 4-kHz analog bandwidth often cited for FM).

5.2 Basic Channel Coding Strategies

The two basic types of channel coding—block codes and convolutional codes—are by now well marked out in the literature.[25] Indeed, in some respects this classical distinction is suspect, for although the two codes are produced in different ways, the trend in code design is toward a blending of the two strategies (as we shall see in Section 5.3). In this section, we shall touch on the basic concepts briefly, followed by a discussion of some of the basic practical issues of decoding solutions.

5.2.1 Block Codes

The first effective channel codes to be developed were developed by expanding "naked" PCM type data words, mapping them into a larger code space— that is, by appending special error correction bits to the so-called "data bits," to create larger codewords that allowed for the detection of errors (Figure 5.18).[26] Because the result of this process was a block of data of fixed length, the term *block code* came into use. Certain mathematical properties of the codes, such as whether two codewords added together (modulo-2) always produce another valid codeword, are sometimes used for designating subcategories, such as *linear* block codes, *cyclic* block codes [21, Chaps. 3 and 4], *group* codes, and *systematic* codes. Other terms are derived from the way in which the codes are generated (the method of mapping data words onto codewords) such as *polynomial* codes [18, Chap. 2]. Specific codes are often designated by the name of their inventors and, in general, all of this terminology constitutes a somewhat imprecise nomenclature that, with one or two exceptions, can be ignored except by specialists.[27]

---

25. There are many references and survey articles. See, for example, [20] for the origins; three fairly representative textbooks are [13, 18, 21].

26. "We will assume that each code word in a group code can be divided into two portions. The first k-symbol portion is always identical to the information sequence to be transmitted. Each of the $n–k$ symbols in the second portion is computed by taking a linear combination of a predetermined subset of information symbols" [18, p. 50].

27. Clark and Cain [18] offer a fair, if unwitting appraisal of the value of the nomenclature per se: "The group codes constitute a vanishingly small percentage of all possible block codes. With very few exceptions, however, they are the only block codes of practical importance. They are often referred to by other names such as linear codes or generalized parity check codes. Within the set of all group codes there is a second major subdivision called polynomial generated codes. Examples of polynomial generated codes are the Bose–Chaudhuri–Hocquenghen (BCH) codes, Reed–Solomon

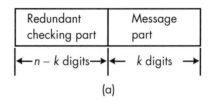

(a)

Messages	Code words
(0 0 0 0)	(0 0 0 0 0 0 0)
(1 0 0 0)	(1 1 0 1 0 0 0)
(0 1 0 0)	(0 1 1 0 1 0 0)
(1 1 0 0)	(1 0 1 1 1 0 0)
(0 0 1 0)	(1 1 1 0 0 1 0)
(1 0 1 0)	(0 0 1 1 0 1 0)
(0 1 1 0)	(1 0 0 0 1 1 0)
(1 1 1 0)	(0 1 0 1 1 1 0)
(0 0 0 1)	(1 0 1 0 0 0 1)
(1 0 0 1)	(0 1 1 1 0 0 1)
(0 1 0 1)	(1 1 0 0 1 0 1)
(1 1 0 1)	(0 0 0 1 1 0 1)
(0 0 1 1)	(0 1 0 0 0 1 1)
(1 0 1 1)	(1 0 0 1 0 1 1)
(0 1 1 1)	(0 0 1 0 1 1 1)
(1 1 1 1)	(1 1 1 1 1 1 1)

(b)

FIGURE 5.18 *Example of block code construction: (a) systematic format of a code-word; and (b) linear block code with k = 4 and n = 7. (From: [21]. © 1983 Prentice-Hall Inc.)*

The first systematic treatment of these sorts of codes was presented by Richard Hamming, one of the most cogent and accessible writers in this end of the business.[28] Part of the importance of Hamming's work was the extension of the geometric metaphor, along the lines suggested above, to enable

codes, generalized Reed–Mueller codes, projective geometry codes, Euclidean geometry codes, and quadratic residue codes. Each of these classes of codes is described by a specific algorithm for constructing the code. The classes form overlapping sets so that a particular code may be a BCH code and also a residue code or it may be a generalized Reed–Mueller code and also a BCH code, etc." [18, p. 49].

28. Hamming's original paper is still a pleasure to read [22]. See also his excellent small book titled *Coding and Information Theory* [12].

researchers to understand precisely the *distance* properties of particular codes, and to design codes to maximize the distance between codewords within a given code space.

Block codes today are somewhat out of favor because of the inability to fully apply soft decision decoding principles. Each block requires a hard decision as to the identity of the transmitted codeword. As we shall see, this imposes a penalty that drives most code designers to choose convolutional solutions whenever they can. There is one important exception: the use of Reed–Solomon coders for a channel characterized by bursts of errors (see Section 5.3).

5.2.2 Convolutional Codes

The other broad class of channel coding techniques is based on a combination of the signal-spreading concept with another, quite different and equally fundamental principle: the convolution of the signal with itself. The concepts and broader significance of convolution as a hardening strategy will be discussed in Section 5.5, but we cannot complete the introduction to channel coding without reference to this approach here.

The original presentation of convolutional coding is usually traced to a paper by Elias [23]. The definition he offers is disarmingly simple: "A convolutional code is defined as one in which check symbols are interspersed with the information symbols ..." [23, p. 51]. Actually, the mechanics of the encoding process are quite different from the block coding approach discussed above. First, a *register* is created with room for a certain number of input symbols. The size of the register is called the *constraint length*, which is normally denominated by the letter k. The stream of data symbols enters the register from one end and passes through and out the other end, so that at each moment in time we can "see" k data symbols in the register. Any symbol entering the register takes k clock cycles to pass entirely through the register (Figure 5.19).

To create the output symbols, the coder takes a snapshot of the contents of the register. It then performs a convolution of some subset of the symbols in the register, combining them together by modulo-2 addition, for example, to produce a new output symbol. Next, *another* convolution is performed on a different subset of the symbols in the same snapshot. Figure 5.20 shows an example of a convolutional coder with a register three symbols long ($k = 3$), and two convolutions being performed on each snapshot of the register (the top convolution combines the first and the last symbols, and the bottom convolution combines all three symbols). Thus, for each

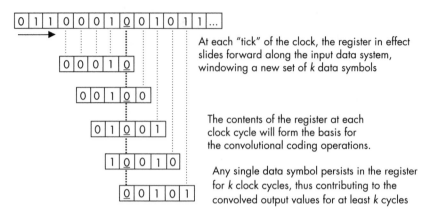

At each "tick" of the clock, the register in effect slides forward along the input data system, windowing a new set of k data symbols

The contents of the register at each clock cycle will form the basis for the convolutional coding operations.

Any single data symbol persists in the register for k clock cycles, thus contributing to the convolved output values for at least k cycles

FIGURE 5.19   *State-by-state mechanics of a register for convolutional coder.*

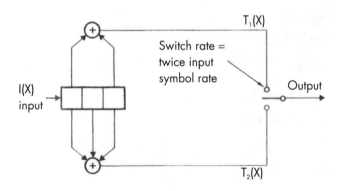

FIGURE 5.20   *Example of a simple convolutional coder. (From: [18]. © 1981 Plenum Press.)*

snapshot of the register, two output symbols are created. Finally, the clock ticks forward one cycle, and the symbols all advance by one space. Note that all of the symbols stay in the register except the oldest one, which exits (to the right) and is replaced by one new symbol coming in (from the left). The coder takes a new snapshot and generates two more output symbols and so forth. Of course, we could have shorter or longer registers, and more than two convolutions, and we can have multiple registers with convolutions across registers. Figure 5.21 shows two versions of a coder with two registers of two symbols each and three convolutions being performed. In principle, the options are limitless.

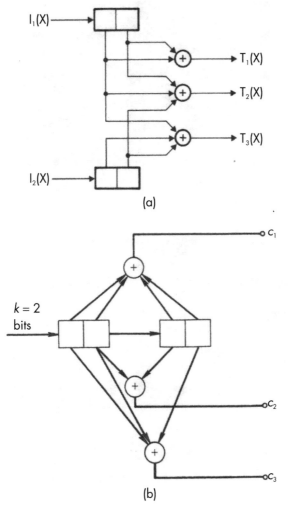

FIGURE 5.21 *More complex convolutional coders: (a) encoder for rate 2/3, k = 4 code (From: [18]. © 1981 Plenum Press.); and (b) rate 2/3 convolutional encoder for encoding both information bits (From: [11]. © 1989 McGraw-Hill Inc.).*

Consider how this process transforms the signal:

1. The output symbol rate is double the input symbol rate (in our example, higher in other examples as a function of the number of convolutions per cycle). Thus, the signal is expanded, just as with block codes.

2. Unlike block codes, in convolutional codes *none* of the input symbols actually survives intact to be transmitted. (This is what Elias [23] meant with his laconic reference to interspersing the parity symbols with the data symbols.) The output symbols are created by combining subsets of the input symbols together and rounding off.

3. Each input symbol has some influence over a long string of output symbols. In our example in Figure 5.21, each input symbol contributes something to the creation of six output symbols (two per cycle, for three cycles). This is another, deeper level of signal spreading, in a sense. The information carried by one input symbol is spread over six output symbols (in this example).

4. Each input symbol is commingled with a number of preceding and following input symbols. This is the heart of the idea of convolution, which comes from the Latin root meaning "to braid" or "to twist together."

5. The process is continuous. It does not produce blocks of symbols of fixed length. As one author has written, "the transmitted sequence is actually a single 'semi-infinite' codeword" [18, p. 227].

Compared to block coding, two things are clear right away. First, the mechanics of the convolutional process are extremely simple (compared to some of the more powerful block codes that require complex mathematical operations to generate the signal expansion). Indeed, whereas the development of new block codes has drawn on advanced algebraic theory and is pursued in a theoretical manner, the discovery of good convolutional codes has been a more trial-and-error process, aided by computerized searches of large numbers of such codes [18, p. 227].[29]

On the other hand, we are struck by the profound nature of the transformation wrought by these codes on the input data stream. The information is decimated, mixed, hammered out, homogenized, distilled, and reblended in a new code stream that has layers of subtle, built-in structure [19, pp. 220ff].

Less obvious aspects of convolutional coding technology have emerged through many years of application. For one thing, the problem of finding an

---

29. In addressing the need for a similar formal treatment of convolutional codes, Forney [24] has called attention to "the difficulty of analyzing convolutional codes, as compared to block codes" [24, p. 720].

easy way to *decode* convolutional expansions took some time to solve (see Section 5.2.3).

More critically, certain kinds of errors can be *fatal* to certain convolutional decoders; they can propagate and throw the decoder completely off-track such that it cannot recover. In block codes, the hard decoding decisions and the strict boundaries between blocks limit the effects of an error to a given block. With convolutional codes, this is not necessarily true.[30] And even when a convolutional decoder is not susceptible to fatal, unrecoverable error propagation, any single error that does occur will generally give rise to strings or bursts of errors.[31] This can be problematic for certain kinds of signals.

Nevertheless, in general, convolutional coding has become the popular choice for many applications, in wireless communications and elsewhere.

> At least for very noisy channels, a convolutional code performs much better ... than the corresponding block code of the same order of complexity ... [such that for a particular example] to achieve equivalent performance asymptotically the block length must be over five times the constraint length of the convolutional code. [25][32]

Most of this relative advantage has to do with the fact that convolutional codes lend themselves particularly well to *soft decision decoding* techniques, whereas block codes often do not.[33] As we shall see below, soft

---

30. "There is a nonzero probability that when a codeword is transmitted over a noisy channel, a finite number of channel errors can change all the nonzero encoded digits into zeroes, thereby making the decoder generate an all-zero output at the receiver. Thus, a finite number of channel errors can cause an infinite number of decoding errors. This situation, clearly undesirable, is known as *catastrophic error propagation* ..." [17, p. 59]. See also [24, pp. 727ff].

31. "When a decoding error occurs, the [convolutional] decoder will necessarily follow this with additional decoding errors. A dependent sequence of decoding errors is called an error event. Not only does one want error events to be infrequent, but one wants their duration to be short when they occur" [19, pp. 231–232].

32. See also [24, p. 720]: "Convolutional codes have proved to be equal or superior to block codes in nearly every type of practical application, and are generally simpler than comparable block codes in implementation," as well as, more recently, [26]: "Generally, anything that can be achieved with a block code can be achieved with somewhat greater simplicity with a convolutional code." One may say that this has achieved the status of conventional wisdom. The issue is less clear once we consider the specifics of the fading channel (see Section 5.3.2).

33. "The use of reliability information or so-called 'soft decisions' is applicable to both block codes and tree codes [this text's term for convolutional codes]. Most of the al-

decisions yield hard performance gains that often outweigh other considerations in the selecting a code for a commercial application.[34]

5.2.3 Decoding Algorithms

Before discussing the payoffs from channel coding, we should touch on what turns out to be the most important implementation challenge, and often acts as the chief constraint on the commercial deployment of a particular coding technique: finding an efficient *decoding* algorithm.

The decoding process is usually far more complex than the encoding process. In wireless applications, it is not uncommon for the processing burden of the decoding operation to be 10 times as much as the processing required for the encoding operation. "In practice," as one textbook notes, "one must often resort to a nonoptimal [decoding] technique" [18, p. 14]. In fact, most of the basic code structures in use today had all been discovered by about 1960, but most of them were impractical, because good decoding procedures did not yet exist.

For example, when Elias originally presented convolutional codes in his 1955 paper, he admitted in his concluding paragraph that "The decoding operation will ... take a great deal of time in interesting cases. No decoding procedure that replaces this operation by a small amount of computing has yet been discovered" [23, p. 52]. The surge in popularity of convolutional coders did not take place until a dozen years later, when Viterbi identified the applicability of his eponymous algorithm to a simplified and reasonably optimal decoding of convolutional systems.

Similarly, the value of Reed–Solomon codes—among the most powerful, and certainly the most popular block codes in commercial use today—lay dormant for nearly 10 years after their first publication, until the breakthrough discovery of an efficient decoding algorithm. A recent retrospective volume summarized this process, which is worth citing at length because it conveys the dynamic of the coding business quite well:

---

gorithms for decoding tree codes make use of this information in a straightforward manner. The use of 'soft decisions' in block codes is somewhat more involved and generally requires significant changes in the algorithms" [18, p. 15].

34. Commenting on Reed–Solomon codes—the commercially most popular type of block code in use today—its proponents admit that "there is no known practical way to 'soft-decode' RS codes, and the 2 dB loss resulting from hard quantization prior to decoding is intolerable in all but a few applications" [27, p. 26].

After the discovery of Reed–Solomon codes [in 1959], a search began for an efficient decoding algorithm. None of the standard decoding techniques used at the time were very helpful. For example, some simple codes can be decoded through the use of a syndrome look-up table.... Unfortunately this approach is out of the question for all but the most trivial Reed–Solomon codes.... Reed and Solomon [later] proposed a decoding algorithm based on the solution of sets of simultaneous equations. Though much more efficient than a look-up table, Reed and Solomon's [decoding] algorithm is still useful only for the smallest Reed–Solomon codes.... In 1960 Peterson['s] ... direct solution algorithm [proved] useful for correcting small numbers of errors but becomes computationally intractable as the number of errors increases. Peterson's algorithm was improved and extended [by various others] ... but Reed–Solomon codes capable of correcting more than six or seven errors still could not be used in an efficient manner. Detractors of coding research in the early 1960s had used the lack of an efficient decoding algorithm to claim that Reed–Solomon codes were nothing more than a curiosity.... The breakthrough came in 1967 when Berlekamp demonstrated his efficient decoding algorithm ... [allowing] for the efficient decoding of dozens of errors at a time using very powerful Reed–Solomon codes ... [27, pp. 12–13]

The history of the field is marked by discoveries of interesting codes that tend to remain "curiosities" during long periods of relative stasis, until breakthroughs in the form of new decoding solutions emerge to enable commercially feasible applications [27, Chaps. 3–7].

A complete survey of decoding techniques would again take us beyond the scope of interest of this volume, but it is useful to maintain the proper orientation toward this branch of signal hardening technology. In some sense, the mathematical basis of channel coding has been well explored, and the onus is on finding commercially engineerable implementations. This is why decoding algorithms must be considered one of the dominant factors. Technical optimality and commercial feasibility interact to select certain codes and advance their popularity during certain periods. Today, it is probably true to say that the two most popular channel coding techniques are, on the one hand, convolutional coding using Viterbi decoders and soft decision processing, and Reed–Solomon block coders for burst-error correction on the other hand. However, many other coding solutions are available, should the technical/commercial balance shift.

One important determinant of this balance is the availability of better computing hardware. Decoding strategies may emerge into a more favorable light as signal processing and computational hardware steadily advances. New-old decoding strategies, based on using previously unpromising algorithms, may become more feasible as the relative cost of MIPS and memory improves.[35]

5.2.4 Performance: Coding Gain

It is immediately clear that channel coding pays off. Figure 5.22 illustrates a comparison of a very basic block-coded system with its uncoded

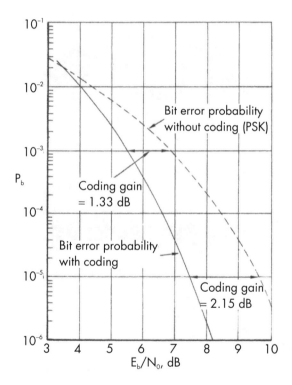

FIGURE 5.22   *Coded versus uncoded system performance. (From: [18]. © 1981 Plenum Press.)*

<hr/>

35. "It is necessary to clearly state that almost all of coding theory was developed in the days when computing was relatively cheap and storage was more expensive. In the year 1985 [!] we are seeing a great decrease in the cost of storage, and hence there is a need to reconsider all the theory we have developed in the past. Methods of design,

counterpart, using one of the simplest digital modulation schemes (BPSK) [19, p. 237]. The channel code in this example is a three-error-correcting Golay code that was discovered more than 50 years ago [28]. The use of this simple block code provides an improvement of 2.1 dB in the required E_b/N_0 for achieving an error rate of 1 in 100,000 transmitted bits.[36] In practical terms, for a wireless engineer, this could mean an extension of the effective range in a system of more than 50%, which would mean fewer cells, less investment, fewer handoffs, and so on. Alternatively, we could spend the gain in signal hardening to allow the system to operate at a higher average noise/interference level, keeping the cell placement the same and allowing a better reuse factor.

The Golay example involves a very simple code. More sophisticated codes can easily achieve coding gains of 6–8 dB, as shown in Figure 5.23. Indeed, as we shall see in the following section, realizable coding gains from the most recent channel coding techniques are now approaching the hard theoretical limits laid down by Shannon himself.

Such improvements have enormous implications for wireless system design. Six decibels of gain represents a fourfold improvement in any of a number of relevant dimensions of the system. Today (unlike even 10 years ago) probably no one would set out to design a wireless system without setting the signal hardening strategy at the top of the list of architectural considerations.

Coding technique	Coding gain (dB) at 10^{-5}	Coding gain (dB) at 10^{-8}	Data rate capability
Concatenated (RS and Viterbi)	6.5–7.5	8.5–9.5	Moderate
Sequential decoding (soft decisions)	6.0–7.0	8.0–9.0	Moderate
Concatenated (RS and biorthogonal)	5.0–7.0	7.0–9.0	Moderate
Block codes (soft decisions)	5.0–6.0	6.5–7.5	Moderate
Concatenated (RS and short block)	4.5–5.5	6.5–7.5	Very high
Viterbi decoding	4.0–5.5	5.0–6.5	High
Sequential decoding (hard decisions)	4.0–5.0	6.0–7.0	High
Block codes (hard decisions)	3.0–4.0	4.5–5.5	High
Block codes–threshold decoding	2.0–4.0	3.5–5.5	High
Convolutional codes–threshold decoding	1.5–3.0	2.5–4.0	Very high
Convolutional codes–table look-up decoding	1.0–2.0	1.5–2.5	High

FIGURE 5.23 *Comparison of major coding techniques. (From: [18]. © 1981 Plenum Press.)*

We should observe, in passing, that the more coding we lay on, the more nonlinear the system becomes. The knee effect, and the potential for sudden deterioration, exists in almost any coded system. The wrong kind or color of noise can produce unwanted outcomes. Strangely, highly coded systems can show *worse* performance at very high noise levels than uncoded digital systems.

> At sufficiently low signal-to-noise ratios, one may observe that the coding gain actually becomes negative. This thresholding phenomenon is common to all coding schemes. There will always exist a signal-to-noise ratio at which the code loses its effectiveness and actually makes the situation worse. [18, p. 35]

This counterintuitive result should help us keep in mind the non-straightforwardness of the coding process. It really is very interesting, and very subtle technology.

5.3 Advanced Coding Strategies

In this section, we will survey a number of the hot topics related to channel coding strategies in the evolution of second- and third-generation wireless architectures. Some of these derive out of latent issues in Shannon theory, whereas others are purely the result of recent engineering coups.

5.3.1 Interaction Between Channel Coding and Source Compression Strategies

In Chapter 6 we will examine source coding technologies as a subset of signal-shaping strategies. Here we consider briefly how the choice of channel coding strategy and source coding strategy can manifest a certain interdependence.

The classical information theory position, put forward by Shannon, repeated over decades, denies the existence of *any* (desirable)

---

encoding, and decoding which depend more on table look-up and other forms of storage and less on computing need to be researched ..." [12, p. 235].

36. E_b/N_0 (pronounced "eb-no") is a proxy for SNRs that is preferred in coding analyses. "The usual figure of merit for a communication system is the ratio of energy per information symbol to noise spectral density (E_b/N_0) that is required to achieve a given probability of error" [18, p. 35].

interdependence between source coding and channel coding. The two operations are seen as formally separate and distinct:

> The essence of the Shannon theory ...learned and applied over nearly half a century ... [holds that one should] completely separate techniques for digital source compression from those for channel transmission, even though the first removes redundancy and the second inserts it. [29, p. 228; see also 30]

One wonders whether this view can be sustained quite so forcefully these days. The argument for such a strong separation *may* have some validity in the case of pure *recoding* of digital bit streams (i.e., where the original source is discrete).[37] But it seems clear that for *analog* sources such as speech or images, the design of the source coder can often benefit from a knowledge of the nature of the channel and the strengths and weaknesses, and the likely failure modes, of the channel coder selected. Alternatively, a change in the source compression strategy can alter the relative merits of different channel coders. In short, the two can be *jointly optimized* to a higher level of performance.

The story of the first use of Reed–Solomon coding—for transmitting images from deep-space probes of the planetary system—illustrates this rather dramatically. It is a good story, and I will cite the account at some length:

> Multi-error-correcting Reed–Solomon codes were used for the first time in deep space exploration in the spectacularly successful *Voyager* mission

---

37. Frankly, I cannot understand the logic for a categorical assertion of separate (as opposed to joint) optimization of source and channel coding systems. It goes against the broad principles of most system design, which holds that suboptimization is a constant risk. In some cases, the mapping of source bits onto the channel coding alphabet can be tweaked to gain performance. Arguably, Gray coding is a form of joint optimization, and when blended with a judicious mapping of perceptually "heavy" bits onto the more protected modulation elements, clearly some sort of joint design is involved. For example: "It is observed ... that [in a particular Gray code] the MSB and MB positions [most significant bit and middle bit, respectively, in an 8-PSK system] provide equal BER performance, which is better than that of the LSB position. This results from the fact that bit decision boundaries can be drawn for the MSB and the MB which support a greater average Euclidean distance between the relevant signal points. The defined format takes advantage of this characteristic by always mapping the most perceptually significant speech bits into the MSB and MB positions. Likewise, the least important speech bits are always mapped into the LSB position" [31, p. 24].

that began in the summer of 1977 with the launch of twin spacecraft (*Voyager 1* and *Voyager 2*) from Cape Kennedy, towards the outer planets Jupiter and Saturn. Earlier deep space missions like *Pioneer, Mariner, Viking* and indeed *Voyager* itself to Jupiter and Saturn, used sophisticated error correction but had no need for Reed–Solomon codes because their digital images were not *compressed* prior to transmission. At Uranus and Neptune, however, *Voyager* transmitted some (though not all) of its images in compressed format, which made RS coding essential. Let's see why this was so.

A *Voyager* full-color image is digitized by the spacecraft's imaging hardware into three 800×800 arrays of 8-bit pixels, or $3 \times 800 \times 800 \times 8 = 15,360,000$ bits. In an uncompressed spacecraft telecommunication system, these bits are transmitted, one by one, to earth, where the image is reconstructed. Of course, if some of the received bits are in error, the quality and scientific usefulness of the image is degraded, and early studies by planetary scientists established 5×10^{-3} as the maximum bit error probability acceptable for images from NASA planetary missions. Thus, when telecommunications engineers designed the error control coding for these missions, they invariably sought to maximize the coding gain at a decoded bit error probability of 5×10^{-3}. For example, in the baseline *Voyager* telecommunication system, which uses a $K = 7$, rate 1/2 convolutional code, the coding gain ... is about 3.5 dB....

By concatenating the [convolutional coder] with an outer RS code [see Section 5.3.5 below on concatenating codes] the coding gain for low bit error rates can be improved considerably.... For example, at [a BER of] 10^{-6}, the concatenated system is about 2.5 dB superior which implies that at [a BER of] 10^{-6} the concatenated system can transmit at a 78% higher data rate than the baseline system. However, as we have seen, planetary missions required only 5×10^{-3}, and so these potential gains ... were apparently of no practical value. Shortly [thereafter], however, an important breakthrough in data compression occurred that changed the situation dramatically.

It had been realized since the early 1960s that planetary images are extremely redundant, and that far fewer than 15 million bits should suffice to represent one of them. However, the known techniques for reducing this redundancy were too complex to be implemented onboard a spacecraft. But this situation changed in the early 1970s when Robert Rice at Caltech's Jet Propulsion Laboratory devised a data compression algorithm that typically compressed a planetary image by a factor of 2.5, with no loss in fidelity, and which was simple enough to be implemented in *Voyager's* software. A factor of 2.5 achieved by data compression translates to 4 dB in system gain, a figure that would be difficult to obtain in any other way. Still, conservative spacecraft engineers judged the Rice algorithm too risky for the all-important basic mission to Jupiter and Saturn, although they were willing to include it as part of a backup mission in case the primary communication link failed and as a way of enhancing the hoped-for extended mission to distant Uranus and Neptune. Even so, there was a stumbling block.

The stumbling block was that Rice's *decompression* algorithm, like most decompression algorithms, is quite sensitive to bit errors. If a compressed line contains even one bit error, Rice's algorithm will, as a rule, garble the line beyond recognition. Thus it was determined that the venerable value of [a BER of] 5×10^{-3} was no longer acceptable; a much lower value was required, a value that could be achieved efficiently only by using concatenation with RS codes....[38]

This is a concrete example of the interaction between the choice of the source coder (in this case an image coding technology) and the channel coder. It stands to reason that such potential interactions do exist and can be exploited even more effectively than in this example. Indeed, the subject of trellis coding (see Section 5.3.7) can be seen as an example of the joint optimization of certain aspects of source and channel coding, which has resulted in very significant coding gains leading to commercially important applications (such as high-rate telephone modems). Joint source/channel coding methods are also favored in mobile video transmission, an extremely challenging application, where the need to transcend the classical Shannon paradigm is explicitly understood:

> Joint source-channel coding (JSC) methods ... are motivated precisely by the practical shortcomings of Shannon's separation theorem: 1) the failure to remove all redundancy [*due to the inherent difficulties in defining it for video sources, which leads to the use of lossy compression techniques in the first place*] and 2) the inadequacy of conventional channel coding methods, especially for noisy or fading channels. These relatively new JSC techniques capitalize on 1) and help to mitigate 2). They can provide channel robustness without incurring any increase in rate ... [30]

The key point is that the Shannon separation theorem, to the extent that it has any real validity, applies only to cases where the definition of redundant information is clear and explicit, allowing lossless compression to be used. This rules out, by definition, any *analog* source, such as speech or

---

38. For the rest of this story, and the even more dramatic events surrounding the coding schemes used on the crippled *Galileo* spacecraft, see [32].

image sources. In such cases, the definition of redundant versus essential information is always somewhat uncertain, and compression is always lossy. As well, there is *always* some residual redundancy. It is this residual redundancy that is, so to speak, available for exploitation by joint source/channel coding strategies [30, p. 1738]. At least, that is one interpretation.[39]

5.3.2 Channel Characteristics and the Choice of Coding Schemes: Burst Errors

Communications theorists typically assume that they are dealing with an unusual type of communications channel, called a *memoryless* channel,[40] in which "the noise affects each transmitted symbol independently" [21, p. 11]. In addition, they assume that the noise itself is of an unusual form, called *additive white Gaussian noise* (AWGN), which means basically that the noise has no discernible structure.[41]

There is some support for the position that this model is reasonably accurate for deep-space and some satellite communications [21, p. 11]. It is generally agreed, however, that the AWGN channel is *not* an accurate model of real terrestrial wireless channels. Such channels are characterized by (1) noise and interference that is often highly structured (nonwhite), and (2)

---

39. Another view, less convincing although it may be simply an "external" interpretation of the same argument, is offered by Hagenauer and Stockhammer [33]. They comment: "Supposedly, this two-step method [i.e., separate source and channel coding] is supported by Shannon's famous separation theorem.... This is true in an information theory sense: long blocks of source-encoded source symbols followed by channel coding, which uses a sequence of random block codes with infinite length. Consequently the delay tends to infinity. In a practical situation, usually the scenario is different and Shannon's separation theorem is no longer applicable. We have short blocks and require small delays as well as low-complexity coding schemes" [p. 1764]. This argument seems to focus on implementation issues, whereas the Van Dyck and Miller [30] position is based on the inherent inability of any coding scheme to discriminate and extract fully the redundancy inherent in an analog source.

40. I consider this term one of the least felicitous expressions in this field, but it is well entrenched. Of course, it has nothing to do with memory per se, but with the existence of a correlation between symbols in a sequence. If this is memory, then we might as well say that bow ties and tuxedos have memory because their occurrences are often correlated.

41. Alternately, if you prefer, it is a process "in which the noise disturbance is assumed to be a zero-mean Gaussian random process with a one-sided spectral density N_0" [17, p. 11].

strong correlations between error probabilities of sequentially received symbols. To put it more plainly, signals in these channels are affected by periods of high interference that cause clusters or *bursts* of errors at the decoder.[42]

The mobile channel is a extreme example of a burst-error channel, a channel with memory. The rapid, violent fading producing by multipath-induced self-interference generates a highly clustered error pattern. Add to this the effects of periodic cochannel interference from other transmitters in a densely packed cellular system, and the mobile channel is perhaps the farthest thing from a memoryless AWGN channel that a communications engineer is likely to encounter.[43]

Not surprisingly, coding schemes that work well with memoryless AWGN channels do not perform well on burst-error channels [35–37].[44] Convolutional coders prefer structureless noise [34, p. 370], and naked convolutional coders (without the use of other countermeasures, such as

42. "On channels with memory, the noise is not independent from transmission to transmission. A simplified model of a channel with memory ... contains two states, a 'good state,' in which transmission errors occur infrequently, ... and a 'bad state,' in which transmission errors are highly probable, ... The channel is in a good state most of the time, but on occasion shifts to the bad state due to a change in the transmission characteristic of the channel (e.g., a 'deep fade' caused by multipath transmission). As a consequence, transmission errors occur in clusters or bursts.... Examples of burst-error channels are radio channels, where the error bursts are caused by multipath transmission, wire and cable transmission, which is affected by impulse switching noise and crosstalk, magnetic recording, which is subject to ... surface defects and dust particles" [21, pp. 11–12]. In short, most channels of interest are non-AWGN channels.

43. The preference for "memoryless" channels is motivated principally by the fact that Shannon theory depends on this assumption for some of its most rigorous results, including the crucial calculation of channel capacity C. "When the channel contains memory, [standard capacity equations] are not easily reconciled (if at all) to yield a capacity C that specifies precisely the set of rates for which reliable communication is possible" [34, p. 347].

44. What is striking is how often an author will acknowledge the importance of burst error channels and, in almost the same breath, dismiss them from his field of interest. For example: "Many communication channels in practice are better described as channels with memory.... The main characteristic of channels with memory is statistical dependence between errors. The errors tend to occur in large numbers, i.e., bursts which are interspersed with essentially error-free communication. The error control strategies required to obtain performance improvements on channels with memory are quite different from those needed for memoryless channels. This book does not treat channels with memory any further" [17, p. 11]. Statements of this sort are surprisingly common in the coding literature.

interleaving; see Section 5.4.2) can be disrupted or even disabled by error bursts in fading channels.[45]

> With correlated fading channels, convolutional codes suffer from short constraint length [i.e., the feasible limits of the Viterbi decoding technique on the number of input symbols that can be convolved] unless interleaving is used. Without sufficient interleaving, a deep fade can 'hit' all of the transmitted information influenced by a data bit; however, the worst type of noise for a convolutional code is probably high-power burst interference, e.g., periodic tone jamming [or co-channel interference from adjacent cells in a cellular system]. Unless specifically prevented, it is possible for the soft-decision metrics to be counterproductive, with the interference-afflicted signal samples regarded as higher quality than the unafflicted signals. [34, pp. 372–373]

In fact, convolutional decoders tend in any case to produce bursts of errors at their output, which can be disruptive to source decoding processes downstream, and in burst-error channels this behavior may be accentuated.

The bursty nature of the mobile channel has helped bring to prominence one particular coding technique that has proven to be especially effective against burst errors: the block-coding technique known as Reed–Solomon coding.[46] Reed–Solomon codes are now, along with convolutional and trellis codes, one of the most popular formats in wireless applications.[47] Like other block coders, the output codewords are structured as a set of data bits, with additional error correction bits appended (Figure 5.24).

---

45. See, for example, Jamali and Le-Ngoc [38]: "A feature of [convolutional schemes] is that the transmitted symbol at any given time depends on the previous input symbol sequence. In the decoding process, hence, the decoder has to know the history of the coded sequences before being able to decide a particular symbol. Furthermore, the decoder has to look at the subsequent history of the sequence to examine the total influence of that symbol on the transmitted sequence. Therefore, if the decoder loses or make a mistake ... *errors will propagate* [emphasis in original]. This event can be considered a drawback in some channels such as mobile radio channels ..." [38, pp. 273–274].

46. "The mobile channel is a nasty environment in which to communicate, with deep fades an ever-present phenomenon. Reed–Solomon codes are the single best solution; there is no other error control system that can match their reliability performance in the mobile environment" [27, p. 14].

47. In addition to the volume edited by Wicker and Bhargava [27], the reader is referred to [18, pp. 206–219] for a good articulation of the formal conceptual framework underlying Reed–Solomon.

Information | Parity

```
0 0 0 0 0  0 0
0 0 0 0 1  6 3
0 0 0 0 2  7 6
0 0 0 0 3  1 5
        ⋮
0 0 0 1 0  1 1
0 0 0 1 1  7 2
0 0 0 1 2  6 7
0 0 0 1 3  0 4
        ⋮
0 0 0 7 0  7 7
0 0 0 7 1  1 4
0 0 0 7 2  0 1
0 0 0 7 3  6 2
        ⋮
0 0 1 0 0  7 3
0 0 1 0 1  1 0
0 0 1 0 2  0 5
0 0 1 0 3  6 6
        ⋮
```

FIGURE 5.24   *Reed–Solomon codes. (From: [19]. © 1990 Addison Wesley Inc.)*

The uniqueness of this code lies in the method of generating these error correction bits, which involves a branch of mathematics known as Galois Field arithmetic. One characteristic of Reed–Solomon is that it operates on multibit symbols, which means that if errors are clustered, they may tend to be contained within symbols, and the symbol error rate may be better than the bit error rate (i.e., lots of errors concentrated into a few symbols).[48]

Reed–Solomon codes also generate a very large dictionary of codewords. Figure 5.24 illustrates some of the codewords from a relatively simple (7,5) Reed–Solomon code, in which each digit is a 3-bit octal symbol. The total number of words for this code is around 32,000. In a larger Reed–Solomon code such as a (255,223) Reed–Solomon, "the number of codewords is astronomical" [19, p. 219]. Reed–Solomon codes thus have "good distance properties" [11, p. 429]—they spread the signal space well. In addition, each

---

48. Golomb, Peile, and Scholtz cover this in [34, p. 374].

codeword contains a very large amount of information, which can often be large enough to contain the burst of errors, which enables the powerful error correction properties of the code to be brought to bear effectively.

5.3.3 Soft Decision Techniques

If the bursty nature of the channel errors pushes the designer in one direction, the potential of soft decision decoding tends to pull him back in the other. What exactly *is* a soft decision? The concept is rather broad and variously formulated. It is used, or explained, in two ways: as an averaging process, or as an assisted decision process.

It is easier to start by defining a hard decision. The receiver is called on to examine the received signal samples and make decisions about which symbol or codeword from the list of valid codewords was actually sent by the transmitter. In a hard decision process, the receiver makes its best guess at the proper value for each sample, and then throws that sample away and goes on to the next sample and repeats the process. In so doing, *two types of information are lost*:

1. The wireless physical channel always creates some kind of *correlation* from symbol to symbol, and if we throw away the signal sample used to decide symbol a_1, we are discarding a small amount of information that may be relevant to helping us decide the following symbol, a_2.

2. We have information related to the *quality* of the data underlying the decision; we know something about whether the received signal was very clean or very noisy; we know how *reliable* the decision is.

If we were to wait for several symbols to arrive, and to evaluate the sequence in its entirety, chances are we could make a better, more accurate decision. If we use our reliability information, we can perhaps make a more reliable decision.

These may sound like descriptions of different processes, and to some degree they reflect different implementations of the soft decision concept (e.g., whether it is applied before or after some form of quantization by the receiver).[49] There is, I think, a unifying idea: the idea of *context*, and of using

---

49. A few examples of how "soft decision" processes are defined in the literature: (1) "Soft decision decoding [is when] a received sequence of samples, quantized to give relative reliabilities, is processed to estimate the most likely code word transmitted, assuming

contextual information to help make a decision about what signal was received. Part of the context of a given signal element is the signal elements that precede and follow it. Another part of the context is the state of the channel itself (relatively noise free or quite noise filled). Actually, we employ the contextual strategy constantly in listening to and decoding speech acoustically, especially in noisy environments.

We all have had the experience of trying to understand what someone was saying, on a noisy street or at the proverbial cocktail party, and at some point in the process being a little uncertain about what exactly was being said at a particular moment. Normally, we simply continue listening. After we have collected a larger sample, a few more words, or a full sentence, suddenly we can decode the earlier portions of the signal that were a little fuzzy at first. Psycholinguistics has firmly demonstrated that we use context *very* actively at all levels—phonemic, word level, phrase level, sentence level—in processing

---

equally likely code words" [34, p. 375]. (2) "At one extreme, the demodulator can be used to make firm decisions on whether each coded bit is a 0 or a 1. We say that the demodulator has made a 'hard decision' on each bit ... [and] since the decoder operates on the hard decisions made by the demodulator, the decoding process is termed *hard-decision decoding*. At the other extreme, the analog (unquantized) output from the demodulator can be fed to the decoder. The decoder, in turn, makes use of the additional information contained in the unquantized samples to recover the information sequence with a higher reliability than that achieved with hard decisions. We refer to the resulting decoding as *soft-decision decoding*. The same term is used to describe decoding with quantized outputs from the demodulator, where the number of quantization levels exceeds two" [11, pp. 363–364]. (3) "When a decoder is used with a demodulator for a Gaussian noise channel, the decoder is called *a hard-decision decoder* if the output of the demodulator is simply a demodulated symbol. Sometimes the demodulator passes other data to the decoder such as the digitized output of the matched filters. Then the decoder is called a *soft-decision decoder*" [19, p. 237]. (4) "A soft-decision decoder accepts analog values directly from the channel; the demodulator is not forced to decide which of the q possible symbols a given signal is supposed to represent. The decoder is thus able to make decisions based on the quality of a received signal. For example, the decoder is more willing to assume that a noisy value in indicating an incorrect symbol than that a clean, noise-free signal is doing so. All of the information on the 'noisiness' of a particular received signal is lost when the demodulator assigns a symbol to the signal prior to decoding" [27, p. 14]. (5) "The *soft-decision decoder* (SDD) ... accepts a vector of real samples of the noisy channel output and estimates the vector of channel input symbols that was transmitted. By contrast the hard-decision decoder (HDD) requires that all its input be from the same alphabet as the channel input.... A SDD for a binary code generally presents an estimate or 'guess' of the binary value of each transmitted symbol and the real number expressing the 'goodness' or reliability of the binary estimate" [39, pp. 108–109]. One can sense the organic elephant underlying these diverse and partial descriptions.

speech aurally. Anyone who has ever had to transcribe a tape-recorded sample of speech will have some sense of how soft decision processes work. If we were presented with recorded speech word by word and asked to write down these words one at a time, we can intuitively see that we would make more errors than if we were allowed to listen to complete sentences. For one thing, complete sentences will generally remove ambiguity like homonyms, which cannot be resolved in a word-by-word decoding process.

In any case, soft decision or contextual decoding can improve the accuracy of the receiver, by 2–3 dB over hard decision decoding for the AWGN channel (Figure 5.25). For the fading channel with clustered errors, the advantage of soft decision decoding can increase to 4–6 dB (Figure 5.26) [34, p. 372; 11, p. 758]. This, of course, is a huge advantage, and virtually free for the taking, if the coding scheme is amenable to it. Indeed, the popularity of convolutional coding is based on its suitability to soft decision decoding. By the same token, it is the difficulty of implementing soft

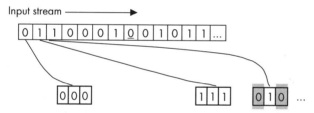

Repetition coding on each bit – relies only on the repetition for redundancy

In bit-by-bit decoding, any two bad bits per triad will result in a decoding error, regardless of positions of those bit-errors. If the probability of a bit-error is 12.5%, then the probability of a decoding error is approximately 4.3% for a best-two-out-of-three repetition code.

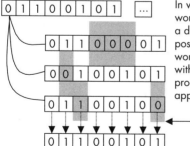

In word-by-word decoding, where each 8-bit word is repeated three times, bit-errors will cause a decoding error only if they occur in the same position in the word. The use of context – the entire word – to assist in decoding the bit means that with the same bit-error probability of 12.5%, the probability of a decoding error is reduced to approximately 1.5%.

Here, the best two-out-of-three comparison of bit positions allows for correct decoding even with a relatively large number of bit-errors.

Repetition coding on 8-bit word – uses context to assist in decoding individual bits

FIGURE 5.25 *Soft decision versus hard decision decoding.*

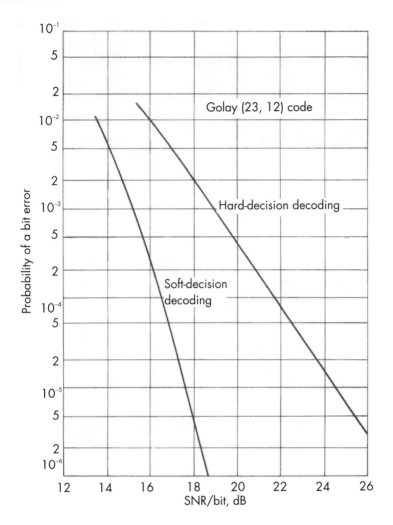

FIGURE 5.26 *Benefits of soft decision decoding for a fading channel. (From: [11]. ©*
1989 McGraw-Hill Inc.)

decision techniques on Reed–Solomon codes that probably prevents them from sweeping the field in mobile communications [27, p.14].

The attraction of soft decision decoding is thus very strong. However, the impact of clustered errors on Viterbi-decoded convolutional codes is also very significant. Figure 5.27 shows one analysis of the relative performance of the two most popular channel coding configurations—Reed–Solomon and convolutional coding—and, as the authors comment: "It is apparent that

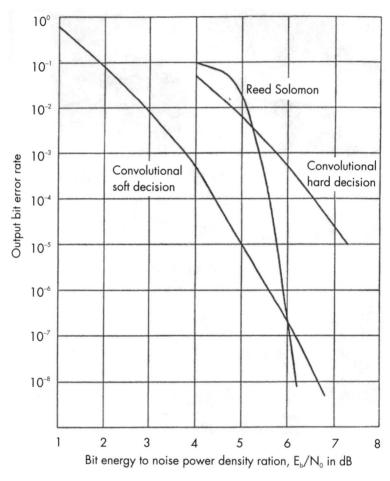

FIGURE 5.27 *Performance of Reed–Solomon block codes versus convolutional codes. (From: [34]. © 1994 Plenum Press.)*

any verdict on which code is better depends on factors external to [this figure]" [34, p. 372]. Several years ago I participated in a heated debate at the beginning of a mobile wireless system development project (which ultimately consumed some $40 million in engineering development alone). On one side, there were several system architects who argued strongly for using a Reed–Solomon code as the cornerstone; on the other side were those who favored an architecture based on convolutional coding. Those supporting the first approach were attracted by the power of Reed–Solomon to handle the highly clustered error patterns of the mobile channel, and they were concerned about the vulnerabilities of convolutional coding in such an environment. The second group felt they had ways to address the vulnerabilities of the convolutional coder to burst errors (principally through heavy use of diversity techniques to be discussed later in this chapter), and that if they did so, they would then reap the benefit of the soft decoding implementation: those 4–6 dB mentioned above. In the end, the second group prevailed, which would be a nice segue to the discussion of diversity techniques in Sections 5.4 through 5.6 if the reader is inclined to page ahead.

5.3.4 Side Information

Closely related to the soft decision concept is the idea of *side information*, the use of extra information about the quality or characteristics of the channel to help the decoder make a better decision. The basic idea is fairly straightforward, and follows the notion of the assisted decision version of soft decision decoding discussed in the previous section.

Let us assume that the receiver is able to make a series of decisions regarding the values of a series of symbols as shown in Figure 5.28. The decisions are accompanied by a second output, which is a reliability metric—an assessment of the channel conditions obtained during the transmission of the symbol in question. Many of the decisions have a high reliability, but a few are passed through with indications of low reliability. Based on experience, we may establish a certain threshold for this metric; if the metric is too low, we assume that the symbol has a high likelihood of being

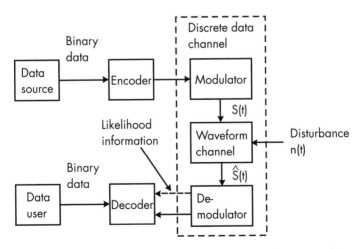

FIGURE 5.28 *Symbol decisions with confidence metrics and erasure. (From: [18].* © *1981 Plenum Press.)*

corrupted due to channel noise or interference, and we cause the symbol to be *erased*. We do so with some confidence, because we have coded the transmission using a channel coder that can correct a high number of erasures (e.g., Reed–Solomon).

The notion of side information can be expanded. The transmitter itself can in some cases determine whether the channel is in a clean or noisy state, and it can transmit information about the channel state along with the user information. This idea was identified by Shannon quite early on as a way to increase capacity [40].

One of the most important applications of side information is in frequency-hopping systems, where on certain hops the desired signal may experience destructive interference. The receiver can detect that such a hit has occurred, and can use this information to erase the corrupted data[50] (Figure 5.29). Pursley has published extensively on the application of side information in frequency-hopping systems, in conjunction with powerful Reed–Solomon codes.[51] He also describes and evaluates different methods

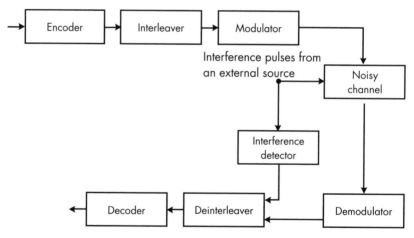

FIGURE 5.29 *The two-detector receiver. (From: [21]. © 1983 Prentice-Hall Inc.)*

---

50. "One possibility [for generating side information] occurs when there is a known interference such as a radar signal whose presence can be determined independently. The reliability information would then be a single symbol that indicates whether the radar is on or off" [18, p. 12]. See also [21, p. 549].

51. Pursley [41] provides a readable summary of much of this research and references to the work in detail.

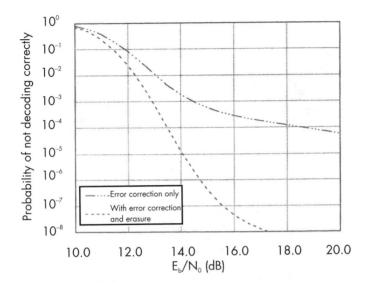

FIGURE 5.30 *Performance with erasure strategy. (From: [27]. © 1994 IEEE. Reprinted with permission.)*

of obtaining side information [41, 42]. Some of the results are amazing: Figure 5.30 shows the performance of a side-information-directed strategy versus a standard Reed–Solomon decoding process without side information. Calibrated at an error rate of 10^{-4}, the results indicate an improvement of approximately 4 dB as a result of the use of side information. This is comparable to the gain from soft decision techniques described in the previous section.

5.3.5 Pilot Signals

One of the ways to generate side information is to embed a special ancillary signal in the transmission, as described by Pursley:

> One method for obtaining side information for frequency-hop communication systems is to transmit special symbols, called *test symbols*, in each dwell interval [i.e., each hop] along with the data symbols. The demodulation of the test symbols from a particular dwell interval provides information on the presence or absence of interference in that dwell interval. [41, p. 162]

This is commonly called a *pilot signal,* and it can be applied for reasons other than supporting side information. For example, in the MIRS (later iDEN) digital mobile radio system developed by Motorola, special pilot signals were used to allow the receiver to correct for signal amplitude alterations caused by multipath in the channel, which in turn allowed the use of amplitude-sensitive modulation schemes like QAM, and a gain of several decibels, in principle, over amplitude-insensitive modulation of similar information-bearing capacity (like PSK).[52]

Another well-known example is the strong pilot signal embedded in the IS-95 (CDMA) downlink channel, which supports a number of functions, including power control and coherent demodulation by the mobile units. The latter, in principle, should provide a 3-dB improvement over noncoherent demodulation—and thus can be viewed as another way in which the signal has been hardened by means of channel coding at the transmitter.[53] A *discontinuous* pilot for phase reference can also be constructed, and will support coherent demodulation as well for frequency-hopped or TDMA systems.[54]

In any case, I expect that we will see the growing use of pilot signals to implement signal hardening strategies. There is a penalty in the use of some portion of the system transmission bandwidth for such pilots, which has probably contributed to a preference for non-bandwidth-consuming implementations where possible. Such blind processes usually require very intensive signal processing capabilities, and at some point the use of a pilot can become an attractive alternative.

5.3.6 Trellis Coding

Most of the basic coding concepts were discovered decades ago, and the history of the field since then has been defined by the challenges of

---

52. The gains of course are mitigated by the loss of capacity needed to support a pilot of sufficient resolution to track the amplitude variations. It is not clear from published material (at least to me) whether in this particular instance the net is a gain or not.

53. Pilot signals are probably not generally accounted for as forms of channel coding per se, because they do not provide error correction for the data bits in the conventional manner. However, the pilot signals used to control erasures in a Reed–Solomon coding scheme certainly would appear to cross that line, and I see no reason—conceptually—not to link pilot signals that enhance the strength of modulation or demodulation processes to the same general strategy of signal hardening.

54. "If several reference symbols of known phase are sent during each hop then the receiver can establish the absolute carrier phase and perform coherent demodulation

implementation, especially in finding efficient decoding algorithms. One area where a real revolution in both theory and practice has occurred more recently (dating effectively from the 1980s) is in the field known as *trellis coding*. Pioneered by Ungerboeck [44], this technique made its way very rapidly into commercial applications such as voiceband telephone modems where it has allowed an increase in signaling rates of almost a factor of 10 compared to rates found in common use at the beginning of the 1980s. I can remember reading statements written at that time which calculated the theoretically maximum rate of data transmission over a standard voice-grade telephone line at around 20–25 Kbps, and I can even recall Shannon's name being invoked to support the idea that this was the limit on capacity.[55] The reason we can now transmit at 56 Kbps over those same lines is due to the tremendous technical success of trellis coding.

Essentially, trellis coding is a technique whereby the coding scheme and the modulation scheme are designed together and jointly optimized to allow the receiver to use the decoding results to assist in the demodulation, and the demodulation results to assist in the decoding. In a sense, the receiver has two discriminations to make from the same signal: It has to discriminate the received signal in the signal space (defined by the physical dimensions of the signal such as phase and amplitude, typically), and it has to discriminate the decoded codeword in the code space (defined by the principles of the channel coding scheme). To make the first discrimination, the receiver would like to see the signal points spread as far apart as possible in the signal space (this is called the *maximum Euclidean distance*). To make the second discrimination, it would like to see the codepoints spread as far apart as possible in the code space (this is called the *Hamming distance*). The problem, as Ungerboeck diagnosed, is a mapping problem (once again):

> Mapping of code symbols of a code optimized for Hamming distance into nonbinary modulation signals does not guarantee that a good Euclidean distance structure is obtained. In fact, generally one cannot even find a monotonic [linear] relationship between Hamming and Euclidean distances, no matter how the code symbols are mapped. For a long time, this has been the main reason for the lack of good codes for multilevel modulation. [44, p. 7]

... The results indicate that if as few as 10 reference symbols/hop are used that near coherent performance is obtained" [43, abstract].

55. I have not been able to put my finger on this reference, and perhaps it is best left uncited, but it was a commonly expressed opinion in those days.

Ungerboeck's key insight was to envision the need to *expand* the signal space—in so-called Euclidean space of phase and amplitude—in a way similar to the expansion of the code space described earlier in the discussion of the underlying concepts of channel coding.

> The essential new concept of TCM ... was to use signal-set expansion to provide redundancy for coding and to design coding and signal-mapping functions jointly so as to maximize directly the ... minimum Euclidean distance between coded signal sequences. [44, p. 5]

The key lies in the *mapping* of one set of values (code values) to the other (physical signal values). Ungerboeck applied a concept called *partitioned-set mapping* (Figure 5.31). By the proper mapping of coded values into the constellation of signal points in the modulation scheme, the so-called Euclidean distance between signal points can be maximized for pairs of codepoints that are separated by the minimum Hamming distance, and vice versa [11, pp. 488ff]. In other words, for two coded values that are close in the code space (in terms of Hamming distance) the corresponding values in the modulation scheme will be far apart. For two modulation values that are adjacent in the phase-amplitude signal space (in terms of Euclidean distance), the corresponding code values will be far apart.

Finally, Ungerboeck extends the concepts of convolutional decoding to the joint demodulation/decoding of received signal sequences. (This is where the trellis comes into play, signifying the decoding tree or trellis used in convolutional decoding algorithms like the Viterbi algorithm.) The power inherent in soft decision decoding is applied across the board.

> The Viterbi algorithm, originally proposed as a ...decoding technique for convolutional codes, can be used to determine the coded signal sequence closest to the unquantized signal sequence.... The notion of error correction is then no longer appropriate, since there are no hard demodulator decisions to be corrected. The decoder determines the most likely coded signal directly from the unquantized channel outputs. [44]

The subject is more complex than we want to get into here [19, pp. 263ff]. Once again, from a practical standpoint, the trellis techniques achieve very impressive gains, from 4–6 dB in many applications.[56] I would say that the most general way to view these techniques is as an extension of several of

---

56. Proakis [11, p. 497] summarizes Ungerboeck's results.

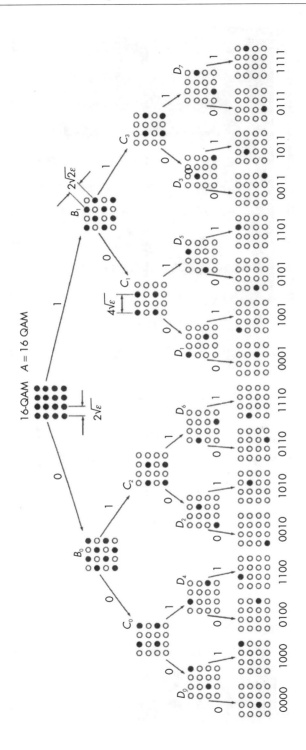

the principles we have covered previously—convolutional decoding, soft decision, signal expansion to achieve redundancy—in an extremely thoroughgoing way to encompass not only the coded information, but the physical characteristics of the signal itself, and to quasi-optimize the design across the entire set of concatenated processes (demodulation, decoding, error correction).

5.3.7 Hierarchical Coding Structures: Concatenated Coding, Turbo Coding, and Parallel Coding

It does not require a great leap of imagination to envision *combining* multiple coding schemes to build even stronger signals, but we must remember that when coding theory was making its pioneering breakthroughs in the 1950s, computers were lumbering behemoths unsuitable for real-time signal processing. Most coding techniques could not be implemented at any cost for real telecommunications systems. Indeed, for the next *40 years*, processing power was the perennial bottleneck for bringing coding applications to market. I can recall that one of the first digital wireless prototype subscriber terminals (TDMA) in the mid-1980s needed *nine* state-of-the-art digital signal processors, for a system that was very lightly coded by today's standards.[57] It was intended for a wireless local loop application, mainly for rural areas. "Keeps your cabin warm, too," the engineers joked. (That is, it used a *lot* of power.) This high cost of signal processing skewed the first- and second-generation wireless architectures, which were designed in the late 1980s, toward modest goals and simple implementations. The signal hardening in today's popular TDMA architectures is quite limited.

The picture is very different now. Moore's Law—the apparently inexorable and exponential growth of computing power available in integrated circuits—has finally begun to catch up to the potential of coding theory. Designers of 3G wireless systems today can much more freely evaluate ambitious signal processing architectures that make use of the latest advances in signal hardening technology. We are not yet at the point where processing is a free resource (from a design standpoint), but we are no longer forced to consider only a small subset of the available alternatives.

This has opened new vistas, leading to hierarchical coding structures such as these:

---

57. This was a TDMA system, using 14-Kbps speech compression and 16-PSK modulation. The production versions used four next-generation DSPs, eventually going to two by the end of the run.

1. *Concatenated coding:* passing the data stream through multiple different coding schemes deployed in series;

2. *Iterative coding:* passing the data through the same coding engine a number of times in succession to asymptotically reduce the error rate;

3. *Parallel coding:* passing the same data through multiple coding engines simultaneously, and then selecting or combining their outputs.

The idea of concatenated coding was put forward by Forney [45] in 1966, and was motivated by the goal of finding a less complex way of achieving greater coding gains. The field at that time was faced with the problem that longer (and more powerful) codes tended to increase the processing requirements exponentially:

> The number of computations required of the decoder would be increasing exponentially with the block length N, so that, while the probability of error would decrease exponentially with N, it would decrease only weakly algebraically with complexity.... One is therefore faced with the problem of finding a code and an associated decoding scheme such that the complexity of the encoder and decoder required to implement the scheme increases much more slowly than exponentially with N. [45, pp. 90–91]

The solution proposed by Forney (Figure 5.32) is to run the signal through a two-stage coding process: an inner code and an outer code. The inner code is generally a convolutional code, and the outer code is almost always a Reed–Solomon block code.[58] This has been called a perfect match because the Viterbi decoder used for the inner code tends naturally to produce bursts of errors at its output (because of the convolutional nature of the

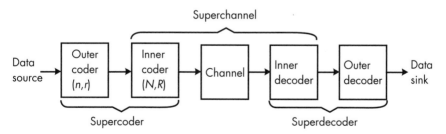

FIGURE 5.32 *Concatenated codes. (From: [45]. © 1974 IEEE. Reprinted with permission.)*

coded data), while Reed–Solomon codes are well suited to handling precisely these burst errors [27, p. 9]. The use of the convolutional coder as the inner code also means that soft decision decoding can be applied to the output of the demodulator, to gain that crucial 2–3 dB of performance. Finally, the inner coder can pass a kind of side information to the outer coder: The Viterbi engine can pass to the Reed–Solomon decoder not only the decoded bit stream, but also an indication of the reliability of each decision, which can assist the Reed–Solomon decoder in deciding when to erase unreliable data, to optimize the power of the Reed–Solomon structure (which can correct twice as many erasures as errors).[59]

> A "best of both worlds" situation can be obtained by combining Reed–Solomon codes with convolutional codes in a concatenated system. The convolutional code (with soft-decision Viterbi decoding) is used to "clean up" the channel for the Reed–Solomon code, which in turn corrects the burst errors emerging from the Viterbi decoder.... Such a system can achieve a bit error rate of 10^{-5} at an E_b/N_0 of 1.7 dB ... [46, p. 243]

Thus, a concatenated code combines almost all of the advanced coding concepts we have touched on: soft decision, burst-error performance, side information. It produces "enormous coding gains" [27, p. 9] with much less complexity than we would need to build a single coder capable of the same performance, at least in certain channels at certain error levels (Figure 5.33).

One of the first uses of concatenated codes was in the *Voyager* and *Galileo* planetary missions (Figure 5.34). The full story of the Galileo mission, where the main spacecraft antenna failed in space, is an engineering

---

58. "Convolutional codes should be used in the first stage of decoding because they can easily accept soft decisions and channel state information from the channel. Reed–Solomon codes are then used to clean up the errors left over by the Viterbi decoder. Indeed Viterbi decoding and the decoding of Reed–Solomon codes complement each other very nicely: the Viterbi decoder has no problem accepting soft decisions from the channel, and it delivers short bursts of errors. Short error bursts do not affect the Reed–Solomon decoder as long as they are within an *a*-bit symbol of the Reed–Solomon code.... [It is possible to] link the two decoders even further [by using] the soft output of a modified Viterbi decoder to generate erasures at the input of the Reed–Solomon decoding process ..." [46].

59. "We consider the possibility of the inner decoder passing along something more to the outer decoder than just its best guess. In particular, we let the inner decoder add to its estimate a real number which indicates how reliable it supposes its estimate to be" [45, p. 91].

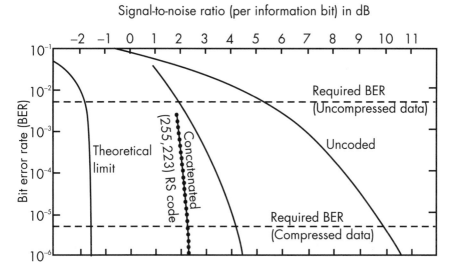

FIGURE 5.33 *Performance of concatenated versus unconcatenated codes. (From: [27]. © 1994 IEEE. Reprinted with permission.)*

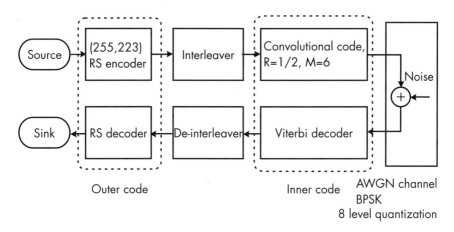

FIGURE 5.34 *Concatenated coding architecture. (From: [27]. © 1994 IEEE. Reprinted with permission.)*

tour de force and mission-rescue epic that rivals, in technical terms, the *Apollo 13* episode [47]. One of the innovations mothered by the necessity of recovering a drastically weakened signal was the idea of repassing the received data through the coding cycle more than once:

In a concatenated system it is possible and desirable to go back to the inner decoding step after the outer decoding step is finished, for we may now have more reliable information about some of the information bits ... [as well] if the decoding of certain information bits is correct ... we might be able to help in the decoding of neighboring bits. [46, p. 255]

This concept of iterative decoding is portrayed (in the *Galileo* implementation) in Figure 5.35. This extension produced a further coding gain of 0.4 dB at a BER of 10^{-5}, which is impressive considering how close to the Shannon limit the system is operating [46, p. 264].

A related idea, which combines concatenation and iteration, is *turbo coding* [48]. This increasingly popular approach is based on concatenating two convolutional coders, and providing a feedback path in the decoder so that the decoded output can be recycled through the decoder engine any number of times (Figure 5.36). The more iterations, the better the results. With 18 iterations through the decoding engine, the turbo code described in the seminal article is able to reach a benchmark BER of 10^{-5} at an E_b/N_0 of only 0.7 dB, which (it is claimed) is within less than 1 dB of the Shannon limit for that type of system (Figure 5.37). (Note by the way the extreme degree of quantization inherent in the performance of the entire system here. With 18 iterations, the system performs in a virtually "on–off" mode:

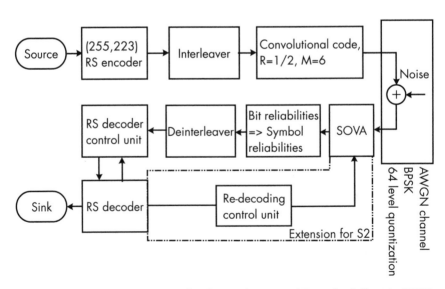

FIGURE 5.35 *Iterative concatenated coding architecture. (From: [27]. © 1994 IEEE. Reprinted with permission.)*

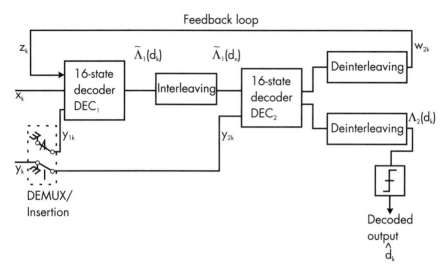

FIGURE 5.36 *Turbo coding. (From: [48]. © 1993 IEEE. Reprinted with permission.)*

At 0.5-dB E_b/N_0 the error rate is catastrophic, and indeed worse than the error rate on the uncoded system, whereas an improvement of only 0.2 dB in E_b/N_0 yields essentially error-free performance. The whole system is now operating like a basic threshold detector.)

Finally, the idea of using multiple independent decoding engines in parallel has been put forward. Basically, this involves a kind of averaging or majority-logic decision process at the global level (global in terms of the coding scheme). Pursley describes one fairly simple design for a parallel Reed–Solomon decoding scheme:

> The simplest parallel decoder consists of an errors-only decoder in parallel with an errors-and-erasure decoder. [As noted, the Reed–Solomon code can support the correction of a certain number of errors, or twice as many erasures. The decoder can be designed to focus on one or the other or both.] Each received word is the input to each decoder, and the results of the two separate decoding attempts are compared. Under fairly general conditions, Reed–Solomon decoders are much more likely to fail to decode [i.e., to output a "cannot decode" signal] than to decode into an incorrect code word; thus it is rare for the two decoders to produce different code words at their output. The most likely situation is that both decoders succeed in decoding and produce the same code word, and the next most likely situation is that one of the decoders fails

Binary
error
rate

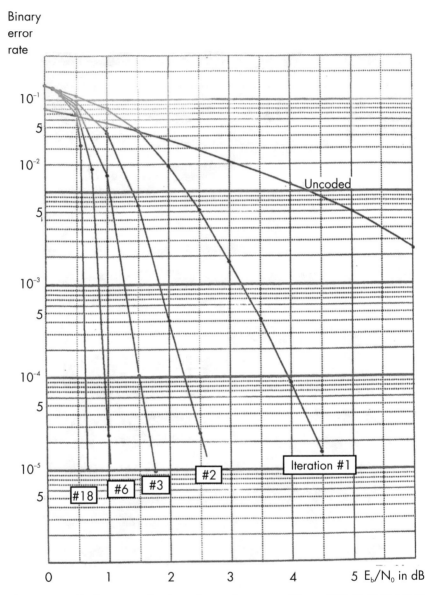

FIGURE 5.37 *Performance of turbo coding architectures. (From: [48]. © 1993 IEEE. Reprinted with permission.)*

to decode and the other produces a code word at its output. The simplest example of an advantage of using parallel decoding is the situation that arises if the side information ... indicates that the number of symbols that should be erased is beyond the erasure correction capability of the Reed–Solomon code. In such a situation ...the system should be designed so that the errors-and-erasures decoder simply defaults to the errors-only decoder. [41, pp. 169–170]

Shannon theory provides (for a given channel model, modulation scheme, and so forth) an asymptotic limit on how well a code can do—no matter how complex or powerful. I think the significance of the recent surge in popularity of these complex, hierarchical coding structures is that the processing power is now available, or at least on the horizon, to allow channel coding schemes to push this strategy of signal hardening nearly to its ultimate potential. It may well be that in this respect coding theory will turn out to be one of those terminal disciplines in the history of science, in the sense that we may soon achieve in practice nearly all the gains that are possible in theory, and research interest will wane. After all, if we know how to approach the Shannon limit within a decibel or less, we have probably entered the region of diminishing returns as far as investing in the search for still more powerful coding solutions.

We are not quite there yet, and undoubtedly there are quite a few implementation challenges ahead of us—although as long as Moore's Law prevails, it should not be long before the glut of processing power will do away with most of the perceived bottlenecks in this field. I would not be surprised to see the industry reach this plateau in the next decade or so. Fortunately for those possessed of the pioneering spirit, there are plenty of other challenging fields in wireless systems design where the potential for dramatic breakthroughs and significant gains in capacity still exist, as we shall see.

5.4 Diversity Techniques

Diversity is not necessarily a helpful term, but we are stuck with it. It encompasses a set of very different techniques that share a common approach to constructing a hardened signal. This approach can be characterized in a number of ways, depending on whether we look at it from the transmitter's perspective, the channel's perspective, or the receiver's perspective. To the transmitter, diversity often appears as a form of signal spreading. From the standpoint of the channel, it can look like an averaging across the relevant dimensions of the channel noise and interference across the physical dimensions of the signal (e.g., S, T, and F). To the receiver, diversity is often implemented by taking multiple samples of the same signal.

Sometimes diversity is constructed actively by the transmitter and coordinated with the receiver (much like channel coding); in other cases, it is a receiver-only function that takes advantage of natural spreading or averaging caused by the physics of the channel. By the same token, sometimes any spreading of the signal (e.g., by a convolutional coder, in the time domain) is referred to as *adding diversity*—and there is some validity to this claim. Yet overall, the breadth and vagueness of the term can be quite confusing. Under the heading of diversity techniques we may find disparate technologies such as frequency hopping, RAKE receivers for CDMA systems, dual antennas, interleaving, wideband transmission, circular polarization, and even soft handoff all referred to on occasion as implementations of diversity.

What do these techniques have in common? I think the best answer is this: Diversity techniques are all focused on *breaking up the structure of an interfering signal* (including the self-interference of multipath) as it impacts the desired signal, rendering it more like white noise. Unlike true interference cancellation (which we will turn to in Chapter 7), diversity does not actually remove or reduce the energy of the interfering signal, but homogenizes it into something much more like white noise.

Let us take as an illustration the most important form of structured interference, and the one against which most diversity techniques are explicitly targeted: multipath fading.

Multipath fading is created by multiple images of the transmitted signal arriving over different reflected paths, with slightly different times of arrival and often rotated in phase. The complex summing of these signals at the receiver's antenna can produce a greater or lesser degree of mutual cancellation of these different images, depending on the phase relationships. Because it is a function of the path geometry, the summed value of these signal images changes as the receiver antenna moves about (and to a lesser extent as the reflectors in the environment may change position as well, although the most important reflectors tend to be stationary). As a mobile or portable cell phone moves through this multipath field, the signal experiences fades of varying depths. The statistics of this fading pattern often fit the so-called Rayleigh distribution (hence, *Rayleigh fading*), which very broadly means that the number of fades (per volume or space, or per second traveled through that space) is inversely related to the depth of the fade. Deeper fades are rarer. Nevertheless, the frequency of fading is great; a single receiving antenna will experience "50–1000 fades per second" [49, p. 189] at cellular frequencies.

Map this fading pattern onto a digital signal and the effect is striking. (Figure 5.38 is a simplified schematic illustrating this idea.) The signal is

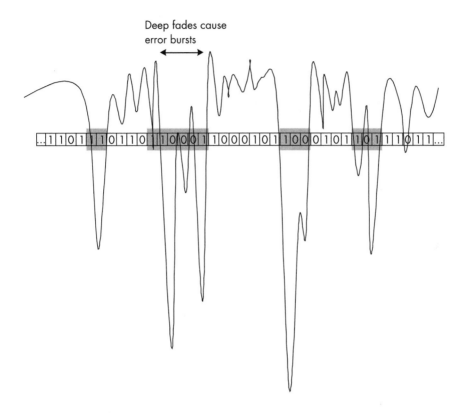

FIGURE 5.38 *Mapping a Rayleigh fading pattern on to a digital transmission in the time domain.*

decimated. During short periods between deep fades, the signal quality and the error rate are good. The fades, especially deep ones, obliterate the signal and cause clusters of errors at the receiver. The interference is not at all noise-like. The average bit error rate over a longish time frame (say, 10 seconds) may appear quite reasonable, but the errors are concentrated in bursts that effectively chop up the transmission and render it unusable. Although some error correcting codes (Reed–Solomon) may be effective to some degree against this type of interference, others, including the popular convolutional coding with soft decision decoding, are undone by it. Even good old robust FM transmission is highly challenged: "This rapid fading alters the signal-to-noise (S/N) performance markedly, washing out the sharp threshold and capture properties of FM" [49, p. 161]. Forget about AM and its cousins (without diversity): "Rapid Rayleigh fading generally has a disastrous effect on single-sideband (SSB) and AM communication systems. Unless corrective measures are taken, the distortion introduced by the fading is larger than the output signal ..." [49, p. 201].

The damage is created largely by the *structure* in the multipath interference. If we could find a way to take the *average* of the losses in signal strength due to fading and spread it evenly over all the digital symbols, we would have a very different situation. Based on the original philosophy of PCM, if the *average* signal loss is modest, the quantized digital signal will be essentially immune to it. If we can convert a situation where we lose every tenth symbol completely into one where we suffer a 10% degradation of each symbol, which the digital threshold detector can handle easily, we will have a much stronger signal *relative to the same level of interference energy*. That is the key: With diversity we are not talking about removing the interfering signal energy (in this case, the multiple images of the desired signal arriving over different paths). We are *redistributing* this energy; we are evening it out.

In terms of the SIR, with multipath fading we may need an extra margin of 20–40 dB above the nominal level required for good reception of an unfaded signal. This value is set by the depth of the worst fades. If we could somehow average out and spread out the effects of this fading more evenly, we might need a margin of only a few decibels. In other words, if we could break up the interference and average it out, we could withstand a much higher level of absolute noise and interference. We can effectively harden the signal by softening the interference.

Though multipath is the most common manifestation of structured interference (because it is not avoidable through buffering), other forms of interference can be analyzed in the same way. Cochannel interference in a TDMA cellular system is also periodic and structured; sometimes it is strongly present, and at other times it may be completely absent. Unaveraged, the margin required to buffer it must be based on the worst case, most of the time it is wasted.

The implementations of diversity are, well, diverse, because it is possible to attack the structure of the interfering signal in any of the three fundamental dimensions of that signal: space, time, or frequency. In the following sections, we will examine some of the techniques applicable in each of these domains. We shall follow the example of multipath fading through each suite of solutions, with occasional comments on other forms of structured interference as well.

5.4.1 Frequency Diversity

Multipath fading is said to be *frequency selective*. That is, for a given point in space, we may find a deep fade at frequency f_1, but if we tune the receiver to another frequency, f_2, we will quite possibly find that a signal received at this new frequency is not in a fade. In fact, as we tune through a wide band of frequencies, without moving the receiver at all, we will find a pattern of fades viewed along the frequency dimension that looks similar to the pattern of fades we saw along the time–space dimension in Figure 5.38. If the total band is divided into a number of narrow channels, we will find that some channels are strongly affected by fades and others are not (Figure 5.39).

How far apart do f_1 and f_2 have to be in order for the fading to be uncorrelated between them?

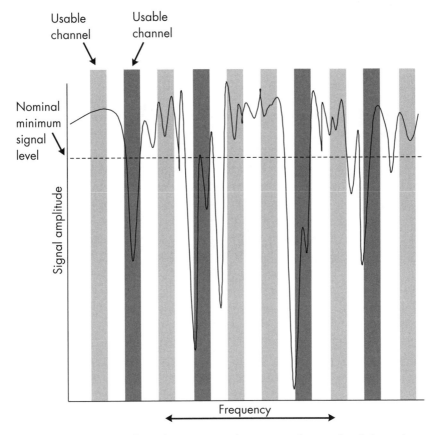

FIGURE 5.39 *Fading in the F-domain mapped onto a set of narrowband channels.*

The answer to this crucial question is a function of the physics of the channel, and the frequency of the transmitted signal (which determines many of the propagation characteristics of the channel, including the multipath geometry). The most important thing, perhaps, is that there *is* an answer: There is a definable separation between two frequencies such that, if f_1 and f_2 are separated by at least that amount, the fading statistics on f_1 and on f_2 will be uncorrelated. This unit of frequency separation is called the *coherence bandwidth*, or *coherence frequency*, F_{coh}. According to Jakes' classic studies of 800-MHz propagation (which formed part of the foundation of the first-generation cellular architectures), the coherence bandwidth for a signal at 836 MHz is approximately 640 kHz [49, p. 51]. He further suggests that to achieve full decorrelation of the fading statistics, two frequency channels should be ideally separated by several times the coherence bandwidth, or as he concludes somewhat vaguely:

> In the mobile radio case, measurements indicate a coherence bandwidth on the order of 500 kHz; thus for frequency diversity the branch separations would have to be at least 1–2 MHz. [49, p. 312]

Alas, the concept is a statistical one, and the vagueness over just how far apart two signal samples need to be to show significant independence has been the source of considerable debate among proponents of different architectures.[60]

But leaving aside the issue of the exact value of F_{coh}, let us establish the concept of frequency diversity. In Figure 5.40, we see a deep multipath fade superimposed on a wider frequency band. There are two ways to achieve diversity in this situation. In Figure 5.40(a), the receiver takes two samples at two different narrowband frequencies that are separated by more than F_{coh}. How is this done? The most common solution is to use frequency hopping, in which a short burst of information is transmitted first on f_1 and then both transmitter and receiver retune, or hop to f_2 for the next burst of information. In principle, if the hopping rate is faster than the fading rate, the effects of the fading will be substantially evened out.[61]

---

60. The IS-95 CDMA channel is "only" 1.25 MHz wide, for reasons that had to do with compatibility with the analog legacy channel allocations in AMPS. Some have argued that this is insufficient to really mitigate the effects of frequency-selective multipath fading, that a channel of 5 MHz (or more) is really required to provide frequency diversity.

61. The interaction of FH diversity and FEC performance is alluded to in [50]: "In general, the hop will be to a frequency beyond the coherence bandwidth of the channel,

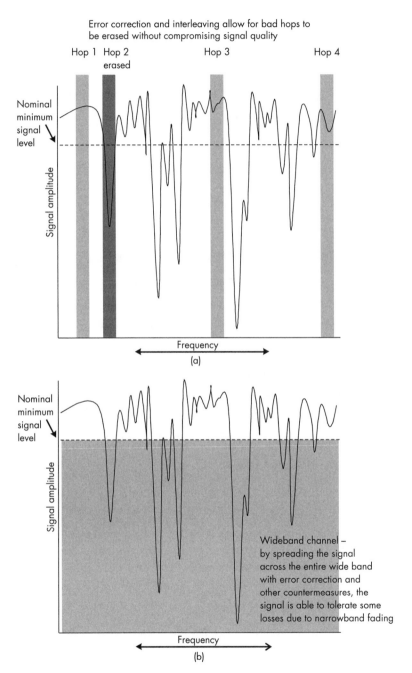

FIGURE 5.40 *Frequency-selective fade imposed on a larger band: (a) frequency hopping strategy to avoid narrowband fades; and (b) wideband signal to counteract narrowband fades.*

Even a slower hopping rate can mitigate some of the effects of structured interference. One of the simplest scenarios involves a mobile antenna that becomes temporarily stationary (for instance, when a car stops at an intersection) and comes to rest in a deep fade. In a nonhopping system, the signal may be completely lost. On the other hand, if we implement even a relatively slow frequency-hopping routine, whereby the transmitter and receiver rotate through a cycle of frequencies that are appropriately separated, the effects of one badly degraded channel will be reduced to a small proportion of the total transmission. If the degraded portion is small enough, the error correction coding may be able to eliminate its effects altogether. (Precisely this type of slow frequency hopping was specified in the extension of the basic GSM architecture, principally for this purpose.)

The second approach, [Figure 5.40(b)], is based on a very different philosophy. Here the transmitted signal itself is expanded, usually by mixing it with a very high frequency noise-like signal called a *direct sequence* or *pseudonoise sequence,* such that the bandwidth of the transmitted signal is wider than F_{coh}. The fade can now be viewed as subtracting only a narrow slice of the total signal. Once again, if the error correction coding is powerful enough, it should be able to handle this partial loss of data much more easily than the perhaps total deterioration of the unexpanded narrowband signal in a deep fade.

The reductions in multipath fading claimed for these two approaches can be quite meaningful, with peak fades reduced from 20–40 dB to only a few decibels.

These techniques can be extended to the management of interference from other users. The use of frequency hopping to average cochannel interference leads to solutions such as slow frequency-hopping TDMA, of which frequency-hopping GSM is an early example. Signal expansion is the basis of wideband, spread spectrum architectures underlying DS/CDMA systems such as IS-95. (Of course, both architectures combine frequency diversity measures with many other hardening techniques, and other signal processing strategies, into very comprehensive and complex systems.)[62]

---

so that the received signal will be faded independently from frame to frame.... This means that FH reduces the burst error length and randomizes errors in a fading channel.... Usually in a fading channel FEC alone provides little BER improvement. This is because errors are not random in a Rayleigh fading environment. When the channel is not in deep fading, few errors occur, hence FEC is not needed. When the channel is in deep fading, too many errors occur, hence FEC is not effective. In an FH system, however, FH randomizes the burst errors, and therefore FEC can work effectively."

5.4.2 Time Diversity

As we have described, multipath fading appears to the mobile receiver as a time-varying phenomenon. In Figure 5.38, we saw how fades produced a pattern of clustering of error bursts in the time domain. If we can smooth out the impact of these hits in the time domain so that the entire signal sees a relatively constant, averaged level of interference, we can improve the performance of the standard error correction coding and reduce the SNR margins required to maintain an adequate threshold for good receiver performance.

Just as there is a coherence bandwidth in the frequency domain, there is a *coherence time*, T_{coh}, in the time domain. Two samples of the signal taken at two instants farther apart in time than T_{coh} will exhibit uncorrelated fading statistics [49, p. 312]. Once again, the value of T_{coh} is based on the physical characteristics of the channel and the signal, but it is also subject to practical interpretation, which renders it somewhat less definite than we might want. Nevertheless, according to Jakes, "for vehicle speeds of 60 mi./hr this time [T_{coh}] is on the order of 0.5 to 5 msec for frequencies in the 1–10 GHz range, respectively" [49, p. 312].

The preferred solution for achieving time averaging of multipath effects is a technique, or family of techniques, known as *interleaving*. Consider a TDMA system that is producing packets on information, to be transmitted on periodic time slots. Let us assume that each packet contains 10 data symbols of information, and let us also assume that instead of transmitting each packet when it is ready, the transmitter accumulates 10 such packets in a buffer. Next, before transmitting any of them, the transmitter interleaves them, or shuffles them together. The interleaver can be visualized as a simple matrix in which the 100 data symbols (10 packets with 10 symbols per packet each) are read into the matrix row by row until it is full, and then read out column by column to create 10 new interleaved packets ready for transmission. Each new packet contains one symbol from each of the original 10 packets. Let us now assume that in the course of transmission over the wireless channel, one of the packets is hit by a fade and the data symbols it contains are corrupted. The other nine packets are received successfully. The ensemble is deinterleaved to recreate the original 10 packets in their original sequence. Now, instead of having lost one packet completely (from which no type of error correction alone could have

---

62. The discussion of system-level architectures like SFH-TDMA and CDMA is beyond the scope of this book, and will be addressed in Volume 2 of this series.

recovered the data), we have lost only one data symbol from each of the 10 original packets. The error correction code (we assume) can correct up to one lost symbol per packet. Thus, by interleaving, we have averaged or spread out the effects of the fading, and (in this example) no data are lost at all. Interleaving is one of the most elegant and straightforward illustrations of the basic idea of diversity: converting the structured interference of multipath fading into nonstructured noise that is far more tractable to our error correction techniques.

Many interleaving techniques have been developed, including block interleaving (as described here), convolutional interleaving (which operates on a continuous stream of data), and pseudorandom interleaving (which removes the inherent periodicity or correlation between the input and the output of the interleaver) [11, p. 440; 18]. Interleaving can be embedded at different levels, either on top of the basic channel coding scheme, or in some cases, within the coding scheme itself. It is easy to implement and it does not require additional bandwidth or add very much complexity at the receiver. The only real design issue is the additional delay created by the interleaver, which depends on its *depth*—that is, the number of input samples that need to be collected before the interleaving function can proceed. The greater the span of the interleaver in time relative to the average fade duration and T_{coh}, the better the averaging and the greater the improvement in performance. It is likely that all third-generation wireless architectures will use interleaving of some sort.

5.4.3 Space Diversity

In addition to T_{coh} and F_{coh}, there is also a *coherence distance* or separation in space between two receiving antennas that ensures uncorrelated fading statistics between the two. The value of S_{coh} is normally given as one-half the wavelength of the radio carrier signal, which is a few inches at the frequencies of current interest for mobile communications [49, p. 311]. A receiver with two antennas separated by at least this distance, which can select the better of the two signals, will show a huge improvement in performance. Lee claims that an 8-dB gain can be achieved in ordinary FM by using two antennas [8]. (To put this in perspective, we should recall that today's most powerful channel coding schemes deliver about the same gain, with much more huffing and puffing, generally). Jakes' extensive analysis would suggest that the gains may be considerably greater, and that "the reduction of cochannel interference [as opposed to multipath] can be the most important advantage of diversity" [49, p. 362]. The various techniques for selecting or combining the different signals from multiple antennas have been well covered in the literature.

5.5   Convolutional Signals

The general goal of all signal hardening strategies is to accentuate and protect the inherent structure of the signal, which carries the information, and to break up the structure of the noise and interference. Diversity techniques work by decorrelating signal and interference.[63] The convolutional strategy works by strengthening the inherent correlations between the elements of the signal, linking them more tightly together by spreading and *overlapping* the information content of those elements.

Consider the classical input signal assumed by Shannon: a series of independent, stand-alone binary digits, or perhaps small blocks of such digits, codewords, each independent of the others. This is the PCM model. If one signal element fails, it fails alone; the others are unaffected. By the same token, the loss of one symbol cannot be recovered by examining the remaining symbols. The information is completely compartmentalized.

Block codes take a step forward by creating correlated blocks of symbols, such that the loss of one symbol in a block can often be recovered by analyzing the surviving symbols. The information of each symbol has been spread among all symbols of the block.

With convolutional coding we found that even stronger correlations could be created by braiding individual symbols together in a continuous strand. Indeed, the metaphor of strands and threads is a useful one: In fabrics like wool and cotton, relatively short fibers are twisted together in a continuous, overlapping fashion to create strands that are much stronger. Actually in convolutional coding, two processes are acting together: the expansion of the signal (expressed as the rate of the code, the ratio of the input symbols to the output symbols) and the interweaving of the signal

---

63. Sometimes diversity seems to imply a conceptual reversal of this idea (of accentuating the signal structure), whereby the signal is scrambled (through interleaving, for example) or multiplied by a noise-like random sequence (as in direct sequence CDMA), but what is really going on is a process of decorrelation: By scrambling the signal *in a deterministic way* (which means that we know exactly how to unscramble it later on), we break up the destructive correlations that exist between the structure of the interference and the structure of the information we want to transmit. In effect, we throw the whole thing into the blender, but we know how to reconstitute the signal, so the interference will remain behind as a homogenized residue.

(expressed as the constraint length, the number of symbols involved in each convolutional cycle to produce a new output symbol).

The convolutional concept also appears in another set of techniques that, when they first emerged in the 1960s, were in some ways rather confounding, because they seemed to many observers to involve the deliberate creation of interference between data symbols, which was (until then) something to be strictly avoided. These techniques were labeled initially *correlative coding* (by Lender, who invented the idea) [51–54], but are more generally known today as *partial response* coding.[64] Among other things, this new signaling method produced the startling result that it became possible to transmit data at speeds *above* the so-called Nyquist rate. This rate sets the maximum rate of transmission of pulses in a channel (without suffering destructive intersymbol interference) and is (approximately) twice the frequency bandwidth of the channel. If we consider a telephone channel with a voice bandwidth of 4,000 Hz, the maximum pulse rate is 8,000 pulses per second, which happens to be the PCM pulse rate.[65]

The Nyquist rate had been enshrined in communications thinking since Nyquist propounded it in his studies of telegraph signaling in the 1920s, to the extent that it had become a kind of limit in the view of many engineers. How then does correlative coding allow us to break the Nyquist signaling barrier? The key lies in the assumption underlying Nyquist's work, which was fully appreciated only after Lender showed how it could be modified, that Nyquist had assumed that *the transmitted pulses must be independent*. What Lender proposed was to allow the pulses to overlap, partially, in a controlled way (Figure 5.41). As we have seen elsewhere, the introduction of a controlled amount of deterministic ISI somehow inoculates the signal and enables it to withstand an even higher level of additional ISI. With the right design, this permits a signaling rate above the Nyquist limit.

In fact, partial response signaling is quite akin to convolutional coding. Because the information content of each input symbol is spread across more than one output symbol, it also can be vulnerable to error propagation, unless certain countermeasures (sometimes called, unhelpfully,

64. "The term *partial response* is used in connection with this technique since the response to an input symbol is spread over more than one [output] symbol interval—the response in a single interval is partial" [54, p. 266].

65. It had become a canonical formula, indeed: "Nyquist's [1924] general result [holds] that in a bandwidth of W hertz, a maximum of kW pulses/sec can be transmitted without intersymbol interference, where $k \leq 2$ is a proportionality factor depending on the pulse shape and the bandwidth" [55, p. 196].

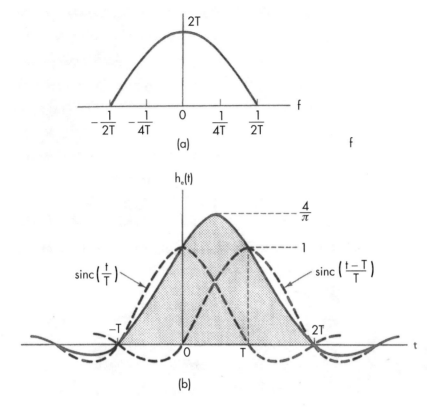

FIGURE 5.41 *Partial response signaling: (a) cosine filter; and (b) impulse response of the cosine filter. (From: [56]. © 1988 Prentice-Hall Inc.)*

precoding) are properly employed [11, p. 540]. It also provides, though this is not its purpose as such, a measure of error control, without using additional bandwidth for redundant symbols.[66] The Viterbi algorithm for soft decoding can also be used with partial response signaling [54, p. 274]. Some authors go so far as to identify partial response as "an example of convolutional coding" [54].[67] However, a convolutional coder produces a digital

66. "Since there is a correlation among the transmitted levels at sampling instants, this correlation, or memory, can be used to detect errors without the insertion of redundant bits at the transmitter" [55, p. 203].

67. Others disagree, sort of: "the duobinary encoder is a convolutional encoder ... [but] the purpose of the encoding is ... different. The purpose is not to suppress noise by

output, while partial response signaling creates an analog signal with convolutional properties—and this is more troubling for many writers, because it is obviously much closer to real interference. How can something be so harmful on the one hand, and so mysteriously beneficial, on the other?

> Intersymbol interference is a harmful mutual interaction of pulses in a transmitted pulse train.... When samples are taken ...they contain unwelcome contributions from neighboring pulses.... The exception to the "unwelcome" statement is for partial response pulses, where correctable intersymbol interference is intentionally inserted ... [54, p. 288]

Partial response signaling is perhaps the most nakedly post-Shannon communications technology in common use. Even though it has been around for more than 30 years, and it is not overly difficult to understand or use, its success is somehow counterintuitive, and a little baffling:

> We come to a remarkable reformulation of the signaling waveform with the property that the unrealizability of the waveform and the unrealizability of the channel can be made to cancel each other leaving a scheme that is realizable in both respects. [19, p. 249]

I would argue that *cancellation*, as a term, should be reserved for a different case, one in which the structure of an interference signal is specifically decoupled from the desired signal and removed. Techniques based on this approach are discussed in Chapter 7. The strategy implicit in partial response is to build correlations in the signal, which allow for a higher level of noise and interference to be tolerated.

Redundancy, diversity, convolution—three broad strategies that share (I claim) a common aim: to harden the signal, and/or to soften the interference, to allow for a communications system to perform well with reduced buffers in space, time, and frequency, and so to gain capacity. In the next chapter, we examine the equally broad range of strategies designed around a complementary objective: to reduce the profile of the transmitted signal, so as to *generate* less interference to other signals, and, by so doing, to allow further reduction of the STF buffering.

---

increasing distance between transmitted waveforms. The purpose of partial response is to manage the spectrum of the transmitted waveform" [19, p. 250].

References

[1]   Oliver, B. M., J. R. Pierce, and C. E. Shannon, "The Philosophy of PCM," *Proc. IRE,* Vol. 36, 1948, pp. 1324–1331. Reprinted in Sloane, N. J., and A. D. Wyner, (Eds.), *Claude Elwood Shannon: Collected Papers,* Piscataway, NJ: IEEE Press, 1993, pp. 151–159.

[2]   Lee, T. H., *The Design of CMOS Radio-Frequency Integrated Circuits,* New York: Cambridge University Press, 1998.

[3]   Calhoun, G., *Digital Cellular Radio,* Norwood, MA: Artech House, 1988.

[4]   Shannon, C. E., "A Mathematical Theory of Communication, Part 2," *Bell System Technical J.,* Vol. 27, October 1948, pp. 623–656. Reprinted in Slepian, D., (Ed.), *Key Papers in the Development of Information Theory,* Piscataway, NJ: IEEE Press, 1974, pp. 19–29.

[5]   Shannon, C. E., "Communications in the Presence of Noise," *Proc. IRE,* Vol. 37, January 1949, pp. 10–21. Reprinted in Slepian, D., (Ed.), *Key Papers in the Development of Information Theory,* Piscataway, NJ: IEEE Press, 1974, pp. 30–41.

[6]   Bellamy, J. C., *Digital Telephony,* New York: Wiley, 1982, pp. 71–72.

[7]   Kucar, A., "Mobile Radio: An Overview," *IEEE Communications Magazine,* November 1991, p. 80.

[8]   Lee, W. C. Y., "Spectrum Efficiency in Cellular," *IEEE Trans. on Vehicular Technology,* Vol. 38, No. 2, May 1989, pp. 69–75.

[9]   Raith, K., and J. Uddenfeldt, "Capacity of Digital Cellular TDMA Systems," *IEEE Trans. on Vehicular Technology,* Vol. 40, No. 2, May 1991, pp. 323–332.

[10] Jayant, N. S., and P. Noll, *Digital Coding of Waveforms,* Englewood Cliffs, NJ: Prentice-Hall, 1984.

[11] Proakis, J. G., *Digital Communications,* 2nd ed., New York: McGraw-Hill, 1989.

[12] Hamming, R., *Coding and Information Theory,* Englewood Cliffs, NJ: Prentice-Hall, 1986.

[13] Gray, R. M., *Source Coding Theory,* Boston, MA and Dordrecht, the Netherlands: Kluwer, 1990.

[14] Sayood, K., *Introduction to Data Compression,* San Francisco, CA: Morgan Kaufmann, 1996

[15] Ifeachor, E. C., and B. W. Jervis, *Digital Signal Processing,* Reading, MA: Addison-Wesley, 1993.

[16] Omura, J. K., and B. K. Levitt, "Coded Error Probability Evaluation for Antijam Communication Systems," *IEEE Trans. on Communications,* Vol. 10, May 1982, pp. 896–903.

[17] Dholakia, A., *Introduction to Convolutional Codes with Applications*, Boston, MA and Dordrecht, the Netherlands: Kluwer, 1994.

[18] Clark, G. C., and J. B. Cain, *Error-Correction Coding for Digital Communications*, New York: Plenum, 1981.

[19] Blahut, R. E., *Digital Transmission of Information*, Reading, MA: Addison-Wesley, 1990.

[20] Berlekamp, E. P., *Key Papers in the Development of Coding Theory*, Piscataway, NJ: IEEE Press, 1974.

[21] Lin, S., and D. J. Costello, *Error Control Coding: Fundamentals and Applications*, Englewood Cliffs, NJ: Prentice-Hall, 1983.

[22] Hamming, R. W., "Error Detecting and Error Correcting Codes," *Bell System Technical J.*, Vol. 29, April 1950,, pp. 147–160. Reprinted in Berlekamp, E. P., (Ed.), *Key Papers in the Development of Coding Theory*, Piscataway, NJ: IEEE Press, 1974, pp. 9–12.

[23] Elias, P., "Coding for Noisy Channels," *IRE Convention Record*, Vol. 3, No. 4, 1955, pp. 37–46. Reprinted in Berlekamp, E. P., (Ed.), *Key Papers in the Development of Coding Theory*, Piscataway, NJ: IEEE Press, 1974, pp. 48–55.

[24] Forney, G. D., "Convolutional Codes I: Algebraic Structure," *IEEE Trans. on Information Theory*, Vol. 16, 1971.

[25] Viterbi, A. J., "Convolutional Codes and Their Performance in Communication Systems," *IEEE Trans. on Communications Technology*, Vol. 19, No. 5, October 1971, p. 767.

[26] Forney, G. D., et al., "Efficient Modulation for Band-Limited Channels," *IEEE J. on Selected Areas of Communication*, Vol. 2, September 1984, pp. 632–647.

[27] Wicker, S. B., and V. Bhargava, (Eds.), *Reed–Solomon Codes and Their Applications*, Piscataway, NJ: IEEE Press, 1994.

[28] Golay, M., "Notes on Digital Coding," *Proc. IRE*, Vol. 37, No. 6, June 1949, p. 657. Reprinted in Berlekamp, E. P., (Ed.), *Key Papers in the Development of Coding Theory*, Piscataway, NJ: IEEE Press, 1974, p. 13.

[29] Viterbi, A. J., "Wireless Digital Communication: A View Based on Lessons Learned," *IEEE Communications Magazine*, Vol. 29, No. 9, September 1991, pp. 33–36. Reprinted in Abramson, N., (Ed.), *Multiple Access Communications*, Piscataway, NJ: IEEE Press, 1993, p. 228.

[30] Van Dyck, R., and D. J. Miller, "Transport of Wireless Video Using Separate, Concatenated, and Joint Source-Channel Coding," *Proc. IEEE*, Vol. 87, No. 10, October 1999, pp. 1735–1750.

[31] Austin, M., et al., "Service and System Enhancements for TDMA Digital Cellular Systems," *IEEE Personal Communications*, Vol. 6, No. 3, June 1999, p. 24.

[32] McEliece, R. J., and L. Swanson, "Reed–Solomon Codes and the Exploration of the Solar System," in Wicker, S. B., and V. Bhargava, (Eds.), *Reed–Solomon Codes and Their Applications,* Piscataway, NJ: IEEE Press, 1994, pp. 25–40.

[33] Hagenauer, J., and T. Stockhammer, "Channel Coding and Transmission Aspects for Wireless Multimedia," *Proc. IEEE,* Vol. 87, No. 10, October 1999, pp. 1764–1777.

[34] Golomb, S., R. E. Peile, and R. A. Scholtz, *Basic Concepts in Coding Theory,* New York: Plenum, 1994.

[35] Drukarev, A. I., and K. P. Yiu, "Performance of Error-Correcting Codes on Channels with Memory," *IEEE Trans. on Communications,* Vol. 34, No. 6, 1986, pp. 513–521.

[36] Kanal, L. N., and A. R. K. Sastry, "Models for Channels with Memory and Their Applications to Error Control," *Proc. IEEE,* Vol. 66, No. 7, 1978, pp. 724–744.

[37] Kohlenberg, A., and G. D. Forney, "Convolutional Coding for Channels with Memory," *IEEE Trans. on Information Theory,* Vol. 14, No. 5, 1968, pp. 618–628.

[38] Jamali, S. H., and T. Le-Ngoc, *Coded-Modulation Techniques for Fading Channels,* Boston, MA and Dordrecht, the Netherlands: Kluwer, 1994.

[39] Cooper, A. B., "Soft-Decision Decoding of Reed–Solomon Codes," in Wicker, S. B., and V. Bhargava, (Eds.), *Reed–Solomon Codes and Their Applications,* Piscataway, NJ: IEEE Press, 1994, pp. 108–109.

[40] Shannon, C. E., "Channels with Side Information at the Transmitter," *IBM J. of Research and Development,* Vol. 2, 1958, pp. 289–293. Reprinted in Sloane, N. J., and A. D. Wyner, (Eds.), *Claude Elwood Shannon: Collected Papers,* Piscataway, NJ: IEEE Press, 1993, pp. 273–278.

[41] Pursley, M. B., "Reed–Solomon Codes in Frequency-Hop Communications," in Wicker, S. B., and V. Bhargava, (Eds.), *Reed–Solomon Codes and Their Applications,* Piscataway, NJ: IEEE Press, 1994, pp. 150–174.

[42] Wang, Q., and Y. Chao, "Frequency-Hopped Multiple Access Communications with Coding and Side Information," *IEEE J. on Selected Areas in Communications,* Vol. 10, No. 2, February 1992, pp. 317–327.

[43] Stadler, J. S., "A Performance Analysis of Coherently Demodulated PSK Using a Digital Phase Estimate in Frequency Hopped Systems," *Proc. IEEE MILCOM '95 Conf. Record,* November 8, 1995, pp. 107–112.

[44] Ungerboeck, G., "Trellis-Coded Modulation with Redundant Signal Sets—Part I," *IEEE Communications Magazine,* Vol. 25, No. 2, February 1987, pp. 5–11.

[45] Forney, G. D., "Performance of Concatenated Codes," in Berlekamp, E. P., (Ed.), *Key Papers in the Development of Coding Theory,* Piscataway, NJ: IEEE Press, 1974, pp. 90–94.

[46]   Hagenauer, J., E. Offer, and L. Papke, "Matching Viterbi Decoders and Reed–Solomon Decoders in a Concatenated System," in Wicker, S. B., and V. Bhargava (Eds.), *Reed–Solomon Codes and Their Applications*, Piscataway, NJ: IEEE Press, 1994, pp. 243–244.

[47]   McEliece, R. J., and L. Swanson, "Reed–Solomon Codes and the Exploration of the Solar System," in Wicker, S. B., and V. Bhargava, (Eds.), *Reed–Solomon Codes and Their Applications*, Piscataway, NJ: IEEE Press, 1994, pp. 25–40.

[48]   Berrou, C., A. Glavieux, and P. Thitimajashima, "Near Shannon Limit Error-Correcting Coding and Decoding: Turbo-Codes," in Rappaport, T., (Ed.), *Cellular Radio and Personal Communications*, Piscataway, NJ: IEEE Press, 1996, pp. 387–393.

[49]   Jakes, W. C., (Ed.), *Microwave Mobile Communications*, New York: Wiley, 1974.

[50]   Guo, Y., and K. Feher, "Power and Spectrally Efficient SFH-FQPSK for PCS Applications," *IEEE Trans. on Vehicular Technology*, Vol. 43, No. 3, August 1994, pp. 795–800.

[51]   Lender, A., "The Duobinary Technique for High Speed Data Transmission," *IEEE Trans. on Communications Electronics*, Vol. 82, May 1963, pp. 214–218.

[52]   Lender, A., "Correlative Digital Communications Techniques," *IEEE Trans. on Communications Technology*, Vol. 12, December 1964, pp. 128–135.

[53]   Lender, A., "Correlative Level Coding for Binary Data Transmission," *IEEE Spectrum*, Vol. 3, February 1966, pp. 104–115.

[54]   Gitlin, R. D., J. F. Hayes, and S. B. Weinstein, *Data Communications Principles*, New York: Plenum, 1992.

[55]   Gibson, J. D., *Principles of Digital and Analog Communications*, 2nd ed., New York: Macmillan, 1993.

[56]   Sklar, B., *Digital Communications*, Englewood Cliffs, NJ: Prentice-Hall, 1988.

6

SIGNAL SHAPING TECHNIQUES (TRANSMITTER-ORIENTED STRATEGIES)

6.1 Concepts of Efficient Transmission: Compression and Shaping

The signal hardening strategies described in Chapter 5 are link oriented. That is, they involve paired, complementary signal processing stages at the transmitter and receiver—coding and decoding, interleaving and deinterleaving, and so forth. In this chapter, we review another set of signal processing techniques that are implemented only, or principally, in the transmitter. Once again, this chapter will bring together a diverse range of technologies that are not normally presented together or analyzed within the same framework, yet it is our contention here that they do share a common goal.

That common goal is *to reduce the amount of physical energy required to transmit the real information content of a signal from point A to point B.* In effect, there is a ratio between the signal *energy* and the signal *information.* There is always a *surplus* of signal energy beyond what is needed to actually accomplish the communications process (i.e., to allow the receiver to create a reasonably accurate copy of the source message). In the case of a single transmitter–receiver link, the surplus transmitted power is of no particular concern. Once we take up the problem of multiple access and the sharing of a limited channel resource (a finite STF communications space) by more than one user, then the surplus becomes a potential problem. Surplus, unnecessary signal power created by each transmitter becomes a source of interference for other users.

Thus, if we can reduce this ratio, reduce the surplus, we should be able to reduce the total amount of interference in the STF communications space. The techniques discussed here all involve strategies for reducing this ratio. We can refer to this idea as *efficient transmission*.

This seemingly straightforward idea combines, and potentially confuses, several concepts. First, there is the notion of the "real information content" of a signal. In *analog* transmission, it is possible to view the communications problem (albeit somewhat naïvely) as a matter of recreating the source signal itself at the receiver. This reduces the problem of fidelity to a comparison of physical parameters of transmitted and received signal copies. Once we cross the line to *digital* transmission, however, we confront the problem of defining and measuring the information content. The original signal is no longer even available to the receiver. Instead, we have a coded message that purports to contain the real information content of the original signal. This content is defined variously, for different types of sources, and it is not always subject to a precise specification of accuracy or fidelity criteria.

What, then, does real information content mean? It implies, for example, that at least two kinds of information (in the pure Shannon theoretic sense) are contained in the original signal. There is information that is important, or relevant, or meaningful, or necessary, or desirable—and there is other information that we can, in principle, do without. As discussed in the previous chapter, every digitalization scheme involves an *explicit* distinction between what is kept and what is discarded (e.g., quantization error), which is in turn based on an *implicit* distinction between what is important, meaningful, and so forth, and what is not. The design of quantization and coding schemes *always involves a theory of meaning* (notwithstanding the many declarations to the contrary to be found in the introductory paragraphs of the foundational documents of the field[1]).

This often vague, undeveloped, and sometimes even unrecognized semantic level in the communications process forms the basis for the

---

1. For example: "Nothing is said [in Information Theory] about the symbols themselves, nor their possible meanings. All that is assumed is that they can be uniquely recognized. We cannot define what we mean by a symbol" [1, p. 1]; or, from Shannon's ur-text itself: "Frequently the messages have *meaning*; that is, they refer to or are correlated according to some system with certain physical or conceptual entities. These semantic aspects of communication are irrelevant to the engineering problem" [2, p. 5].

concept of *compression*. In effect, we are positing a *second* ratio: between meaningful information and unimportant information, which are blended in the original source. If we can find a way to squeeze out the unimportant information while retaining the important information, we can reduce the amount of signal energy required to transmit the message.

Compression is a very important strategy for post-Shannon communications. By reducing the transmitted message to a more concentrated payload of meaningful information, we reduce the amount of signal energy required to transmit this information successfully. To give some idea of the potential gains, consider that in the past 15 years or so the standard for acceptable transmission of digitized voice has dropped by a factor of 10 (in terms of bit rate): from 64,000 bps of traditional *pulse code modulation* (PCM) to around 4,000–8,000 bps today. Low-bit-rate coders have been one of the most important sources of capacity gains in the second-generation wireless architectures, and continue to play a significant role in strategies for third-generation systems.

Compression schemes are further divided into *lossy compression,* in which the criterion for meaningful versus nonmeaningful information is difficult to specify, and *lossless compression*, in which, in principle, no meaningful information is sacrificed. In other words, in lossy systems the compression process involves the elimination of nonmeaningful information along with at least some information that may be potentially meaningful, whereas lossless systems claim that only nonmeaningful or redundant information is eliminated, leaving purely meaningful information. (We examine these concepts further in Section 6.3.)

The second important concept embedded in the idea of efficient transmission is the idea of *surplus* signal energy. What exactly does this mean?

Consider the classical link budget, which specifies the relationship between the signal power required at the transmitter in order to achieve a specified level of signal power at the receiver that is sufficient for the receiver to operate and recreate the signal. To build the link budget, we start with a transmitter power of typically a watt or more. We add and subtract from this value, to account for the transmission through various stages of the end-to-end channel. Mostly, we subtract. The most important factor that reduces the energy value of the signal is the free-space transmission channel itself. From a level of, say, 1W of power at the output of the transmitter antenna, we find that the actual received signal at the receiver antenna is reduced by an enormous factor—up to 100 dB or more.

In other words, the receiver actually needs only a few billionths or even a few trillionths of the original transmission energy in order to do its job. To overcome the losses of the channel, the transmitter has to emit an enormous amount of energy, only a very small amount of which is actually used by the intended receiver; the surplus is radiated into the communications space where it creates interference for other users. If there were a way to precisely shape and target the transmitted signal so that it only delivered the required amount to the receiver, without creating all of the surplus that leads to interference, the overall interference levels in the system could be dramatically improved. The buffers could be reduced, the system could be packed more densely in the available STF communications space, and we could gain capacity.

This signal shaping can take place at different stages: Some operate at baseband, and other techniques can be used to shape the RF waveforms. *Inherent* shaping techniques change the way the signal is generated, such as modulation schemes that reduce the F-interference from unwanted sidebands (Section 6.5). *External* shaping techniques, like filtering, modify the signal after it has been generated to reduce the unwanted outputs. Such shaping can be performed in the S-domain, the F-domain, or the T-domain. It can involve a simple component-level implementation (better filters) or a system-level architectural concept (e.g., adaptive system concepts described in Section 6.7).

These two ratios—the meaningful information versus unimportant information, and the necessary signal energy versus the surplus signal energy (Figure 6.1)—thus suggest a categorization of transmission strategies:

1. *Compression:* These strategies are designed to improve the first ratio, between meaningful information and unimportant information in the payload; this is often called source coding, which can be further divided into (a) *lossy compression,* in which the criterion for distinguishing between the two is imprecise and (b) *lossless compression,* in which meaningful information and unimportant or redundant information can be clearly distinguished

2. *Signal shaping:* These strategies are designed to improve the second ratio, that between necessary signal energy and surplus signal energy; which can be divided into (a) *S-shaping strategies,* which attempt to shape the spatial characteristics of the transmitted signal, to reduce S-interference; (b) *F-shaping strategies,* which focus on shaping the frequency characteristics of the transmitted signal, to reduce F-interference; and (c) *T-shaping strategies,* which address the time-domain characteristics of the transmitted signal, to reduce all forms of interference.[2]

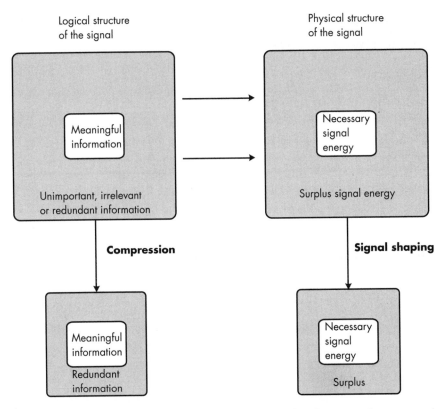

FIGURE 6.1 *Two ratios: meaningful to nonmeaningful information (compression) versus necessary signal energy to surplus signal energy.*

This conceptual scheme is offered in order to connect with the framework elaborated on in the previous chapters, but is not applied rigorously to the following discussions. Instead we divide the techniques according to the more conventional disciplinary boundaries: source coding (Section 6.3), shaping strategies implemented at baseband (Section 6.4), strategies implemented at radio frequencies (Section 6.5), and smart antenna technologies, especially those focusing on the transmitter antenna configuration (Section 6.6).

Before turning to specific techniques, however, we need to touch on another fundamental conceptual aspect of the problem of transmitter

---

2. As we shall see in Section 6.4, the most important time-domain approaches yield benefits in the form of reduced interference in all three dimensions of the communications space.

interference reduction that cuts across all of these boundaries: the problem of handling nonlinear signals.

6.2 Signal Nonlinearities: A Conundrum

The manipulation of communications signals is an intricate art. Starting with a source signal, many different kinds of operations can be performed, such as amplification, filtering, or sampling and quantization. Each of these processes can be viewed as a *mapping* of an input signal to an output signal. If we analyze the relationship between input and output, we find that sometimes the relationship is smooth, or *linear*. That is, specific input–output pairings can be graphed, like *x,y* coordinates, and the resulting plot will be—in some sense—"straight" (Figure 6.2).[3]

On the other hand, some of the processes applied to a signal will produce *x,y* plots that exhibit curved or broken or squared-off shapes. These processes are then said to be *nonlinear*. Such nonlinearities introduce a number of problems for system designers. For example, in the amplification of a weak signal to produce a strong signal, we generally want to make sure that the translation is as linear as possible. Nonlinearities in the amplifier can show up as distortion. In general, any nonlinearity creates potential difficulties. (Of course, if the distortion or nonlinear translation affects a signal parameter that we are not concerned about, we may not notice that particular form of nonlinearity. For example, some modulation schemes are insensitive to changes in signal amplitude. FM is a good example. Thus, FM is not significantly affected by amplitude nonlinearities in the amplification process, which means that one can use cheaper, less linear power amplifiers. It is important to keep in mind, however, that these nonlinearities may still occur; it is just that they affect properties of the signal which do not, in a particular architecture, happen to carry information. They may still contribute to the problem of interference.)

A major problem is that nonlinearities in the *input* signal, or events in which an abrupt change occurs in an important signal parameter, tend to

---

3. The definitions of linearity can be forbiddingly dense, for example: "A linear channel is one that satisfies the superposition principle: if input $c(t)$ causes output $v(t)$ and input $c'(t)$ causes output $v'(t)$, then for any real numbers a and b, input $ac(t) + bc'(t)$ causes output $av(t) + bv'(t)$" [3, p. 21]. I must ask: Why is there this penchant for obscurely couched presentations of fundamentally simple ideas? It pervades the entire literature.

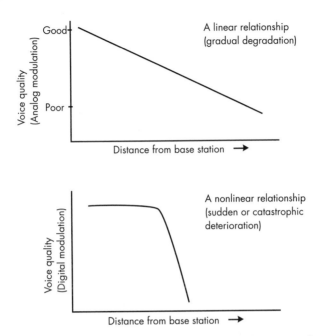

FIGURE 6.2 *Input-to-output signal mappings, with linear and nonlinear relationships.*

lead to an *expansion* of the *output* signal bandwidth. Why? Because sudden changes in the signal inherently embody a high-frequency component, and high-frequency components will expand the signal in the frequency domain.[4] Indeed, it is often said that all real signals (i.e., those with no constraints on the inputs) will produce outputs that extend *infinitely* in frequency.[5] If we try to contain this frequency spreading by filtering the signal—band-limiting it in the frequency domain—we find that it will tend to spread in the time domain—and again this expansion is, in principle, *infinite*.[6] This translation between frequency and time domains is one of the foundational principles of signal processing.[7] For example, Figure 6.3 shows the time–frequency transforms (Fourier) for several simple types of signal elements, or pulses.

---

4. "[Any signal which] consists of abrupt changes ... gives rise to spectral components at high frequencies" [4, p. 287].

5. "Rather generally, waveforms ... have spectral components which extend, at least in principle, to infinite frequency" [4, p. 25].

6. "Strictly [frequency] band-limited signals are not realizable since they imply signals with infinite duration" [5, p. 43].

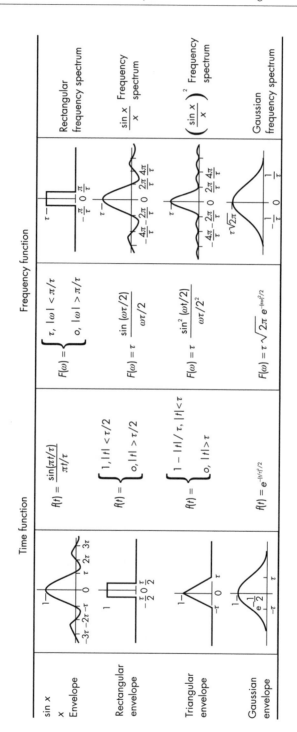

FIGURE 6.3 *Fourier transforms. (From: [6]. © 1984 John Wiley & Sons Inc.)*

Let us review this more closely. Examine the Fourier transform for the time and frequency representations of the two most basic signal elements: a perfectly square pulse and a sinusoidal pulse (Figure 6.3). A pulse that is sharply confined in the time domain will have a quasi-infinite representation in the frequency domain. That is, if we use pulses that are precisely controlled in time, the extremely abrupt edges will generate infinitely high-frequency components, and the waveform will spread significantly in terms of its occupied frequency bandwidth. If, on the other hand, we apply perfect filters in the frequency domain to confine the signal energy to a precisely defined bandwidth, the signal will automatically spread in time. In a digital system, for example, the sharper the filter, the more intersymbol interference is created (by the filter itself, quite apart from any ISI created by the channel).[8]

This is a profound and somewhat mysterious relationship. That a signal should show such strict transformations between its frequency characteristics and its time characteristics is in a way surprising. That the confinement of the signal in one domain automatically creates a quasi-infinite expansion of the signal in the other domain is not necessarily an intuitive result, considered from a physical perspective. It is a foundational premise of communications engineering that is widely recognized, but often passed over as though the statement of mathematical formulas were sufficient explanation. Why, for example, should a bandpass filter per se create a time delay? What is the exact underlying physical mechanism? (Engineers seemingly do not always need explanations for the results they obtain, as long as they can reliably obtain them.)

In any case, the communications engineer faces a dilemma. If the system introduces, or is required to process, nonlinear signals, *the output signal will tend to expand*. He or she can take steps to control the spreading in the frequency domain, but the signal will then spread in the time domain, and countermeasures, like equalization, will be needed. He or she can try to

---

7. Indeed, many communications textbooks begin with a presentation of the basic notions of spectral analysis and time–frequency transforms. "We most often see signals presented in the time domain (that is, as functions of time). Any signal, however, can also be presented in the frequency domain, and transforms (mathematical operators) are available for converting frequency- or time-domain functions from one domain to the other and back again" [6, p. 1].

8. "We might try to alleviate this difficulty [frequency spreading due to nonlinearities] by passing the baseband signal through a ... filter to suppress the many side lobes. Such filtering will cause intersymbol interference" [4, p. 287].

control the time domain—to reduce ISI—and the signal will expand in its frequency bandwidth and require a larger (frequency) channel allocation.

This has led sometimes to a mind-set that views nonlinear processes as inherently risky and problematic.[9] The more linear we can make the system, the fewer abrupt changes or breaks in the signal, the better. A great deal of effort has been devoted to finding ways to shape signals to avoid or reduce such effects. Much of the rest of this chapter is devoted to signal shaping and smoothing techniques that seek to avoid undesired signal expansion by eliminating nonlinearities as much as possible, at all stages of the signal processing chain.

And, yet, there is a conceptual dilemma here. The very foundation of modern spectrally efficient, post-Shannon communications, is digitalization, *which is the most nonlinear of all processes we can imagine.* Indeed, it is precisely the nonlinearity of the digitalization process that creates its value. The idea of a threshold, which is the basis of quantization and its signal-to-noise performance, is a pure instance of a nonlinear input–output relationship. Once again, Shannon put his finger on this matter right from the start:

> There will be a certain threshold effect when we perturb the message. As we change the message a small amount, the corresponding signal will change a small amount, until some critical value is reached. At this point the signal will undergo a considerable change. In topology it is shown that it is not possible to map a region of higher dimension into a region of lower dimension continuously. It is the necessary discontinuity which produces the threshold effects....
>
> It is [therefore] evident that any system, either to compress [the signal—i.e., gain efficiency] or to expand it and make full use of the additional volume, must be highly nonlinear in character. [8, pp. 165–166]

Shannon is not simply saying that we can tolerate nonlinearity; he is asserting that the system *must* be highly nonlinear, that we should actively design nonlinearity into the system. We have seen in Chapter 5 that as coding schemes become more and more powerful, moving closer to the Shannon theoretical limits, the whole characteristic of the system (and not just the individual quantization elements) becomes more nonlinear.

---

9. "Abrupt changes can be problematic since they theoretically require an infinite bandwidth, and ways are sought to avoid them" [7, p. 3].

By its very nature decoding [i.e., any quantized process] is nonlinear. In fact, it is almost the opposite of a linear operation: small disturbances of the decoder input signal should not affect the decoder output signal at all. [9, p. 99]

As we noted in our discussions of noise and interference in the previous chapters, it appears that the post-Shannon framework has inverted the traditional valuations, and has revalued nonlinearity from a bad or problematic system characteristic to a constructive and desirable one.

This is not a matter of either/or. Nonlinearity is another of the complex, multifaceted conceptual components of the Shannon revolution, and it does cut both ways. Essentially, it is profoundly true that—like noise, and interference—what we find we want is *designed* nonlinearity, or "good" nonlinearity; and what we want to suppress or avoid are those "bad" nonlinearities that come in several varieties. In Chapter 5, we looked at the ways of designing good nonlinearity. Another way of construing the subject of the signal shaping in this chapter, particularly Sections 6.5 and 6.6, is as a survey of some of the techniques for avoiding or reducing bad nonlinearity.

6.3 Compression: Post-Shannon Source Coding Strategies

Many different kinds of sources are available for communications signals, including text, speech, audio (music and so forth), still images, and video. In one sense, the miracle of digital communications lies in its ability to reduce all of these sources to a homogeneous stream of 1's and 0's, which means that the same sort of physical layer protocol and the same physical channel can handle, in principle, any type of signal. We must also recognize that productive *compression* strategies are closely tied to the nature of the source, as well as the nature of the receiver (especially for sound and images). In another sense, therefore, not all bit streams are truly alike. Processing techniques that may be acceptable for compressing one kind of digitized source may not be viable at all for another. Specialization in the source coding field is intense, for reasons described below, such that there is very little overlap between work on text compression and speech compression, for example. For these reasons, the technology of source coding has become so vast that even a proper survey is not less than a book-length undertaking.

The more limited goal of this section will be to touch on several of the most important meta-concepts in compression technology that are most relevant to the post-Shannon wireless revolution. These are the core ideas that, in some cases, divide the field into subdisciplines, while in others they provide a degree of technical commonality across these specialized fields.

It may be useful to first develop some of these ideas intuitively. Let us begin with a sample of live human speech, presented to the input of a communications system. In its initial analog form, the amount of information this sample contains is quasi-infinite—or at least very large. In reality, information theory cannot really be applied to analog signals per se.[10] Not, that is, until we have tamed the analog signal in certain ways, which we shall get to in a minute. So we start with a quasi-infinite bit rate, so to speak.

On the other hand, the actual rate of linguistic information in human speech, corresponding to the information that is captured in a textual representation of that speech, is quite low. Shannon cites a figure on the order of 15 bps.[11]

This is quite a gap. Yet if the essential linguistic information is occupying a bandwidth of only 15 bps, what is the rest of that quasi-infinite information rate of the analog signal all about? The basic answer is that it provides the information that creates the tone, timbre, inflection, and so forth of the speaker's voice—all the things that go under the heading of *voice quality*.

We may next ask: How much information is really required to carry this voice quality information? Presumably, there is a specifiable upper limit, based on the sensitivity of the ear and the number of nerve synapses or hair cells in the cochlea of the ear. The very best audio quality speech recording uses several hundred thousand bits per second.

Next, do we really need to transmit this much voice quality, or can we reduce the amount of information and still maintain an acceptable signal? The answer is yes, of course. The telephone industry is still based, for the most part, on the standard rate of PCM of 64,000 bps, which in turn derives from the historical transmission characteristics of analog telephone wires. This yields telephone quality voice, which is by no means as good as real audio quality (think of the sound of telephone conversations played on radio stations, compared to the higher quality of the voices of the announcers). Yet telephone quality has become an accepted standard, at least for the time being, for telecommunications purposes.

---

10. "A continuously variable quantity [i.e., an analog signal] ... requires an infinite number of binary digits for exact specification" [2, p. 73].

11. "Taking 100 words per minute as a reasonable rate of speaking, we obtain 15 bits per second as an estimate of the rate of producing information in English speech when intelligibility is the only fidelity requirement" [10, p. 193].

There is still a lot of apparently extra information in the voice signal—several thousand times more than what would appear to be necessary to convey the real linguistic information. How much further can the payload be reduced? How much of this extra information is truly necessary?

In today's second-generation wireless systems, acceptable telephone quality voice transmission is being achieved at rates lower than standard PCM by up to a factor of 10 or more. Indeed, this reduction of the payload has been probably the single most important source of capacity improvement in second-generation architectures. As we shall see, some of the more exotic post-Shannon architectures in the marketplace today really rely on voice compression to achieve most of their capacity gains.[12] Clearly then, speech compression plays an important role in the overall framework of high-capacity wireless systems.

Let us consider a second example: the transmission of a short sample of English text, such as an e-mail message. Immediately, it is apparent that this source is quite different. For one thing, we can precisely specify how much information (in information theoretic terms) the message contains. In addition, we can directly and unambiguously analyze the structure of the information contained in the message, and we can determine that some of the information is, so to speak, *redundant*. In principle, we do not need to transmit all of the message in order for the receiver to be able to recreate it perfectly. We can leave some message elements out or repackage them more compactly.

If we look closer, we see that there are apparently different types of redundancy. For example, some of the message elements, like the *u* following a *q* in standard English, are purely redundant and can be dropped. All we need is a rule at the receiver to reinsert a *u* after every *q* and we can eliminate the *u*'s from the signal and reduce the bit rate, however slightly. In other cases, it appears that we can also eliminate some message elements that are not as strictly redundant. For example, it is sometimes possible to drop vowels from the text. If we transmit the sequence *txt* we can assume that the receiver should be able to recreate the word *text* because there is no other obvious reconstruction in English. In this case, we are assuming that the receiver not only possesses certain rules to reconstruct the message, but that it also has an English dictionary.

---

12. It is arguable that almost the entire purported advantage of today's CDMA systems over their TDMA counterparts is a result of the implementation of voice coders in CDMA that can take advantage of speech activity patterns to reduce the average bit rate of the transmission.

Other tricks of a similar sort are used. For example, in the case of Morse code, we replace all the letters of the English alphabet with sequences of dots and dashes. As is generally well known, the most commonly occurring letters, like *e* (the most common letter in English), are represented by the shortest sequences. Rare letters use longer sequences. By weighting the substitution rules in this way, we can reduce the bit rate required to transmit a text message.

What is clear is that there is quite a bit we can do to compress the information produced by a transmitting source. The tricks and technologies, and the trade-offs, depend a great deal on the type of source. If we can reduce the size of the payload, we need less signal energy to carry it, which means less interference and denser packing of the signals in the STF communications space.

The preceding paragraphs have touched on some of the important concepts of source coding and compression, which we will now outline more systematically.

6.3.1 Lossless Compression

We have shown that intuitively a distinction can be conceived (though not always drawn) between *essential* information and *nonessential* information, which are blended together in the uncompressed source. Sometimes, it is possible to separate the information unambiguously into essential and redundant information. The redundant information can be eliminated without affecting the essential information at all. Some forms of payload compression can be implemented without losing any of the essential information.

The key to all truly lossless compression is that the process is *reversible*. The information that has been removed by the transmitter can be recovered and reinserted by the receiver [11, p. 16].

Some lossless compression can normally be achieved with *any* source. The applicability of lossless compression is strongly related to the statistical characteristics of the data produced by the source. If the statistics are strongly skewed—that is, if some message elements occur with a much higher probability than others—it may be possible to implement lossless compression and achieve meaningful gains in capacity.

Some of the techniques for lossless compression include run-length coding, speech activity (or channel activity) coding, so-called entropy coding (Huffman coding and related techniques), and data compaction techniques.

Run-Length Coding

Consider (for simplicity) a source that produces two kinds of message elements, such as black (b) pixels and white (w) pixels (or, indeed, 0's and 1's) and that tends to produce long runs of identical message elements. We can conceptualize such a source as a two-state Markov model, as shown in Figure 6.4.[13] The probabilities for w/w and b/b are very high, and the probabilities of a transition (either w/b or b/w) are small.

> The highly skewed nature of the probabilities ... says that once a pixel takes on a particular color (black or white), it is highly likely that the following pixels will be of the same color. So, rather than code the color of each pixel separately, we can simply code the run lengths of each color. For example, if we had 190 white pixels followed by 30 black pixels, followed by another 210 white pixels, instead of coding the 430 pixels individually, we would code the sequence 190,30,210. [13, p. 123][14]

We find such a source in certain kinds of images. For example, facsimile images of documents usually fit this model, and run-length coding has been built into the Group 3 facsimile standard.

The use of run-length encoding is well advanced; because of the inherent simplicity, such schemes have been embedded in many communications standards. For the appropriate sources, compression rates of up to a factor of

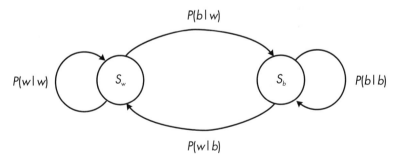

FIGURE 6.4 *Two-state Markov model (Capon model). The two-state Markov model represents a source of two-valued signals suitable for run-length encoding, such as a black-and-white fax message. (From: [13]. © 1996 Morgan Kaufmann Inc.)*

---

13. This model was first developed by Capon [12].

14. See also [11, Chap. 10].

several hundred can be obtained (although it is highly dependent on the actual statistics of the source document, for fax, as shown in Figure 6.5) [14].

Voice Activity Compression

The same principle can be applied to speech activity, which also exhibits strongly skewed statistics. Speech is very much a bursty or on–off process. That is, a voice channel is normally only occupied by speech data for a portion of the total time it is in use. For example, when a user is listening to the other person talk, he or she is not generating speech information to transmit. The gross channel occupancy rates for voice transmissions, the so-called "on–off rate," have been reported as low as 35% to 40%.[15] This is certainly intuitively reasonable, if we assume that each speaker is speaking about half the time and listening have the time, with a little additional off time when neither is speaking.

By analyzing the speech transmission from an individual user with even finer resolution, it is discovered that even *within* a speech burst that is seemingly continuous, only 65% to 75% of the channel time is actually used to carry speech data. If we sample the speech source finely, we discover that microscopic silent periods can be identified, corresponding to short pauses for breathing, intersyllabic gaps, and normal hesitations associated with thought processes [18].

In principle, if we can identify and code the silences in a manner similar to run-length coding, we can compress a speech signal by a factor equal to the inverse of these macroscopic and microscopic activity factors, that is, typically two or three times.

Source description	Original size (pixels)	MH (bytes)	MR (bytes)	MMR (bytes)	JBIG (bytes)
Letter	4352 × 3072	20,605	14,290	8,531	6,682
Sparse text	4352 × 3072	26,155	16,676	9,956	7,696
Dense text	4352 × 3072	135,705	105,684	92,100	70,703

MH, MR, MMR, and JBIG are different facsimile coding algorithms, all based on run-length encoding principles. Significant compression can be obtained.

FIGURE 6.5 *Run-length encoding gains. (From: [13]. © 1996 Morgan Kaufmann Inc.)*

---

15. The original studies of speech activity statistics are found in [15–17].

We can take advantage of this compression opportunity in any of several ways. From the simplest to the more complex, these include the following:

1. *Discontinuous transmission (DTX):* This involves simply turning the transmitter off when there are no speech data to transmit. This typically reduces the overall interference in the system by approximately a factor of 2, or a 3-dB reduction in cochannel and adjacent-channel interference. In terms of system impact, DTX is one of the easiest ways to gain potential capacity—if the traffic is conventional voice traffic. This gain comes from reduced interference loads, allowing better reuse factors and smaller S-buffers generally.

2. *Digital speech interpolation (DSI):* This is a multiple access architecture, in which silent periods from any given channel are returned to a pool of available channel resources and may be reallocated to other users. DSI is widely used on certain kinds of wireline channels, and has been used in second-generation wireless systems on a limited basis.[16] Direct capacity gains of around 2 have been observed.

3. *Packetized architectures:* The exploitation of true packetized transmission for voice is still a relative novelty, certainly in the wireless arena. However, some architectures based on exploiting the activity factor in this way have been proposed.[17]

4. *Variable rate coding:* The use of a variable rate coder, which adjusts between higher rates for periods of active speech and much lower rates for silent periods (and indeed, may involve multiple steps based on the putative information density of the signal at any given moment), is another way of exploiting the compression opportunity. Such a coder has been implemented in the U.S. CDMA standard (IS-95), and it is claimed that the resulting interference reduction of 3 dB leads directly to system capacity gains.[18]

---

16. In the late 1980s and early 1990s, Hughes Network Systems employed DSI in its E-TDMA cellular system, as an extension of IS–54 (the U.S. TDMA standard). I believe that Hughes still uses the E-TDMA approach in some of its wireless local loop systems.

17. A notable example is *packet reservation multiple access* (PRMA), proposed by Goodman et al. [19].

18. See the references for Chapter 6 of [20].

Entropy Coding

Assume that the source produces symbols drawn from a known, finite alphabet (like the English alphabet), and that we know a priori the probability of seeing any given symbol. A very basic notion, intuitively stated, is that the probability of seeing a symbol is inversely related to the amount of information which that symbol contains. The less frequently we see a given symbol, the more its occurrence surprises us. This is sometimes called the *self-information* of the symbol [11, p. 617].

The *average* amount of self-information produced by the source, as it emits various symbols over time, is called the *entropy* of the source [11, p. 617; 21, p. 9]. In effect, the entropy is a measure of the statistical skewedness of the source.

Without delving more deeply into the challenging conceptual treatments of entropy that have motivated much of the broader interest in information theory, we can connect it to an important coding and compression concept: If we map the source alphabet onto a second channel alphabet of message elements or codewords in such a way that the most frequent source symbols are matched with the shortest message elements, and the less frequent symbols are represented by longer codewords, we can achieve a significant improvement in the transmission efficiency.

> Entropy coding is obtained by means of a variable length coding procedure which assigns codewords of variable lengths to the possible [source symbols] such that highly probable outcomes are assigned shorter codewords, and vice versa. [11, p. 147]

In other words, the self-information of each symbol is approximately matched to the length of the codeword. Why does this yield a compression? If we were to use the same amount of binary data to transmit every symbol in the source alphabet, then most of the time we would be using more binary data than necessary to transmit *frequently* observed symbols. By matching the statistics of the source with the code lengths in this way, we can bring the bit rate much closer to the optimum rate. In many ways, this is a natural process that we observe everywhere, at all levels of the composition of the message. For example, even the process of *abbreviation* of common words and phrases (using "U.S." for "United States") is a kind of entropy coding procedure. One of the handiest examples is the Morse code, in which the most frequent letters like *e* are represented by the shortest message elements.

Designing good entropy codes also requires paying attention to the decoding process. The receiver should be able to uniquely determine which symbol has been transmitted without having to perform extensive analysis or code look-ups. In other words, the code should be self-punctuating [3, p. 296]. One of the earliest and best solutions for generating such a code was created by David Huffman, and has become known as Huffman coding [22]. It is a simple technique for generating a uniquely decodable variable-length codebook and assigning the codewords appropriately [1, pp. 63ff].

Blahut [3] presented a standard example of a Huffman code, as shown in Figure 6.6, which provides the probabilities of source symbols, matched

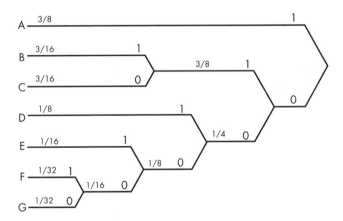

Source symbol	Probability	Codeword	Blocklength
A	$\frac{3}{8}$	1	1
B	$\frac{3}{16}$	011	3
C	$\frac{3}{16}$	010	3
D	$\frac{1}{8}$	001	3
E	$\frac{1}{16}$	0001	4
F	$\frac{1}{32}$	00001	5
G	$\frac{1}{32}$	00000	5

FIGURE 6.6 *Huffman coding. The basic principle in Huffman coding is to match the shortest symbol elements to the most frequent (highest a priori probability) message elements. (From: [3]. © 1990 Addison Wesley Inc.)*

with codewords that are derived from the tree structure shown in the example and are self-punctuating. If we used fixed-length codewords for each of these seven symbols, we would need at least eight of them, and the average length of the codeword would be 3 bits. By using Huffman coding, the average block length in this example is reduced to 2.44 bits per source symbol, which is very close to the measurable entropy of the source, 2.37 bits per source symbol. This Huffman code is close to the optimum. It will save almost 19% of the channel capacity.

Huffman coding is a very general procedure that can be applied to almost any data stream from any type of source. It may be applied to the data in a number of different ways. Sayood discusses the applications of Huffman coding to different kinds of sources, and gives a range of results for the compression that can be obtained [13, Chap. 3].

We should bear in mind that entropy coding works best when the source produces a highly skewed statistical distribution: "It is only when the probabilities of the source symbols of the message are very different that we get a significant economy from the Huffman encoding process" [13, p. 69]. That is to say, if the source produces highly structured data, with skewed statistics, then entropy coding is productive. The more noise-like or randomized the input data, the less we can gain from this approach.

Data Compaction

The classical Huffman concept is based on a priori knowledge of the source statistics. What if we do not know the statistics ahead of time?

One perhaps obvious solution is to modify the Huffman procedure to work adaptively. It becomes "a two-pass procedure: the statistics are collected in the first pass, and the source is encoded in the second pass" [13, p. 43]. Indeed, coding ingenuity soon finds a way to accomplish the code development in a single pass, generating and modifying the Huffman codebook "on the fly."

Another approach is to construct a true dictionary for the source as we go along. The value of a dictionary for compression purposes is that it allows us to identify an entry by an index or a pointer. Instead of transmitting the actual data, we transmit the index. In adaptive dictionary compression schemes, both the transmitter and the receiver build identical dictionaries as they process the data streams (Figure 6.7).

The most popular adaptive dictionary techniques are based on two papers by Jacob Ziv and Abraham Lempel, published in 1977 and 1978. In one version of this type of compression algorithm, the encoder keeps a copy of the data that has been already encoded, and it searches this buffer as each

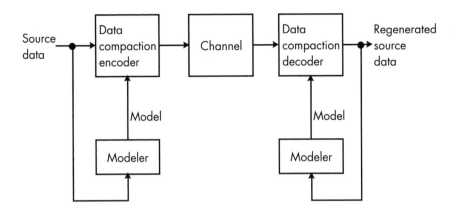

Modelers at both the transmitter and the receiver are actively creating and updating their dictionaries of transmitted and received sequences.

FIGURE 6.7 *Data compaction at TX and RX. (From: [3]. © 1990 Addison Wesley Inc.)*

new substring of data is encountered, identifies the match—that is, it looks up each substring in its dictionary—and then encodes not the substring itself, but a pointer (index) and a length value (Figure 6.8) [3, p. 315].

The Ziv-Lempel algorithms have become mainstays of file compression for all sorts of computer applications and are widely used to reduce the payload of transmitted signals involving text files, as well as previously quantized image files [13, pp. 100ff].

6.3.2 Lossy Compression

For many signals, such as speech, the essential information and nonessential information are so blended that they cannot be unambiguously separated. Different users evaluate differently which information, and how much, is really essential. A military radio system operator may be quite happy if his system produces intelligible speech, even if much of the color and character of the speaker's voice is lost. A radio station cannot adopt the same attitude with regard to its audio broadcast quality.

In such cases, compression is achieved by the judicious *destruction* (euphemistically called a "loss") of some of the information contained in the source signal. This sets a hard ceiling on the quality of the transmitted signal. No matter what, the information that is destroyed in this kind of

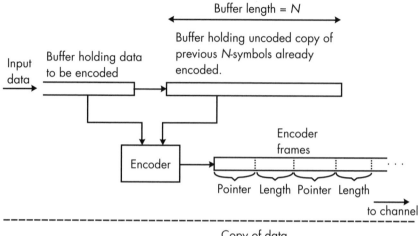

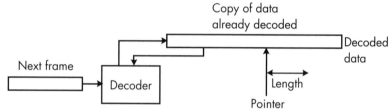

FIGURE 6.8 *Ziv-Lempel encoder/decoder. (From: [3]. © 1990 Addison Wesley Inc.)*

compression cannot be recovered or accurately recreated, although some-times it may be simulated.[19] Lossy compression is *irreversible*.

One's attitude toward the partial destruction of the signal (for the sake of efficient transmission) depends on one's point of departure. For some, the result is tainted, no matter its practical value: "A data compression code [means] that the original data cannot be cleanly recovered; distortion is implicit in the code" [3, p. 364]. For Shannon, who was concerned with finding a way to handle intractable analog signals (like speech) within the framework of his new theory, the issue seemed minor. Indeed, since "exact transmission [of an analog signal] is impossible," then

---

19. An ironic example of such simulation is the creation of artificial background noise, called *comfort noise,* which is sometimes inserted on top of speech in digital circuits that have become *too* quiet. Human listeners apparently rely on low-level background noise to provide a sort of channel status indicator, and will sometimes assume that a channel that is too quiet may have been disconnected. The insertion of comfort noise is intended to alleviate this problem.

[t]he real issue ... practically [is] that we are not interested in exact transmission when we have a continuous source, but only in transmission to a certain tolerance. [21, p. 26]

His position is principled: Because perfect transmission is theoretically unachievable, there is by definition a choice, for the system designer, regarding what sort of signal degradation to tolerate. For Shannon, the controlled reduction of signal quality achieved through quantization is clearly preferable to the uncontrolled effects of analog noise.

Other writers have attempted to gloss over the process as merely the elimination of irrelevant information (as compared with the elimination of redundant information by lossless compression procedures).

Quantizers [the most basic of lossy compression processes, of course] are devices that remove the *irrelevancy* in the signal; they accomplish this through an irreversible, information-lossy procedure. [11, p. 16]

This will not do, clearly. The selective elimination of portions of the signal is not based here on anything as neutral or as cleanly characterizable as redundancy. It *is* a matter of tolerating a real reduction in signal quality.

Implicit in any lossy compression process, therefore, is a *criterion* of quality (also called a *rate distortion measure*) that specifies an acceptable level of information loss. This criterion is typically expressed as a quantitative measure of the amount of difference between the source signal and the compressed version of the signal. Yet it indicates that a *subjective* element has entered the calculation, for the basis of the quality assessment is eventually a subjective judgment of signal quality. In the early development of PCM, it was possible to sidestep this issue, in a sense, by employing a simple metric like mean squared error (the absolute value of the difference between the two signals). As lossy compression schemes have become more and more aggressive, the use of simple metrics has become insufficient. For example, a slight rotational shift in a digitized image will produce a large mean squared error, but may be perceptually negligible. On the other hand, a bright horizontal streak may show a much smaller measure of error, but is quite objectionable.[20]

---

20. "A standard objective measure of coded waveform quality is the ratio of signal variance to reconstruction error variance, referred to for historical reasons as the signal-to-noise ratio (SNR) ... We shall note the inadequacy of SNR as a perceptually meaningful measure of digitized waveform quality. Very briefly, the inadequacy of

Nevertheless, any distortion measure must serve at least two masters: the human recipient of the message who will evaluate its quality and acceptability, and the communications engineer who needs a simple, relatively repeatable way of benchmarking the performance of his system.

> The selection of a distortion measure for a particular problem can be a difficult and often controversial problem. Ideally, the distortion should quantify to subjective quality, e.g., a small distortion in an image coding system should mean a good looking image, large distortion a bad looking image. If one is to have a useful theory, the distortion measure should be amenable to mathematical analysis.... If the distortion measure is to be useful for practice, one should be able to actually measure it, that is, it should be computable. Unfortunately, one rarely has a distortion measure which has all three properties of subjective meaningfulness, mathematical tractability, and computability. In fact, the first property is often at odds with the other two since a distortion measure might need to be extremely complicated to even approximate subjective feelings of quality. [23, p. 39]

In truth, the quality criterion always implicitly embodies a model of *human* signal processing—specifically, how we hear and see, and what really matters, and what does not matter so much. As the compression objectives have become more aggressive in the post-Shannon era, the strategies for deciding *how* to compress the signal have grown much more sophisticated, and we have entered more deeply into the disciplines that are starting to be known as *perceptual coding*.

6.3.3 Perceptual Coding

The newer approach to signal quality assessment and lossy compression in general is based on the idea of matching the coding output to the characteristics of the human receiver, especially to the human ear and eye.

> It is imperative that we design the coding (or compression) algorithm to minimize a perceptually meaningful measure of signal distortion, rather

---

SNR has to do with the fact that reconstruction error sequences in general do not have the character of signal-independent additive noise; and that therefore the seriousness of the impairment cannot be measured by a simple power measurement" [11, pp. 6–7].

than more traditional and more tractable criteria such as mean squared difference between the waveforms at the input and output of the coding system. [24, p. 1385]

At one level, the idea that human beings can tolerate certain kinds of noise better than others is perhaps intuitively obvious, and was commented on by Shannon in the earliest papers:

> In the case of speech, the ear is insensitive to a certain amount of phase distortion. Messages differing only in the phases of their components (to a limited extent) sound the same. This may have the effect of reducing the number of essential dimensions in the message space [i.e., allow compression].... If the ear were completely insensitive to phase, then the number of dimensions would be reduced by one-half due to this cause alone ... [25, pp. 33–34]

Indeed, the shape of human acoustic response patterns was implicit in many of the strategies that were developed in earlier systems. For example, Shannon also noted that the ear is relatively insensitive to higher frequencies. Instead of digitizing all frequency bands at the same rate, we can delete more information from the higher frequencies without impairing the subjective quality of the compressed signal. This is the foundation for sub-band coding and related techniques [26]. A similar, more formal strategy has been the use of *nonuniform quantization*, especially logarithmic quantization scales, to allow for finer discrimination of relatively quiet signals and coarser quantization of very loud signals [11, p. 141].

The emerging view is more detailed. Speech, as a source, deviates from the typical model of classical source coding theory in a number of ways, and the ear does not respond to all forms of distortion equally. Perceptual coding is based on the idea of identifying these structural characteristics and marshaling them in support of coding strategies that "minimize the perceptibility of the distortion by shaping it advantageously in time or frequency, so that as many of its components as possible are masked by the input signal itself" [24, p. 1393].

The concept of *masking* is central to achieving significant compression. Masking refers to "the inability of the human perceptual mechanism to distinguish between two signal components (one belonging to the signal, one belonging to the noise) in the same spectral, temporal or spatial locality" [24, p. 1385]. In effect, there are opportunities to hide the noise within the signal itself, in such a way that it becomes completely imperceptible to human observers "even if the objectively measured local SNR is modest or low" (Figure 6.9) [24, p. 1385]. The results can be striking:

> In one example, with a signal which consists of a male *a capella* chorus, the coder is perceptually transparent at a 13 dB overall SNR, when the noise is inserted according to psychoacoustic principles. On the other hand, the coder is nontransparent to additive white noise at an SNR as high as 60 dB, and nontransparent to locally adjusted white noise at an SNR of 45 dB. With a 1-kHz tone input, transparency with additive white noise requires an SNR of 90 dB. With ideally shaped noise, transparency occurs at an SNR of 25 dB. [24, p. 1409]

Perceptual coding is in its early stages, theoretically. As the authors of a recent pioneering article have noted, "A biologically correct and complete model of the human perceptual system would incorporate descriptions of several physical phenomena including peripheral as well as higher level effects, feedback from higher to lower levels in perception, interactions between audio and visual channels, as well as elaborate descriptions of time–frequency processing and nonlinear behavior" [24, p. 1395]. Phew! Now there's a research agenda.

6.3.4 Correlative Quantization

Another group of related strategies involves the development of more sophisticated quantization techniques that achieve compression by exploiting the

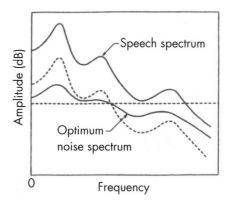

FIGURE 6.9 *Perceptual masking: matching the noise spectrum to the signal. The optimum spectrum is weighted toward an improved SNR at low frequencies. (From: [24]. © 1983 IEEE. Reprinted with permission.)*

patterning or structure of the source data, the inherent correlation between successive samples or message elements. These techniques are often fairly easy to implement and tend to find their way into many of the wireless 2G and 3G compression technologies, usually in combination with more explicit perceptual coding techniques and sometimes with source modeling techniques (Section 6.3.5) as well. As illustrations of this approach, we will look at four such strategies: difference encoding, adaptive quantization, vector quantization, and codebooks.

Difference Encoding

Standard quantization techniques encode each sample independently. Yet there is almost always a high degree of correlation from sample to sample. If we can capture the thread of this correlation, so to speak, we can compress the amount of information that has to be transmitted to reconstruct the signal. As a very basic illustration, nothing is simpler than the idea of difference encoding. In lossless compression, difference encoding means sending only the quantized difference between a sample and the previous sample. In other words, we encode only the new information, or the changes observed since the previous sample. Because signals tend to change slowly relative to the sample rate, the dynamic range, so to speak, of the difference is usually much smaller than the range of the entire signal and can thus be encoded with fewer bits.

In the context of lossy compression, difference encoding is usually referred to as delta modulation.[21] Here, in the simplest form, the encoder only determines if the new sample has increased or decreased in amplitude value compared to the previous signal, and it sends a signal bit to indicate track up or track down (Figure 6.10).

As Figure 6.10 shows, delta modulation generates two different kinds of error, really two different quantization effects, which represent the loss of information in the compression process. A steady or slowly changing signal will produce a jittery back-and-forth tracking behavior in the delta modulation coder, which is often called *granular noise*. A rapidly changing signal, on the other hand, outraces the tracking signal and produces a gap sometimes

---

21. Delta modulation has to be one of the best, or worst, examples of obfuscatory engineering terminology in the field of communications theory. The term *delta* refers to the old mathematical symbolism that uses *delta* to stand for "difference." The use of *modulation* here is simply a well-entrenched inaccuracy. The term *deltamod* has nothing to do with modulation; it is simply a quantization rule. For a good presentation of this technique, see [11, Chap. 8; 13, Chap. 10].

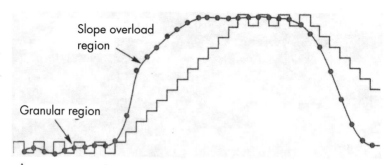

FIGURE 6.10   *Delta modulation. Note: Step size is uniform, which reduces the effectiveness of tracking a rapidly changing input signal. (From: [13]. © 1996 Morgan Kaufmann Inc.)*

called *slope overload*. It is not a big leap to the next improvement: a variable step size that tracks fast moving signals with large steps, and slow moving signals with smaller steps (Figure 6.11).

Intuitively, we are now transmitting samples with a single bit instead of a multibit word. However, the apparent gains from difference encoding are somewhat offset by the need to maintain a rapid bit rate to track a fast changing signal. Nevertheless, many applications have shown that modest compression—up to a factor of 2 or so—can be achieved by delta modulation-type techniques that are extremely simple to implement. In the days when processing power was still relatively expensive, delta modulation

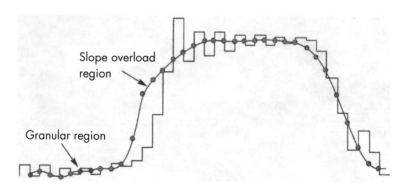

FIGURE 6.11   *Adaptive delta modulation. Note: Step size varies—increasing when the input signal is changing rapidly, and decreasing when the input signal is stable—which allows for a better overall matching of the input to the output. (From: [13]. © 1996 Morgan Kaufmann Inc.)*

was a popular compression strategy, and it can still be employed to advantage (especially in lossless implementations, such as for further compression of previously quantized images[22]).

Adaptive Quantization

These methods work, in information theoretic terms, by exploiting the fact that samples are not independent, but correlated; the true information in one sample overlaps with adjacent symbols, in the original source. Mechanistic quantization (like PCM) destroys this correlative information, which means that we need to recreate the meaningful information in each sample from scratch. Delta modulation takes advantage of the fact that if we know sample t_1, we can use this knowledge to reduce the amount of information that we have to send about sample t_2.

Difference encoding can be viewed as a very simple form of adaptive quantization. It uses only the correlative properties that exist between two immediately adjacent samples. Yet the structure in the signal and the correlative properties in the sampled, quantized representation of the signal extend far beyond a mere two-symbol span. Consider the fact that a delta modulation system as described earlier does not know whether the next sample will track up or down. Yet it is possible, conceptually, to take a longer look at the data and actually try to predict whether the next sample with be up or down (Figure 6.12).

In fact, we can recover even more of the correlative information in the signal by looking at longer sequences and trying to develop better and better predictions of the following values. We can then, for example, send only the *difference* between the *predicted* value and the *actual* value, which should be even smaller in its dynamic range than pure difference encoding without prediction. Indeed, there are many techniques, labeled in the literature with words like *adaptive* and *predictive* that use the longer term structure of the signal to reduce the amount of information that has to be sent for any particular sample.

In general, these approaches "adapt the quantizer to the statistics of the input" [13, p. 185]. This is important, conceptually: These approaches are statistical, not deterministic. In Section 6.3.5 we discuss approaches in which the source is actually modeled, and deterministic predictions may be derived. Here, we are talking about the much simpler concepts of tracking

---

22. See [13, pp. 53ff] for examples of actual coding gains using difference encoding versus full-sample encoding of images.

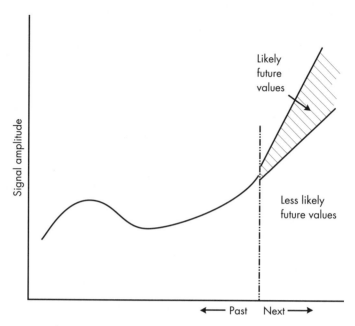

FIGURE 6.12 *Prediction of future signal values.*

the large (and slow) patterns in the data that can be used to provide a statistical prediction, which can allow for a reduction in the rate of information to be transmitted.

The type of adaptation or statistical tracking can vary [11, pp. 188ff]. The adaptation can be performed at the encoder and transmitted as a kind of side information (similar to forward error correction) or it can be regenerated by the receiver (Figure 6.13) [11, p. 192; 13, p. 186].

Vector Quantization

The classical approach to quantization, as noted often in this discussion, is to quantize each sample independently. This is called (not altogether helpfully) *scalar quantization* [3, p. 374]. As we have noted, one result of scalar quantization is the destruction of correlative information about the signal.[23]

If, on the other hand, the individual samples are grouped in blocks, and the blocks are coded—that is, the codeword that is transmitted

---

23. Of course, one can always recode the quantized output, and in principle recover some of these correlation properties. That is, in a way, one idea of vector quantization.

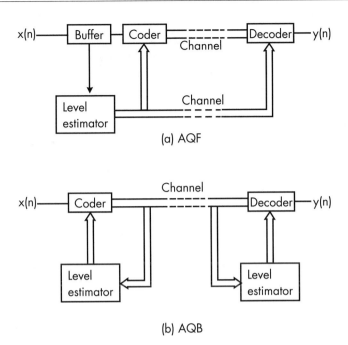

FIGURE 6.13 *Adaptive quantization with (a) forward estimation of input level (AQF), and (b) backward estimation of input level (AQB). (From: [11]. © 1984 Prentice-Hall Inc.)*

represents a block of data symbols—the correlative properties within that block are conserved. We have seen block codes used for signal hardening (error correction) in Chapter 5. This is one way to exploit the correlation properties. Another angle, so to speak, on the same phenomenon is to use block codes to achieve compression, in which case the most common term is *vector quantization* (VQ). Indeed, this aspect of block coding—the ability to optimize capacity through compression—is itself another angle on Shannon's basic argument or existence proof of optimal codes.[24] Still another conceptual link is with *soft decision* ideas, as described in Chapter 5. In effect, all of these are facets of a single, central phenomenon: the use of longer blocks, larger samples, and more time to generate more optimal codes.

24. "The idea that encoding sequences of outputs can provide an advantage over the encoding of individual samples was first put forward by Shannon, and the basic results of information theory were all proved by taking longer and longer sequences of inputs" [13, p. 213].

VQ has become one of the most popular and successful approaches to data compression. The literature is vast, and interesting.[25] Again, VQ can be applied as a lossless compression technique. Sayood develops a very simple example to show why this can work. It is worthwhile quoting at length:

> In [Figure 6.14] we introduce a source that generates the height and weight of individuals. Suppose that the height of these individuals varies uniformly between 40 and 80 inches, and the weight varies uniformly between 40 and 240 pounds. Suppose we allow a total of six bits to represent each pair of values. We could use 3 bits to quantize the height and 3 bits to quantize the weight.... When we look at the representation of height and weight separately, this approach seems reasonable.
>
> But let's look at the quantization scheme in two dimensions. We will plot the height values along the x-axis and the weight values along the y-axis. Note that we are not changing anything about the quantization process [i.e., there is no loss of information here]....
>
> From the figure [Figure 6.14] we can see that we effectively have a quantizer output for a person who is 80 inches (6 feet 8 inches) tall and weighs 40 pounds, as well as a quantizer output for an individual whose height is 42 inches but weighs more than 200 pounds. Obviously, these outputs will never be used ... [13, p. 216]

If we redesign the quantizer so that it codes for pairs of the height/weight values, the so-called (h,w) vectors, the overall code space can be compressed to fit the region shown in Figure 6.14(b). What we are taking advantage of is the fact that height and weight are correlated. As Sayood comments, "looking at long sequences of inputs brings out the structure in the source ... [and] this structure can then be used to provide more efficient representations" [13, p. 218].

In passing, we may note that it has also been argued that even apparently structureless or uncorrelated data can be compressed with VQ techniques, although the argument is not entirely clear to me.[26]

---

25. See [27] for an excellent collection of the basic papers surrounding the tremendous surge in the popularity of this approach in the 1980s, in connection with wireless 2G architectures.

26. According to Blahut [3], "It is easy to surmise wrongly that when the source is memoryless, there is no benefit in quantizing in blocks. Actually one can usually do a little better by quantizing even a uniform memoryless source in blocks" [p. 383]. Blahut draws out a lengthy argument to this effect, which I have trouble following. The reader is referred to the horse's mouth on this one. A somewhat more lucid

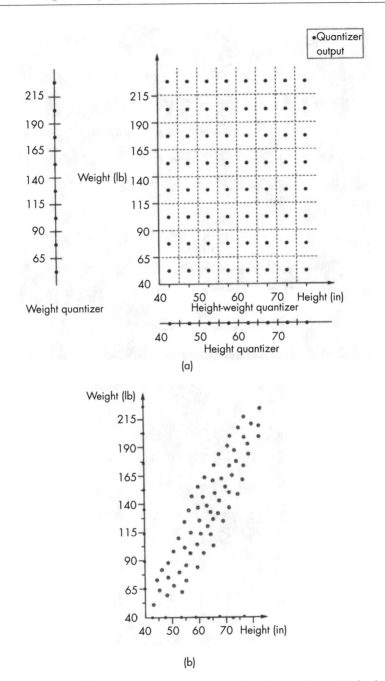

FIGURE 6.14 *The VQ concept, exploiting correlations in the data: (a) the height-weight scalar quantizers when viewed in two dimensions; and (b) the height-weight VQ. (From: [13]. © 1996 Morgan Kaufmann Inc.)*

Codebooks

The most successful uses of VQ, however, are those that employ a further compression technique to create a very powerful *lossy* compression concept. The idea is to perform a second-level quantization or, as it is sometimes referred to, a *pattern matching* on the VQ data. Each vector is compared to a preexisting list of likely or expected vectors, the so-called "codebook," and the nearest match is found. Then the index of this matching codeword is transmitted. This combination of VQ with codebooks is so powerful that it is common to identify VQ with codebook-based approaches, although as we have seen, the two ideas are conceptually distinct. Jayant and Noll describe VQ in these terms:

> Consider an R-ary coder sequence of length N and rate R bits/sample. This implies a total of $J = (2^R)^N = 2^{NR}$ unique *codewords*.... A codebook coder (also called a vector quantizer) accepts a block of N input samples, searches through the codebook with $J = 2^{NR}$ entries, finds the output sequence best matching the input block [there has thus been a reduction from the original J to something less than J entries—this is the implication of the phrase best matching], and transmits the corresponding codeword index ... [11, p. 432]

VQ is, as noted, similar to block encoding. It preserves the correlation properties. Combined with a codebook, this can achieve large compression gains. The onus, intuitively, shifts to the development of the codebooks. Good codebooks make use of both raw statistical properties and known perceptual phenomena, where appropriate. Codebooks can also be designed to

---

comment is offered by Proakis [28]: "A fundamental result of rate distortion theory is that better performance can be achieved by quantizing vectors instead of scalars, *even if the continuous amplitude source is memoryless* [emphasis added]" [p. 120]. In the next breath, however, Proakis adds that "basically, vector quantization of blocks of data may be viewed as a pattern recognition problem"—which clearly focuses on the correlative properties of the data, no longer "memoryless." He pursues a lengthy example [28, pp. 125–126] and notes that when an operation is effected to "render the two random variables statistically independent," then "scalar and vector quantization achieve the same efficiency." He finally concludes that "consequently, vector quantization will always equal or exceed the performance of scalar quantization." I must confess that I cannot see why so-called memoryless data—that is, completely random data showing no structure whatsoever—would yield any compression gains from the use of lossless vector quantization. This is clearly a footnote controversy, at best, since, practically speaking, there is no such thing as a memoryless source.

be adaptive, built up (similar to the data compaction algorithms discussed earlier) from an ongoing analysis of the data stream.

Indeed, the VQ-plus-codebook combination is analogous to sampling plus quantization. All of the same issues, trade-offs, and benefits accrue, but the power of handling larger blocks of data, on the one hand, and the complexity of the processes of designing, searching, and adaptively rebuilding codebooks, on the other hand, push the art far beyond the simplicity of scalar quantization techniques.

VQ techniques (with codebooks) are now the mainstay of some of the best voice compression algorithms in second-generation systems, and the application of VQ can be expected to penetrate into more areas as processing power (Moore's Law) allows for the practical implementation of ever more aggressive and complex VQ strategies.

6.3.5 Source Modeling

The last major approach to compression involves the use of more explicit modeling techniques to create not just a statistical analysis of the patterns in the data to be transmitted, but (at least in principle) a partially deterministic model of the source itself. The idea is to build identical models of the source for both the transmitter and the receiver. If this can be accomplished—or, to say it differently, *to the degree* that it can be accomplished –we can move away from transmitting the actual data and simply transmit certain information about the settings and operation of the model.[27] The receiver then uses these settings to calibrate its model of the source, and regenerates or synthesizes the signal which corresponds to the original source data.

The classical example of this approach is in voice coding models based on analyzing the speech production process [29]. In the words of one text, "the mechanics of speech production impose a structure on speech" [13, p. 3] and if we can recreate (model) the mechanism whereby speech is produced, we can recreate the structure of the signal. Speech, simplified, is viewed as a sound source, called the *excitation*, and a sound channel defined by the vocal tract, which is viewed as a *filter* (Figure 6.15). The nature of the excitation changes, especially between two states: voiced (corresponding

---

27. Blahut [3] says, "Another method of data compression is to fit a parameterized model to the source. Each realization of the source output data is described by adjusting the parameters of the model. Then the model is digitized by quantizing the model parameters. The decoder receives the digital parameters, constructs the source model, and regenerates the source output from the source model" [3, p. 392].

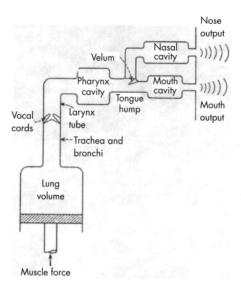

FIGURE 6.15 *Modeling the vocal tract. (From: [29]. © 1972 Springer-Verlag. Reprinted with permission.)*

generally to activity from the vocal cords) and unvoiced (corresponding to noise produced by constricting the airflow, associated generally with certain consonants like *s*). Also, the shape of the sound channel changes, as the tongue, lips, mouth, and throat are altered to produce different types of sounds. This system can be modeled mathematically.

One approach is called *linear prediction,* in which the excitation source is fed through a digital filter that is modified to reflect different configurations of the vocal tract. In operation as a coder, a source sample is provided to the vocal tract model, and the model is converged through feedback until it approximates the input (minimizes the error or residual). The parameters of this filter, plus the excitation, the pitch, and the gain (amplitude) can all be extracted. Then, together with some information about the residual (usually, for this is where much of the speech coloration and quality are contained), these parameters are encoded and sent to the receiver.

These techniques are increasingly used in combination. The voice coders in today's second-generation wireless systems are based on vector quantization, codebooks, speech synthesis, and perceptual coding principles, combined in highly complex and artful technologies. They are capable of yielding acceptable voice quality at bit rates in the range of 4–8 Kbps, which is a huge compression gain compared to 64-Kbps PCM. It is likely, in fact, that compression of speech signals has reached its practical limits, or nearly so. Another reduction by half may be about all that we can hope for. The emphasis in third-generation systems may shift to other signal shaping strategies.

6.4 Baseband Signal Shaping

After the information payload has been fully compressed, the focus shifts from the information domain to the domain of the physical signal (or from coding to modulation, as these terms are best understood). Given the payload, how can we reduce the signal energy required to transmit it from A to B? In particular, how can we reduce the energy that tends to spill over into adjacent STF cells in the communications space?

The signal starts to spread once we introduce nonlinearities into it, which is to say, once we quantize it. The creation of a train of baseband pulses (0's and 1's) is the beginning of trouble in this regard. Telegraphers learned long ago that a square input pulse produces a spreading signal. This leads to intersymbol interference (T-interference), which sharply limits the practical signaling rate and channel capacity. Unless we ensure that the pulse shape is such that we can determine there will always be sampling times in which the spectrum of the preceding pulse(s) will be at a zero-crossing instant. In other words, the spreading pulse will show a dampening oscillation such that at time T (or $-T$) it will be crossing the zero-energy level; this will be followed by another zero-crossing at $2T$, another at $3T$, and so forth (Figure 6.16). Thus, if the train of pulses can be arranged such that the peak of the next pulse occurs precisely at time T, and the next at time $2T$, and so forth, we can achieve zero intersymbol interference even with spreading signals (Figure 6.17).

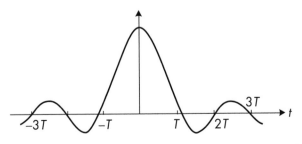

FIGURE 6.16 *Zero-crossing waveform. (From: [3]. © 1990 Addison Wesley Inc.)*

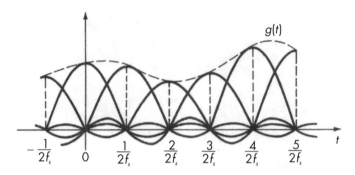

FIGURE 6.17   *Nyquist signaling. (From: [31]. © 1993 Macmillian.)*

Any pulse that satisfies this property is called a *Nyquist pulse,* and the appropriate signaling rate is the *Nyquist rate* (after Harry Nyquist, a pioneer in the information theory field and Shannon's greatest precursor) [3, p. 29; 30, p. 617]. Yet there are many pulse shapes that meet the Nyquist criterion. All are alike in creating an exploitable pattern of zero-crossings, but some generate much more energy in their tails than others. If the timing of the samples is imperfect (as it is, by definition, in real systems), we are therefore concerned with ensuring that the amount of energy contained in the spreading signal is minimized.

> We would prefer that the tails of the time-domain pulse shape decay as rapidly as possible, so that the jitter in the pulse sequence or sampling times will not cause significant intersymbol interference. This last requirement is a very pragmatic one, but it is necessary since any data transmission system will have some timing jitter, which will cause pulse sampling to occur at times other than the exact center of each pulse interval. If the pulse tails are not sufficiently damped, sampling in the presence of timing jitter can cause substantial intersymbol interference. [31, p. 191]

Once again, this entails the *avoidance of nonlinearities* in the pulse shape. "To design practical pulses that have no intersymbol interference, one must design Nyquist pulses who time-domain sidelobes fall off rather quickly [and] this means that the spectrum should not have sharp edges" [31, p. 31]. A square pulse produces large sidelobes; an appropriately rounded shape in the initial pulse can reduce the sidelobes significantly (Figure 6.18).

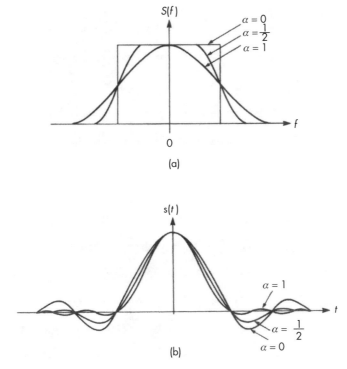

FIGURE 6.18 *Nyquist pulse shaping: (a) spectra with various roll-off factors; and (b) roll-off affects the amount of energy in the "tails." (From: [3]. © 1990 Addison Wesley Inc.)*

Nyquist pulse shaping is part of the classical foundation of the field. Modern research is taking the idea of pulse shaping in the direction of reintroducing various degrees and types of correlation. For example, a modulation scheme called *quadrature pulse overlapping modulation* (QPOM) is described by Jamali and Le-Ngoc [32, Chap. 10]. The pulse train is divided into two separate streams, and odd and even pulses are shaped according to different equations, a time delay is inserted in one of the streams, and they are then recombined in an output stream with controlled overlapping (controlled ISI). Depending on the parameters, a tradeoff is obtained between the width of the main lobe of the modulated spectrum (in the frequency domain) and the energy in the sidelobes (Figure 6.19). Indeed, the family of quadrature amplitude modulation techniques generally is motivated to exploit pulse shaping as an inherent part of the generation of the physical signal. In such systems, the data stream is divided, offset (normally), and then a pulse-shaping stage process follows, as in Figure 6.20:

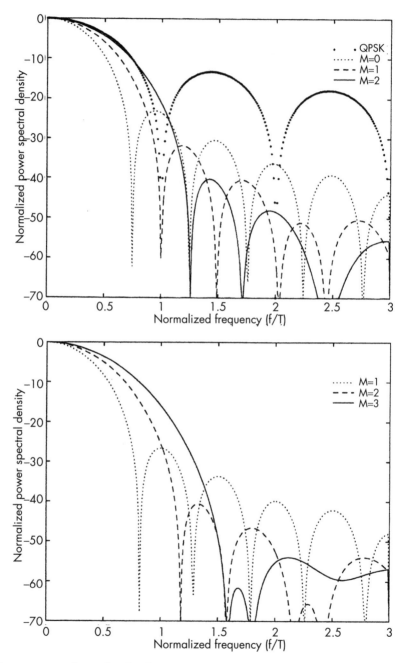

FIGURE 6.19 *Advanced pulse shaping. Normalized power spectral density of QPOM signals with double interval pulse. (From: [32]. © 1994 Kluwer Academic Publishers.)*

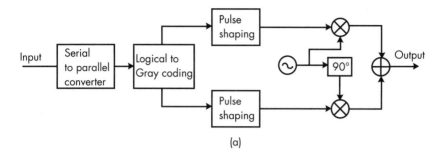

(a)

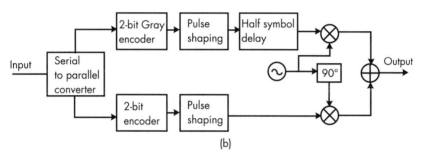

(b)

FIGURE 6.20 *QAM with pulse shaping.: (a) standard QAM modulator; and (b) offset QAM. (From: [7]. © 1994 IEEE. Reprinted with permission.)*

D/A conversion is ... followed by pulse-shaping, before upconverting the signal to the carrier frequency. It is often advantageous to perform the pulse-shaping prior to the D/A conversion by oversampling the transmitted data and using further look-up tables containing the pulse-shaping law in order to produce a digital representation of the smoother signal. [7, pp. 115ff]

Even more generally, *partial-response signaling*, which is closely related to correlative coding (discussed in Chapter 5) is based on the idea of allowing the pulse train to blend together and overlap in a controlled manner. It is really the same process, but viewed in terms of a different objective. The purpose of correlative coding per se is to strengthen the signal in the presence of noise, whereas "the purpose of partial-response signaling is to manage the spectrum of the transmitted waveform" [3, p. 250][28]—a distinction perhaps without a difference.

---

28. See also [31, pp. 198ff].

Thus, the common theme of post-Shannon technology reappears: The pre-Shannon notion of keeping signal elements carefully separate has moved toward a certain optimum, and then beyond it—in the post-Shannon framework—to the reintroduction of controlled self-interference. Properly handled, this can reduce the amount of surplus signal energy that is needed to carry the payload.

6.5 RF Signal Shaping

6.5.1 Bandwidth-Efficient Modulation

The prior example involving modulation and pulse shaping illustrates the straightforward conceptual transition to signal shaping strategies that are implemented at the RF level. The goal is basically the same—*how to pre-shape the input signal so that the output signal spreads as little as possible*—except that it is implemented at a higher frequency, and the most important form of signal spreading is in the *frequency* domain, rather than the time domain [7, Chaps. 3 and 4]. The generation of modulation pulses can create surplus energy in the radio signal, most often measured in the frequency domain as sidelobes in the power spectrum (Figure 6.21). The larger the sidelobes, the more interference is created for users in adjacent F-subspaces (adjacent frequency channels).

Once again, the main strategy is *to identify and eliminate sources of nonlinearity* in the modulated signal. Some standard modulation techniques, such as QPSK, involve abrupt transitions in the modulated waveforms and zero crossings, which create high-frequency components (sidelobes) in the RF signal. Figure 6.22 shows the pattern of transitions in the QPSK waveform, as well as two stages in the direction of mitigating these effects by offset-QPSK and so-called π/4-QPSK. Both of these modifications lead to smoother waveforms with fewer and less abrupt transitions. The next step is to eliminate abrupt phase shifts altogether, and to impose a requirement for *continuous phase* transitions.[29]

---

29. "Abrupt switching from one oscillator output to another in successive signaling intervals results in relatively large spectral side lobes outside of the main spectral band of the signal and ... requires a large frequency band for transmission of the signal. To avoid the use of signals having large spectral side lobes, the information-bearing signal frequency modulates a single carrier whose frequency is changed continuously. The resulting frequency-modulated signal is phase continuous and, hence, it is called continuous-phase FSK (CPFSK)" [28, p. 172].

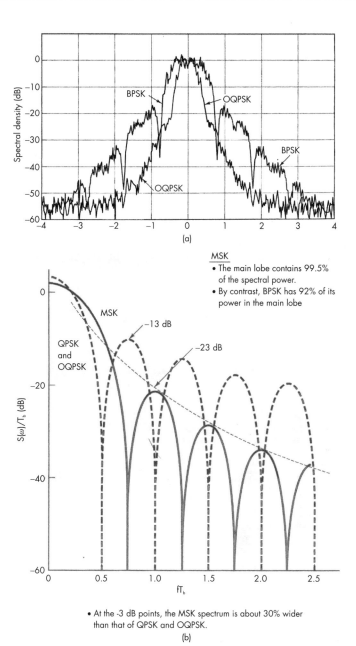

FIGURE 6.21 *Power spectrum sidelobes for various modulation formats: spectral side-lobe regrowth of BPSK and absence of regrowth of OQPSK after hard-limiting amplification; and (b) power spectral density of QPSK, OQPSK, and MSK. (From: [33]. © 2000 Prentice-Hall Inc.)*

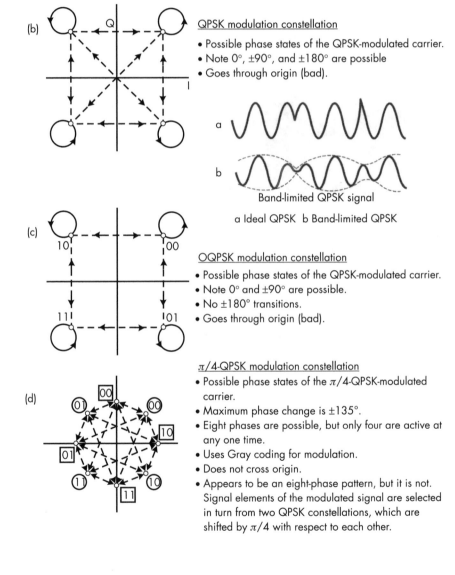

(a) BPSK modulation constellation

(b) QPSK modulation constellation
• Possible phase states of the QPSK-modulated carrier.
• Note 0°, ±90°, and ±180° are possible
• Goes through origin (bad).

a Ideal QPSK b Band-limited QPSK

Band-limited QPSK signal

(c) OQPSK modulation constellation
• Possible phase states of the QPSK-modulated carrier.
• Note 0° and ±90° are possible.
• No ±180° transitions.
• Goes through origin (bad).

(d) π/4-QPSK modulation constellation
• Possible phase states of the π/4-QPSK-modulated carrier.
• Maximum phase change is ±135°.
• Eight phases are possible, but only four are active at any one time.
• Uses Gray coding for modulation.
• Does not cross origin.
• Appears to be an eight-phase pattern, but it is not. Signal elements of the modulated signal are selected in turn from two QPSK constellations, which are shifted by π/4 with respect to each other.

FIGURE 6.22 *Spectrum shaping versions of QPSK signaling, modulation constellations for: (a) BPSK, (b) QPSK, (c) OQPSK, and (d) π/4-QPSK. (From: [33]. © 2000 Prentice-Hall Inc.)*

The most popular form of continuous phase modulation is *minimum shift keying* (MSK). MSK is really just a form of frequency shift keying, alternating bit values between two frequencies chosen such that the transition between the waveforms is smooth and continuous (Figure 6.23) [3, pp. 52ff].

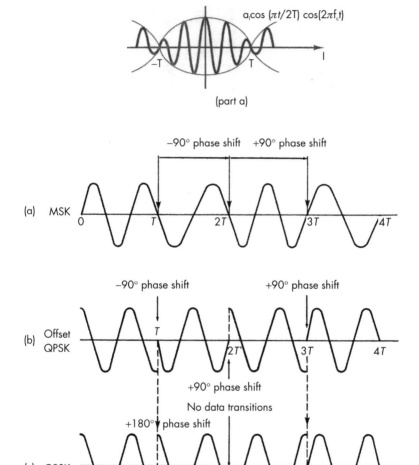

FIGURE 6.23 *MSK modulation: (part a) generation of MSK via quadrature modulation (From: [33]. © 2000 Prentice-Hall Inc.); and (part b) signal waveforms for MSK, OQPSK, and conventional QPSK (From: [34]. © 1976 IEEE. Reprinted with permission.).*

The elimination of nonlinearities reins in the spreading of the signal. The main lobe of an MSK waveform is a little broader than a QPSK waveform, but the sidelobes (the main source of F-interference) are much reduced (Figure 6.24). The MSK principle can be generalized under the category of *continuous phase modulation* (CPM) as a family of methods that shares these characteristics [3, pp. 259ff; 28, pp. 171ff; 31, pp. 387ff].

The interrelationship of these shaping concepts is evident in the literature. The root of the similarity of pulse shaping, correlative coding, partial response, and continuous phase modulation methods lies in the blending of different data elements (symbols, pulses, modulation states) to create more analog-like signals, convolutional signals, and purposeful interference, which in turn display various positive properties such as signal hardening, compression, and reduced interference. In a way, all of these processes are aimed at restoring something of the native robustness and threaded structure of the analog source, which was lost when the source signal was disarticulated through the quantization process. We create nonlinearities, and then we seek to ameliorate them.

A great deal of subtlety surrounds this nexus of connected topics and effects. It points to a certain unsteadiness in some of the most fundamental theoretical concepts of information. We recall that Shannon viewed the analog mode as intractable and reduced it by the legerdemain of rate distortion and fidelity criteria to the digital or discrete case. He tolerated signal expansion as a reasonable price to pay. In the post-Shannon framework, where capacity really does come at a premium price, the trend is toward undoing the effects of signal expansion by reversing the original bargain, in part, and with great care. It is odd, at least, to find so many of the latest techniques trending toward the reestablishment of the smooth and continuous contours of analog signaling.

6.5.2 Linearized RF Systems

The creation of nonlinear signals—that is, signals carrying significant embedded nonlinearities, such as digital transitions—in turn creates a need for the transmission system to behave linearly.

This seeming paradox is the fulcrum of many design problems and architectural choices facing the wireless engineer. The more inherently linear the signal—the smoother its transitions, the more analog it appears—the more forgiving it is of nonlinearities in various processing stages. As highly nonlinear signals are constructed—signals that are heavily coded, precisely

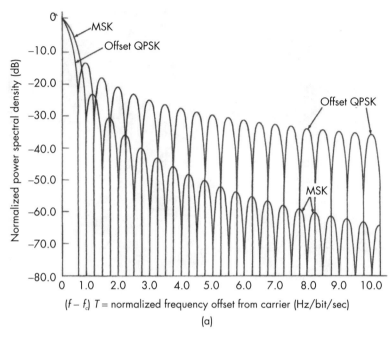

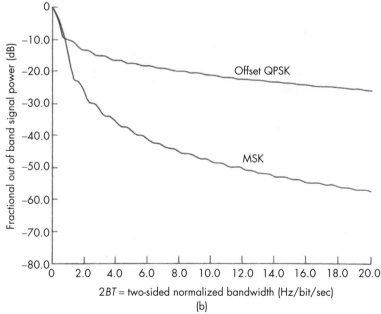

FIGURE 6.24   *Sidelobe energy of MSK and QPSK: (a) power spectra density of MSK and QPSK and (b) out-of-band power (normalized two-sided bandwidth = 2BT). (From: [34]. © 1976 IEEE. Reprinted with permission.)*

defined—the more we need very high-fidelity components that handle them very precisely and accurately.

One of the most troublesome stages in the transmission process is the power amplifier. The information-bearing signal is created, manipulated (coded, modulated), upconverted, and made ready for transmission—all of which takes place at relatively low power levels suitable for conventional electronic circuitry. The last stage is to amplify the signal, to boost the power by many orders of magnitude, in order to survive the passage through the radio channel itself. This amplification must be accomplished without introducing too much distortion, which is difficult. The design of power amplifiers is one of the most problematic fields in communications engineering generally, and as second-generation technologies have revolutionized the air interface, the industry has been forced to look for more and more linear power amplifiers. In particular, the choice of spectrum-efficient modulation techniques—that is, those with narrow main lobes in their power spectrum—has driven the industry toward more demanding and expensive power amplifier specifications:

> Other modulation schemes require linear amplification [because they introduce nonlinearities in the signal to be amplified] ... $\pi/4$-DQPSK modulation, used with a raised-cosine pulse-shaping filter to reduce signal bandwidth without causing intersymbol interference, results in a nonconstant envelope modulated signal, and requires linear power amplifiers. [The $\pi/4$-DQPSK is used in the IS136 TDMA standard, for example.] Linear power amplifiers tend to have much lower power efficiency than nonlinear amplifiers.... These variations in the envelope, due to band-limiting of this signal performed by the pulse-shaping filter, need to be preserved in the transmitted signal in order to retain its band-limited spectrum. [35, p. 94]

This requirement is especially crucial at the base station receiver, where many signals of disparate characteristics are handled together.

> Spectral regrowth is a major concern in personal communications since transmission often occurs on a channel adjacent to one in which reception of a much weaker distant signal may be taking place. To ensure freedom from interference, transmitter intermodulation distortion products must be below the carrier-to-interference ratio by 35 to 65 dB, depending on the application. [36, p. 23]

Linear amplifier technologies have thus become a key enabling technology for the second-generation architectures and are likely to attract increased attention in third-generation systems. Three concepts for achieving linearity are feedforward linearization, feedback linearization, and predistortion linearization.

Feedforward Linearization

This involves the subtraction of the known input signal from the output signal, which leaves as a remainder the distortion introduced by the amplifier. Thus isolated, this distortion can be itself subtracted from a delayed second copy of the output signal, leaving, in principle, the undistorted input signal (amplified) (Figure 6.25).

Feedback Linearization

"In these approaches, an amplifier's input and output signals are detected and the resulting baseband signals are compared. The error signal is used to modify the amplifier's characteristics so as to minimize distortion ... " [36, p. 34] (Figure 6.26). The shape of the distortion is in a generic sense inverted and fed back into the input stream so that its effects are self-canceled.

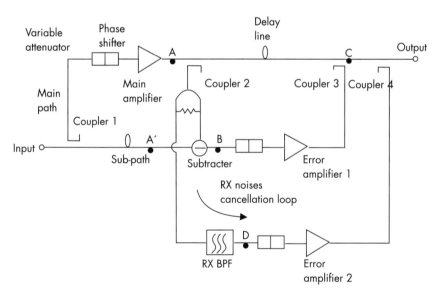

FIGURE 6.25 *Feedforward linearized power amplifier block diagram. This is a representative use of feedforward techniques in the design of an amplifier. In this case, there are two separate feedforward loops—one designed to correct intermodulation distortion and one targeted at receiver noise. The correction signals are added to the transmitter output. (From: [37]. © 2002 Horizon House Publications Inc. Reprinted with permission.)*

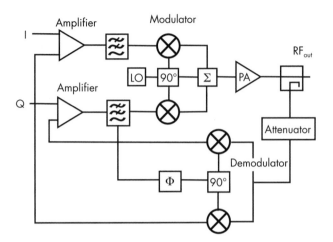

FIGURE 6.26 *Feedback linearized power amplifier block diagram. The Cartesian loop is a representative use of feedback techniques in the design of an amplifier. In this case, there are two separate feedback loops for the I and Q components of the signal. The correction signals are inverted and fed back to the input of the amplifier. (From: [36]. © 1999 Horizon House Publications Inc. Reprinted with permission.)*

Predistortion Linearization

The concept of predistortion is related to other ideas we have touched on, such as pulse shaping and equalization. Essentially, if we know the type of distortion produced by the amplifier, we can predistort the input signal with an inverted image of the expected distortion pattern of the amplifier, so that when it passes through the amplifier the predistortion and the added distortion will cancel each other (Figure 6.27). Once again, we come across the strange notion that by adding distortion to the signal we can achieve a positive benefit:

> An alternate way of thinking of a PD [predistortion] linearizer is to view the device as a generator of [intermodulation distortion, IMD]. If the IMD components produced by the linearizer are made equal in amplitude and 180° out of phase with the IMD products generated by the amplifier, the IMD will cancel. [36, p. 36]

All of these techniques can be substantially elaborated and combined, and in many cases are amenable to either digital or analog implementations.[30] Note also that these concepts are quite general. Feedforward correction, feedback correction, and predistortion can be applied at many stages of

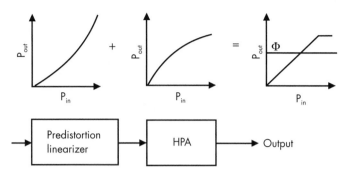

FIGURE 6.27 *Predistortion. (From: [36]. © 1999 Horizon House Publications Inc. Reprinted with permission.)*

the signal processing chain to counteract the effects of nonlinearities that have been either intentionally introduced or are the result of uncontrollable factors, such as the channel itself. They share the underlying strategy of attempting to detect the structure of the undesired distortion and then subtracting or canceling it deterministically.

6.6 Smart Antenna Technologies (Transmission)

Spatial shaping of the signal is another vast field, involving smart antennas, which combine often elaborate hardware arrays with extremely intensive signal processing engines. These are used in various ways to focus the transmitted or received signal. The smart transmitter process is easily conceptualized: Instead of radiating the signal in all directions, the transmitter locates the receiver within a much narrower beam and directs most of the signal energy in that more or less precise direction (Figure 6.28). It is like using a searchlight, rather than a floodlight, to find a target in the darkness.

For various reasons, it has proved easier in mobile applications to manage beam forming by the receiver rather than the transmitter. This may seem counterintuitive; after all, it would seem to be easier to direct a beam in a particular manner than to find the unknown direction of a distant transmitter. Yet tracking a mobile unit with a smart antenna beam, which can be either a transmitting path or a receiving path, requires knowledge of the

---

30. For a recent, somewhat more thorough, yet still brief survey of PA linearization techniques, see [7, pp. 128ff].

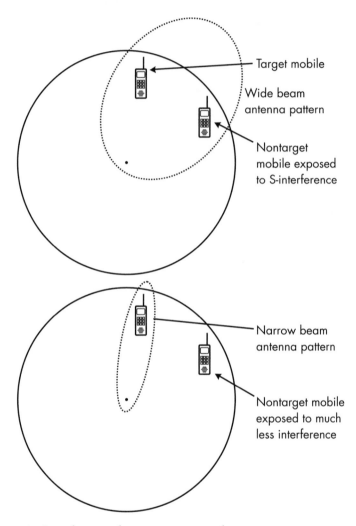

FIGURE 6.28 *Omnidirectional transmission versus beam transmission.*

specific, time-sensitive conditions of the channel. Essentially, in many architectures the transmitter does not have good information about the channel, because the signal has not yet traversed it, and cannot achieve effective focusing and tracking in a mobile, multipath environment [38]. Thus, most of the work on smart antennas has focused on receiver implementations, where the receiver can extract information about the channel from the received signal and use these to help determine the direction of the transmitter.[31] We consider these receiver implementations in the following chapter, so we postpone a more detailed presentation of the fundamental beam-forming principles until then.

Nevertheless, the use of smart antenna techniques by the *transmitter* is possible, and thus depends on developing an effective way to estimate the channel characteristics so that the beam that the transmitter emits can be steered and shaped appropriately. The channel is changing very rapidly (due mainly to the movement of the mobile unit and the rapidly shifting multipath patterns), and the changes are also quite frequency sensitive. Essentially, drawing on the discussion in Chapter 5, the beam-forming transmitter needs channel information that falls within the coherence time (T_{coh}) and the coherence bandwidth (F_{coh}) of the actual signal in order to be confident in its ability to characterize the spatial properties of the signal. This may be possible in a *time-division duplex architecture*, where the same frequency channel is shared back and forth by the two transmitters at the base and mobile. If the time switching is fast enough, the transmitter can use its own receiving side channel information to shape the transmit beam [39, p. 68]. In a frequency-division duplex system, where the uplink and the downlink are widely separated in frequency (and thus do not lie within the critical F_{coh} bandwidth), the challenge is much greater.

In reality, therefore, most implementations require a feedback-based or closed-loop algorithm (similar to what we shall see for power control algorithms in Section 6.7). Using either a training sequence or a pilot signal (two names for the same idea, more or less), the transmitter enables the receiver to detect the relevant channel characteristics and then to issue control information back to the transmitter:

> A direct approach to estimating the forward channel is to feedback the signal from the mobile unit and then use either a blind or nonblind method for estimating the channel. To illustrate ... assume that we transmit a training signal through one antenna at a time [in a multiele-ment antenna array; see Chapter 7] from the base station to the mobile. We can then ... estimate the channel from each base station antenna to

---

31. "In the forward link, the space–time processing [for beam forming] is carried out prior to transmission and therefore before the signal encounters the channel. This is very different from the reverse link where the space–time processing is carried out after the channel has affected the signal. Therefore, while in the reverse link it is possible to use blind or nonblind techniques on the received signal to either estimate the channel or data, in the forward link, the base station needs feedback of the signal received at the mobile station for analogous processing" [39, p. 68].

the mobile. Once we determine the channels for each antenna, we can write down the total channel. [39, p. 69]

The awkwardness of this procedure and the need for substantial overhead to be invested in supporting the closed-loop approach militate against spatial processing on the transmitting side. As we shall see in the next section, the unavoidable need to solve a similar problem in power control has become a huge implementation challenge for CDMA-based architectures.

6.7 Adaptive Link Technologies

The use of the word *adaptive* to describe various wireless solutions generally refers to the incorporation of technologies that can modify their transmission parameters to better match the changes in the channel characteristics over *time*. Of course, the same concept of self-modification can be developed for frequency-adaptive and spatially adaptive solutions. For example, the use of adaptive frequency hopping in the so-called *cellular digital packet data* (CDPD) cellular data transmission standard is a good exploitation of adaptive principles in the frequency domain, where small data packets are transmitted during idle periods on a set of intermittently active cellular voice channels. The use of adaptive spatial tracking-beam systems is discussed in the next chapter. Most of the important examples of adaptive techniques focus on the time-varying nature of the channel; or, to say it a little differently, they track the changes as measured in the time domain, and adjust the system transmission parameters according to some optimization process.

Adaptiveness in this sense is an old and familiar idea. The thermostat that controls the furnace in a house is a device that tracks the variations in the external temperature over time and adjusts the heating system to maintain a more or less constant temperature. Control theory, systems theory, and many other by now venerable disciplines have embedded this sort of adaptive optimization in many fields of engineering design.

Strangely, adaptive techniques are still quite new to communications. Most communications systems are built around fairly fixed performance benchmarks, which are designed to provide at all times a surplus signal (over and above the minimum that would be required for adequate communications at any given moment). In the wireless channel, where the basic characteristics of the channel fluctuate over a wide range, this means that there is almost always a large surplus of unnecessary transmission energy being radiated into the communications space. It is not uncommon to design for a

10-dB margin—to transmit enough signal energy such that the received signal will always be at least 10 dB about the threshold required by the receiver, even in the worst case (statistically). This means, effectively, *that 90% or more of the transmitted signal energy is unnecessary* for good communications and serves no purpose except to create a huge amount of interference. This in turn drives the need for STF buffers, described in earlier chapters, and reduces usable system capacity.

Running the transmitter at a constant power in a time-varying channel is like turning the furnace on full blast and consoling ourselves with the thought that at least we won't freeze to death. If we can find ways to track the relevant changes in the channel, we can begin to implement various homeostatic algorithms to reduce the amount of interference we generate.

There are many ways to design an adaptive signal. Adaptive techniques can be divided, perhaps usefully, into two groups:

1. Techniques that try to adapt to the setup conditions of the channel and to tune the system parameters based on the actual needs of the specific link rather than on the precalculated statistical requirements for the channel; this allows, in some cases, the design of the system to the *average case* conditions, instead of the worst-case conditions.

2. Techniques that track the rapid fluctuations in the signal in real time.

In the following sections, we will look at three representative techniques that illustrate the range of adaptive solutions that are being developed for third-generation wireless systems: adaptive power control, adaptive time alignment, and adaptive modulation.

6.7.1 Adaptive Power Control

One of the most straightforward adaptive solutions for reducing unnecessary interference is the use of adaptive power control. The needs of the receiver are usually defined in terms of a fixed threshold for the received signal level. If the received signal is above this threshold, the signal quality will be deemed to be acceptable. As noted earlier, traditional nonadaptive solutions simply make sure that the transmitter is capable of delivering the required threshold level, plus a margin, under all channel conditions that we design for. This last phrase implies a statistical distribution, because the relationship between transmitting power and received signal level is not fixed, but is a so-called "random" variable. If the transmitter power is fixed, the received signal will fluctuate based on two factors: (1) the distance from the transmitter, or free-space loss, with which we may combine general terrain features that are fixed in nature and may cause further signal attenuation, and (2) fast fading due to self-interference (multipath) (Figure 6.29).

In the early days of radio transmission, systems were noise limited and the excess transmitted energy did not constitute a problem per se. When cellular architectures were introduced, the systems became interference limited [26]. The excess energy meant that the channels could not be reused efficiently. This led to the development of basic power control algorithms, probably the first use of adaptive principles in wireless design. The initial power control solutions utilized a slowly time-varying measure of received signal level, which was retransmitted to the base station from the mobile unit (or vice versa), to control the transmit power. These solutions were able to reduce interference by several decibels in system-wide implementations. Unfortunately, a relatively large margin was still required because the first-generation power control algorithms did not track the fast fading of the signal. The margin for multipath was still needed.

With the emergence of second- and third-generation wireless systems, the issue of power control has become much more central. Indeed, for the

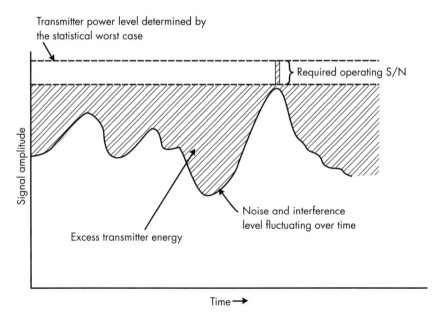

FIGURE 6.29 *Fluctuating RSSL versus receiver threshold.*

successful implementation of spread spectrum CDMA architectures, power control emerges as the single most important technical challenge.[32] To handle the need of CDMA for extremely accurate power control—to track the signal within 1 dB or even less—adaptive solutions have been developed that control the required power on a symbol-by-symbol basis, fast enough to track the multipath statistics of the mobile channel.

The importance of power control for CDMA is based on the signal architecture. In a CDMA system, all users share the entire available communications space. Any surplus signal contributed by any one mobile unit impacts the performance of all other users in the cell. In particular, CDMA suffers from the well-known near–far problem: A user close to the base station will tend to jam a user farther away from the base unless all users are power equalized to within a very narrow range. Imperfect power control, and the resulting interference, has a direct effect of reducing system capacity, often dramatically. Various simulations and measurements of the effects of imperfect power control have been published, with estimates of as much as 50% to 60% loss in cell capacity from as little as a 1-dB deviation from perfect power control tracking (Figure 6.30) [41, p. 72].[33]

Implementation of effective power control solutions has proven to be the breakthrough for enabling the development of feasible CDMA systems. Today's standard CDMA (IS-95) incorporates both open-loop and closed-loop power control algorithms.

> Power control can be established by letting the base station continuously transmit a (wideband) pilot signal that is monitored by all mobile terminals. According to the power level detected by the mobile, the mobile adjusts its transmission power. Hence mobiles near the cell boundary transmit at a lower power level than mobiles located close to the base station. This is open loop power control. It is also possible [indeed, necessary] to use closed loop power control.... In this case the base station measures the energy received from a mobile and controls

---

32. "Power control is recognized as the most important system requirement for cellular CDMA" [40, p. 515].

33. Other studies have generated lower estimates of the capacity impact. Cameron and Woerner cite a figure of only 15% capacity reduction at a 1-dB deviation, declining however rather steeply to 30% at 1.4 dB and 60% at 2 dB [42, p. 780]. Delli Priscoli and Sestini find a loss of only 30% capacity at a 2-dB deviation [43, p. 1817]. Finally, Viterbi and Viterbi argue that a 2.5-dB deviation will cause a loss of only 20% [44]. This has obviously been a highly contentious matter in the by-now moot arguments over the merits of CDMA versus other architectures.

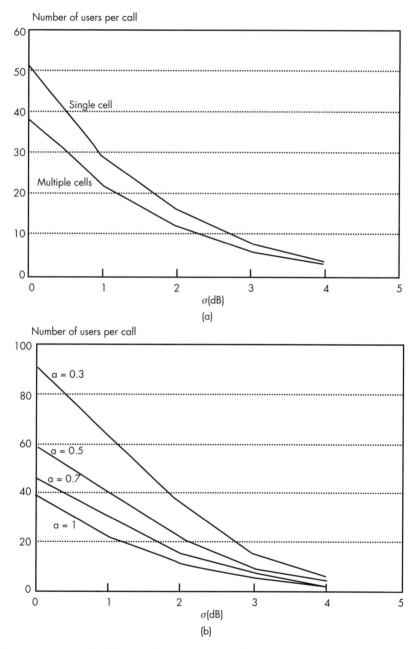

FIGURE 6.30 *Penalty for imperfect power control; capacity as a function of power control imperfection for: (a) a single cell and multiple cell system and (b) several values of the voice activity factor. (From: [41]. © 1995 IEEE. Reprinted with permission.)*

the transmit power of the mobile by sending a command over a (low data rate) command channel. [41, p. 68]

The full description of power control solutions in CDMA is beyond our scope of interest here. Suffice it to say that multiple-algorithm or layered-algorithmic approaches are needed to ensure that all units are controlled to within such narrow tolerances. This indicates how extensive the effort to implement an adaptive approach can become, and the fruits it can bear—in this case, not merely a reduction of interference, but the enabling of a broad new architectural approach.

Outside of the CDMA environment, power control may seem today a rather mundane technique—and yet there are still significant gains available from better power control in channelized systems. Better forward-link power control has been studied for use in third-generation TDMA, and the prospective gains are surprising:

> New slot formats contain at least one reserved bit for power control....
> Given that the bits occur once every 20 ms [in IS-136], power commands are possible 50 times a second. In comparison with the power control available today which is once every few seconds, this is quite a bit faster; however, it is still not fast enough to track the Rayleigh [multipath] fading, and improvements are thus confined to reducing the shadow variations. Nevertheless, [studies] have shown that the performance can be significantly improved (3–4 dB in average C/I) despite even a 10 percent error rate in the power control bit. [45, p. 27]

If we consider that the entire CDMA standard was promoted on essentially a claim of gaining 3–4 dB over TDMA, the possibility of gaining another similar amount from such a simple signal-shaping technique (far simpler than the multilayered power control system required to make CDMA operable at all) is certainly attractive.

6.7.2 Adaptive Time Alignment

In a time-division multiplexed system, two users occupying adjacent slots on the uplink may shift in time relative to one another because of differences in propagation delay (Figure 6.31). The signal received from a mobile unit transmitting from the edge of a large cell will be delayed relative to a nearby mobile, and the two time slots will overlap. If we want to avoid losing data from both users, we must provide a guardband in the time domain that is large enough to accommodate the maximum possible difference between

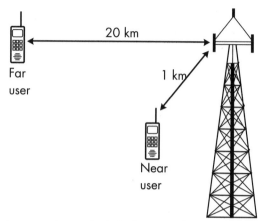

Actual received signals shifted in time due to propagation delay

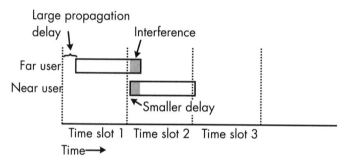

FIGURE 6.31 *TDMA guardband, with different propagation delays creating overlap.*

two such users (Figure 6.32). If the time buffer is set according to the worst-case scenario, this can require a significant overhead (up to 10% of the bandwidth in some systems, which must be left vacant or filled with dummy symbols).

By using an adaptive solution we can recover most of this lost capacity. The base station measures the delay shift during the first few cycles and determines how far away the mobile unit is. The base then transmits to the mobile a transmission offset value. If the mobile unit is quite close to the base, it will be instructed to delay its transmission slightly so that the arrival times are equalized with respect to more distant users on adjacent time slots (Figure 6.33). The time alignment is monitored and continuously adjusted (although the tracking of this very slowly changing parameter is much simpler than the dynamic power control described in the previous section).

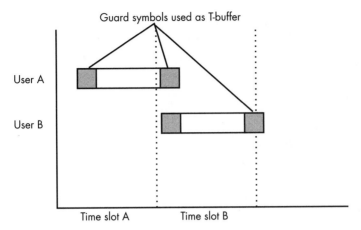

FIGURE 6.32 *Fixed temporal guardbands.*

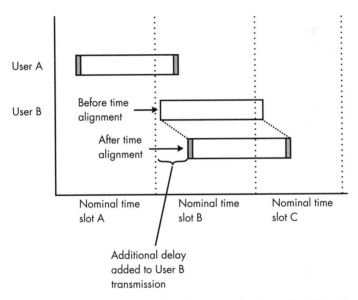

FIGURE 6.33 *Adaptive temporal guardbands: overhead allocated to T-buffering can be reduced.*

Allowing for slight imperfections, it is possible in principle to reduce the guardband to a single guard symbol. Of course, a small amount of bandwidth is required to implement the time alignment algorithm, but overall there is usually a significant capacity gain.

6.7.3 Adaptive Modulation

An even more fundamental adaptation involves the modification of the bit rate to fit changing channel characteristics. That is, with a good SNR or signal-to-interference ratio, we can transmit at a higher rate; as the SNR degrades, we can reduce the bit rate. This is the solution employed by the current wireline fax standards, for example, and implicitly, by many asynchronous protocols (such as TCP/IP). Similar solutions are proposed for third-generation wireless data services. One way to achieve this is by using higher level modulation (more bits/symbol) for good channel conditions, and allowing the system to fall back to more basic modulation schemes under conditions of heavy interference. This can further be combined with adaptive channel coding—stronger codes for more challenged channels. For example, the enhanced data service for GSM evolution (EDGE) standard is based on adaptive modulation and adaptive coding to achieve a range of bit rates depending on channel characteristics and traffic loads (Figure 6.34):

> This system will adapt between Gaussian minimum shift keying (GMSK) and 8-PSK modulation with up to eight different channel coding rates ... to support packet data communications at high speeds on low-interference/noise channels and at lower speeds on channels with heavy interference/noise. EDGE can be deployed with the conventional 4/12 reuse used for GSM. However ... a combination of adaptive modulation/coding, partial loading, and efficient automatic repeat request (ARQ) permits operation with very low reuse factors [as low as 1/3]. [46, p. 16]

Figure 6.35 shows a set of simplified throughput performance curves for three different modulation schemes considered for enhancement of the

Scheme	Modulation	Max. rate (kb/s)	Code rate
MCS-8	8PSK	59.2	1.0
MCS-7	8PSK	44.8	0.76
MCS-6	8PSK	29.6	0.49
MCS-5	8PSK	22.4	0.37
MCS-4	GMSK	17.6	1.0
MCS-3	GMSK	14.8	0.8
MCS-2	GMSK	11.2	0.66
MCS-1	GMSK	8.8	0.53

FIGURE 6.34 *Adaptive coding and modulation. (From: [46]. © 1999 IEEE. Reprinted with permission.)*

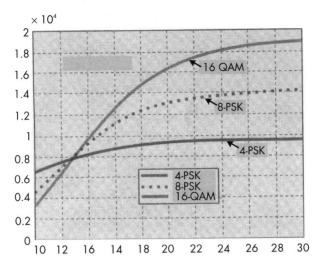

FIGURE 6.35 *Performance benefits from adaptive signal design. (From: [46]. © 1999 IEEE. Reprinted with permission.)*

IS-136 TDMA standard: 4-PSK (2 bits/symbol), 8-PSK (3 bits/symbol) and 16-QAM (4 bits/symbol). Obviously, if the channel SNR is good, the system can gain significant capacity by using the higher level modulation:

> As can be seen, it is advantageous to use 16-QAM over a wide range of SNRs. Below 18 dB, the performance of 16-QAM as well as 8-PSK deteriorates rapidly, and at 10 dB 4-PSK offers three times more throughput than 8-PSK or 16-QAM. In the presence of imperfect channel estimation, it is expected that 16-QAM and 8-PSK will be worse that 4-PSK at even higher SNRs. Furthermore, 8-PSK and 16-QAM will be progressively less robust to noise, interference, and delay spread compared to QPSK. Thus, mode adaptation ... which falls back to π/4-PSK under difficult channel conditions [is needed]. [46, p. 12][34]

Again, this type of adaptation can be performed on the fly or at setup. For voice services, where a constant bit rate is often desirable, the adaptation may be instituted at the time the call is created; in this case, the bit rate is constant, but the different modulation options result in different channel

---

34. See also [45, pp. 20–33]; indeed, all the articles in this special issue devoted to TDMA 3G evolution touch on adaptive modulation and adaptive link design to some degree.

allocations. In the 1980s, the author was involved in the development of a wireless local loop system, in which calls could be established in either 16-PSK (4 bits/symbol) for good conditions, typically close to the base station, or 4-PSK (2 bits/symbol) for the fringes of the coverage area. If 16-PSK was used, the TDMA structure was a four-slot format. In the QPSK mode, a two-slot format was used. This allowed for higher capacity than if the system had been designed, nonadaptively, to support the worst-case link conditions (the fringe).[35]

In general, the third generation architectures—especially those based on TDMA—will increasingly move to embrace the idea of *link adaptation*. This encompasses on-the-fly adaptation of both the modulation scheme, the user payload (e.g., variable voice compression schemes), and channel coding to match the transmission format as closely as possible to the time-varying characteristics of the channel (reflecting both natural effects, such as the multipath configuration, as well as the effects of changing system traffic loads and cochannel and adjacent channel interference levels). By allowing all three transmission shaping parameters to vary as a function of the channel conditions, the system can adapt to handle much higher levels of traffic.

Adaptation is indeed the meta-concept underlying the various technologies of compression and signal shaping. The uniqueness of the wireless channel, as discussed in Chapter 3, derives in part from its unengineerability. The channel is subject to large natural and human-made variations, which cause large and rapid changes in all relevant signal parameters. Traditionally, in nonadaptive architectures, this variability has been handled by building in huge margins in the transmission end of the link budget, to make sure that even under the worst conditions a sufficient received signal level would be maintained. This means, however, that traditional systems are like furnaces without thermostats: They radiate enormous amounts of excess, unnecessary energy, which in the communications context translates into interference, which means lost capacity.

The eventual solution to reduction of interference—that is, lowering the amount of unnecessary energy that must be transmitted to accomplish the communications objective—will involve the ever more comprehensive application of the idea of adaptation, at all levels of transmission signal processing (source coding, channel coding, modulation, spatial directivity, time alignment, and other timing processes).

---

35. I have recounted this story at some length in an earlier book [20].

REFERENCES

[1]   Hamming, R. W., *Coding and Information Theory*, 2nd ed., Englewood Cliffs, NJ: Prentice-Hall, 1986.

[2]   Shannon, C. E., "A Mathematical Theory of Communication," 1948. Reprinted in Sloane, N. J., and A. D. Wyner, (Eds.), *Claude Elwood Shannon: Collected Papers*, Piscataway, NJ: IEEE Press, 1993, pp. 5–83.

[3]   Blahut, R. E., *Digital Transmission of Information*, Reading, MA: Addison-Wesley, 1990.

[4]   Taub, H., and D. L. Schilling, *Principles of Communications Systems*, 2nd ed. New York: McGraw-Hill, 1986.

[5]   Sklar, B., *Digital Communications*, Englewood Cliffs, NJ: Prentice-Hall, 1988.

[6]   Dixon, R. C., *Spread Spectrum Systems*, New York: Wiley, 1984.

[7]   Webb, W. T., and L. Hanzo, *Modern Quadrature Amplitude Modulation*, Piscataway, NJ: IEEE Press, 1994.

[8]   Shannon, C. E., "Communications in the Presence of Noise," 1948. Reprinted in Sloane, N. J., and A. D. Wyner, (Eds.), *Claude Elwood Shannon: Collected Papers*, Piscataway, NJ: IEEE Press, 1993, pp. 160–172.

[9]   Loeliger, H.-A., et al., "Decoding in Analog VLSI," *IEEE Communications Magazine*, Vol. 37, No. 4, April 1999, pp. 99–101.

[10] Shannon, C. E., "Recent Developments in Communication Theory," 1950. Reprinted in Sloane, N. J., and A. D. Wyner, (Eds.), *Claude Elwood Shannon: Collected Papers*, Piscataway, NJ: IEEE Press, 1993.

[11] Jayant, N. S., and P. Noll, *Digital Coding of Waveforms*, Englewood Cliffs, NJ: Prentice-Hall, 1984, pp. 190–193.

[12] Capon, J., "A Probabilistic Model for Run-Length Coding of Pictures," *IRE Trans. on Information Theory*, 1959, pp. 157–163.

[13] Sayood, K., *Introduction to Data Compression*, San Francisco, CA: Morgan Kaufmann, 1996

[14] Arps, R. B., and T. K. Truong, "Comparison of International Standards for Lossless Still Image Compression," *Proc. IEEE*, Vol. 82, June 1994, pp. 889–899.

[15] Brady, P. T., "A Technique for Investigating On–Off Patterns of Speech," *Bell System Technical J.*, January 1965, pp. 1–22.

[16] Brady, P. T., "A Statistical Analysis of On–Off Patterns in 16 Conversations," *Bell System Technical J.*, January 1968, pp. 73–91.

[17] Brady, P. T., "A Model for Generating On–Off Speech Patterns in Two-Way Conversations," *Bell System Technical J.*, September 1969.

[18]   *Telecommunications Transmission Engineering, Volume I—Principles,* 3rd ed., Bellcore Technical Publications, 1990, p. 398.

[19]   Goodman, D., et al., "Packet Reservation Multiple Access for Local Wireless Communications," *IEEE Trans. on Communications,* August 1989, pp. 885–890.

[20]   Calhoun, G., *Wireless Access and the Local Telephone Network,* Norwood, MA: Artech House, 1992.

[21]   Shannon, C. E., "A Mathematical Theory of Communication, Part 2," *Bell System Technical J.,* Vol. 27, October 1948, pp. 623–656. Reprinted in Slepian, D., (Ed.), *Key Papers in the Development of Information Theory,* Piscataway, NJ: IEEE Press, 1974, pp. 19–29.

[22]   Huffman, D., "A Method for the Construction of Minimum Redundancy Codes," *Proc. IRE,* September 1952, pp. 1098–1101.

[23]   Gray, R. M., *Source Coding Theory,* Boston, MA and Dordrecht, the Netherlands: Kluwer, 1990.

[24]   Jayant, N., J. Johnston, and R. Safranek, "Signal Compression Based on Models of Human Perception," *Proc. IEEE,* Vol. 81, No. 10, October 1983.

[25]   Shannon, C. E., "Communications in the Presence of Noise," *Proc. IRE,* Vol. 37, January 1949, pp. 10–21. Reprinted in Slepian, D., (Ed.), *Key Papers in the Development of Information Theory,* Piscataway, NJ: IEEE Press, 1974, pp. 30–41.

[26]   Calhoun, G., *Digital Cellular Radio,* Norwood, MA: Artech House, 1988.

[27]   Abut, H., (Ed.), *Vector Quantization,* Piscataway, NJ: IEEE Press, 1990.

[28]   Proakis, J. G., *Digital Communications,* 2nd ed., New York: McGraw-Hill, 1989.

[29]   Flanagan, J. L., *Speech Analysis, Synthesis and Perception,* 2nd ed., Berlin, Germany and New York: Springer-Verlag, 1972.

[30]   Nyquist, H., "Certain Factors Affecting Telegraph Speed," *Bell System Technical J.,* April 1924, pp. 324–339.

[31]   Gibson, J. D., *Principles of Digital and Analog Communications,* 2nd ed., New York: Macmillan, 1993.

[32]   Jamali, S. H., and T. Le-Ngoc, *Coded-Modulation Techniques for Fading Channels,* Boston, MA and Dordrecht, the Netherlands: Kluwer, 1994.

[33]   Pattan, B., *Robust Modulation Methods and Smart Antennas in Wireless Communications,* Upper Saddle River, NJ: Prentice-Hall, 2000.

[34]   Gronemeyer, S. A., and A. L. McBride, "MSK and Offset QPSK Modulation," *IEEE Trans. on Communications,* Vol. 24, August 1976, pp. 809–820.

[35]   Gupta, R., and D. J. Allstot, "Fully Monolithic CMOS RF Power Amplifiers: Recent Advances," *IEEE Communications Magazine,* April 1999.

[36] Katz, A., "SSPA Linearization," *Microwave Journal*, Vol. 42, No. 4, April 1999, pp. 22–44.

[37] Jeong, Y. C., "A Feedforward Power Amplifier with Loops to Reduce RX Band Noise and Intermodulation Distortion," *Microwave Journal*, Vol. 45, No. 1, January 2002, pp. 80–91.

[38] Godara, L., "Applications of Antenna Arrays to Mobile Communications, Part I: Performance Improvements, Feasibility, and System Considerations," *Proc. IEEE*, Vol. 85, No. 7, July 1997, pp. 1031–1060.

[39] Paulraj, A., and C. Papadias, "Space-Time Processing for Wireless Communications," *IEEE Signal Processing Magazine*, Vol. 14, No. 5, November 1997.

[40] Tonguz, O., and M. Wang, "Cellular CDMA Networks Impaired by Rayleigh Fading: System Performance with Power Control," *IEEE Trans. on Vehicular Technology*, Vol. 43, No. 3, August 1994, pp. 515–527.

[41] Jansen, M. G., and R. Prasad, "Capacity, Throughput, and Delay Analysis of a Cellular DS CDMA System with Imperfect Power Control and Imperfect Sectorization," *IEEE Trans. on Vehicular Technology*, Vol. 44, No. 1, February 1995, p. 72.

[42] Cameron, R., and B. Woerner, "Performance Analysis of CDMA with Imperfect Power Control," *IEEE Trans. on Communications*, Vol. 44, No. 7, July 1996, pp. 777–781.

[43] Delli Priscoli, F., and F. Sestini, "Effects of Imperfect Power Control and User Mobility on a CDMA Cellular Network," *IEEE J. on Selected Areas in Communications*, Vol. 14, No. 9, December 1996, pp. 1809–1817.

[44] Viterbi, A. M., and A. J. Viterbi, "Erlang Capacity of a Power Controlled CDMA System," *IEEE J. on Selected Areas in Communications*, Vol. 11, No. 8, August 1993, pp. 892–899.

[45] Austin, M., et al., "Service and System Enhancements for TDMA Digital Cellular Systems," *IEEE Personal Communications Magazine*, June 1999, pp. 20–33.

[46] Sollenberger, N. R., N. Seshadri, and R. Cox, "The Evolution of IS-136 TDMA for Third Generation Wireless Services," *IEEE Communications Magazine*, June 1999.

7

Signal Recovery Techniques (Receiver-Oriented Strategies)

Traditionally, communications engineers have focused on the architecture of the transmitter, and on the detailed construction of the transmitted signal—the coding, hardening, and shaping techniques discussed in the previous chapters. In the classical Shannon framework, both source coding and channel coding are seen as fundamentally transmitter-oriented processes in which the receiver merely unpacks and reassembles what the transmitter has previously encoded. Even where the use of channel state information was considered by Shannon, it was—strangely—seen as a resource solely for the transmitter.[1] The receiver is often assumed to be a fairly simple detection and regeneration device, incapable of adding anything to the signal that is not already there.

Recently, however, receiver architectures have gained new attention in their own right as a source of potential system performance improvements.[2] The theoretical basis of these newer receiver-oriented techniques

---

1. "Channels with feedback from the receiving to the transmitting point are a special case of the situation where there is *additional information available at the transmitter* [emphasis added]" [1, p. 289].

2. The predominance of transmitter-oriented technologies has not been absolute. Some of the key innovations in receiver architecture—such as the RAKE receiver—date back to the 1950s, and at least one of the techniques described in this chapter, equalization, has been used extensively since the 1960s.

rests on an important fact: *Unlike the transmitter, the receiver is in possession of the actual received signal.*

This truism may sound like a mere tautology, until we realize that the received signal is more than just a low-quality replica of the original message. It is actually a rich composite, comprised of some components derived from the transmitted signal but also other components embodying other, very different types of information. These extra elements include both *logical* and *physical* extensions or transformations of the original transmitted signal, which can be exploited to enhance the receiver's performance. The logical extensions include, for example, knowledge about the *reliability* of the receiver's detection decisions. Physical extensions can encompass much more: It turns out that the received signal bears the imprint of the entire physical world through which it has passed.

In short, *the received signal contains much more information than the transmitted signal.* In fact, the amount of information contained in the received signal is arguably *infinite.* How can this be so? It is because the received signal has again become an *analog* phenomenon—the carefully crafted digital signal elements sent forth by the transmitter have again mutated into analog entities. According to Shannon, any analog signal contains an infinite amount of information. To complete its task of detecting and decoding the blurry message, the receiver must once again requantize—redigitize—the analog received signal. As we saw in Chapter 5, quantization involves partitioning this analog entity into two portions: one that is kept (the message), and another that may be discarded (normally considered as noise or interference). Or so the process has traditionally been understood. This is the wheat-and-chaff model, in which the receiver's job is to winnow the useful, valuable information and separate out the useless or harmful chaff.

Here we reach the first crux: We must now begin to consider *the potential information value of this noise and interference.* What we have previously regarded as merely a kind of entropic accretion, or distortion of the true signal, will turn out to be a gold mine of useful information for the well-tuned receiver. This information includes the following:

1. *Quantization by-products:* To complete the detection process, the receiver eventually has to make a hard decision as to what it will register as a detected information symbol. However, we know that quantization of any sort involves the shucking off of a portion of the original signal, and this normally discarded portion of the analog signal contains a great deal of information. This soft information

that would otherwise be lost in the event of a hard detection decision can be saved and used to interpret the remaining data.[3]

2. *Patterns:* The receiver knows what it has seen until now. It has information, if it chooses to use it, about the contextual characteristics of each particular element in the data stream. It has information about the patterns in the data and the statistical interdependencies in the source, as well as information about the reliability of the quantization/detection decisions that it has made previously.

3. *Channel information:* This is the huge trove of information that is created by the physical channel. Potentially, it enables the receiver to model the physical world in which it operates and to use this knowledge to adjust its decisions. We will see in Section 7.2 just how powerful this component of the received signal can be.

4. *Foreign signals:* Information is also present that has been added by specific interferers—other structured signals that are blended with the desired signal. If the receiver is capable of detecting its own desired signal, it may also detect other signals and use this information to enhance its discrimination of the target.

The techniques surveyed in this chapter have to do with the ways in which this surplus information can be exploited by an active receiver. Many of these techniques have mirror-image twins on the transmitting side, some of which have been touched on in earlier chapters, but transmitter-oriented techniques always operate with imperfect information about the actual state of the channel and the real transformation of the signal by the physical environment, typically relying on estimates or delayed feedback from the receiver. Transmitter-generated estimates can never surpass the quality of the actual channel information embedded in the received signal and available, in principle, only to the receiver.[4] Nevertheless, in the historical

---

3. "Soft decisions inherently carry information that enable the assessment of their reliability" [2, p. 158]. This passing remark touches on a truth that runs rather deeper, it seems to me, than the normal superficial presentations of the advantages of soft decision decoding.

4. One recent study has concluded that for certain fading channels, which exhibit *independent and identically distributed* (i.i.d.) fading, the receiver can achieve almost the same capacity gains by processing the received signal as could be achieved with both

development of modern signal processing techniques, there has been a bias toward transmitter-side approaches until relatively recently. It is based perhaps on what we might call the pervasive hubris of digital design—that is, the assumption that designed-in information is somehow more pliable, even if it is by definition imperfect, than the rich but unruly information added by nature.

Following the framework used in the previous chapters, we will initially consider receiver-oriented strategies based on the physical signal dimension on which they primarily operate: space, time, or frequency. In some cases, the techniques operate in only one of these dimensions (e.g., time-domain equalization), whereas some of the interesting newer approaches involve joint optimization in more than one dimension (such as space–time processing).

We will also, however, examine these techniques from a somewhat different perspective based, roughly speaking, on whether they work by *subtraction* or by *addition*. Subtractive techniques focus on analyzing the signal into a desired component (presumptively the part contributed by the transmitted signal) and an undesired component (the distortion introduced by the channel). Once the undesired component has been isolated, so to speak, it can be subtracted from the received signal to recover a better copy of the original. On the other hand, an additive strategy aims at separating the various partial images of the transmitted signal (e.g., different multipath images of a short symbol pulse) in order to be able to combine them constructively, to add them together. Moreover, the signature of multiple images in one or more dimensions can be used to discriminate between different users in a multiuser environment.

Finally, we will look at what may be the ultimate solution (at least from the perspective of the receiver)—a realization of the "interference demon" described in Chapter 4—by means of a family of techniques that may be grouped under the heading of multiuser detection and interference cancellation (Section 7.6). These techniques purport to combine the logical and physical analyses, to detect and subtract the foreign information contained in the received signal.

---

transmitter and receiver cooperating on the same channel side information. That is, even if the transmitter actively attempts to predistort or otherwise adapt the signal to known channel conditions, the receiver can do just about as well acting entirely on its own. In cases of correlated fading, on the other hand, the transmitter–receiver cooperation can outperform the receiver-only processing [3].

7.1 Logical-Level Signal Recovery Strategies: The Active Receiver

An important trend in third-generation wireless technology is to build more intelligence into the receiver to enable it to perform a *logical recovery* of the signal. The receiver possesses, in principle, some knowledge of *actual or likely errors*, information about the *reliability* of its decisions, and other similar factors that are not available to the transmitter. It has an idea of which elements of the received signal are solid and which are doubtful or uncertain. Moreover, the receiver also generally knows something about the *model* of the signal—its statistical characteristics, which patterns to expect, and so on. This can enable it to guess at the true values of some uncertain elements of the received signal. A smart receiver can make use of this extra information to make better decisions than it could make if it relied strictly on the prepackaged error coding provided by the transmitter.

The underlying shift in philosophy involves questioning the assumption we make (usually implicitly) about the receiver's inability to respond independently to the occurrence of a decoding error. In the early days of signal processing, it was normally assumed that the receiver was a passive detection device. That is, the receiver was assumed to be capable of registering a physical event—such as making a measurement of signal amplitude or phase at a given sampling instant—and assigning that event a code value (a 0 or 1, for instance). With respect to the output stream of decoded values, the receiver was assumed to be helpless, or at least completely dependent on the embedded error correction provided by the transmitter. If an error occurred, the receiver could attempt to correct it using this redundant information (if available), but beyond that it was assumed that little could be done.

An Illustration: Image Coding

A recent review of image-coding techniques, which we shall follow at some length, illustrates how this perspective is changing [4].

Images are likely to become a significant part of the traffic flowing through third-generation wireless networks. A significant problem with image transmission over a mobile channel is by now familiar: Source coding techniques developed for engineerable channels (i.e., wire*line* channels), and which have become widely disseminated and standardized, are often grossly inadequate for wireless transmission:

> Typical image compression algorithms produce data streams that require a very reliable *and in fact perfect* [emphasis added] communication

channel—they are not designed for transmission in an environment in which data may be lost or delayed. [4, p. 120][5]

The normal solution for shaky source coding is aggressive signal hardening by adding layers of error correction coding (see Chapter 5). Still, *any* error correction code will on occasion fail, and with standard transmitter-oriented image compression techniques, the receiver is acutely sensitive to uncorrected errors: "Transmission errors can cause catastrophic decoder [i.e., receiver] errors.... Even a single bit error left uncorrected by the channel code can render the remainder of the bitstream useless [!]" [4, p. 122]. This is the paradigm of the passive receiver. The entire responsibility for ensuring the integrity of the signal falls on the transmitter. If somehow an uncorrected error still penetrates the decoder, the receiver is helpless.

The most efficient codes are the most vulnerable: for example, *entropy-coded* bit streams (see Section 6.3.1) produce *variable-length codewords* (VLCs), and the loss of the critical punctuation bit (analogous to a comma) at the end of the VLC—somewhat misleadingly referred to as *loss of synchronization*—can destroy the entire signal. What to do? Unfortunately, all the standard fixes come with a price tag. One technique is to replace variable-length codes with fixed-length block codes, which limits the effect of an error to a single block, while substantially increasing the signal overhead. Another strategy is to add periodic reset commands in the bit stream:

> The simplest technique to deal with errors in variable length code bitstreams is to employ resynchronization flags.... Such flags are called *restart markers* in JPEG [a popular image compression technique] or *synchronizing codewords*.... They can be inserted at user-defined intervals; a shorter interval improves robustness but decreases compression efficiency since the restart markers represent no image data. [4, pp. 122–123]

Transmitter-oriented signal design[6] always seems to run itself into the same corner, trading robustness for efficiency, and then vice versa, calling forth ever more ingenious solutions:

---

5. "The compressed video signal is extremely vulnerable against transmission errors, since low-bit-rate video coding schemes rely on interframe coding for high coding efficiency. Therefore, the loss of information in one frame has considerable impact on the quality of following frames. Since some residual transmission errors inevitably corrupt the video bit-stream, this vulnerability precludes the use of low bit-rate video coding schemes designed for error-free channels without special measures" [5, p. 1707].

More sophisticated techniques to provide robustness [include] *error-resilient entropy code (EREC)* ... in which the input signal is split into blocks that are coded as variable length blocks of data ... [with] negligible overhead. *Reversible variable-length codes* are uniquely decodable both forward and backward and are useful for both error location and maximizing the amount of decoded data.... *Resynchronizing variable-length codes* allow rapid resynchronization following bit or burst errors and are formed by designing a resynchronizing Huffman code and then including a restart marker at the expense of slight non-optimality of the resulting codes ... [4, p. 123]

The thread of transmitter-oriented thinking is perhaps clear enough from this example. Aggressive coding strategies have been the bread and butter of communications engineering for several decades now. Yet, in the same article, another very interesting strategy is hinted at, in the form of a hypothetical *code-free* transmission scheme.

A second hypothetical extreme exists in which ... the uncoded image [i.e., without *channel* coding] is simply transmitted, and the redundancy present in the image is used to compensate for lost data. In this case, raw data can be corrupted, but the uncoded image has sufficient redundancy to allow successful concealment of the errors. [4, p. 120]

What can this mean? If the reader has arrived here at Chapter 7 having traversed Chapters 1 through 6 in a linear sequence, then the paragraph just cited must come as something of a shock. It would have seemed that the benefits of error correction coding have been fully established. The arguments for signal hardening developed at length in Chapter 5 would appear convincing, even decisive. Generations of researchers, starting with Shannon himself, have embraced the idea that signal construction, and code design, are the way to cope with a noisy real-world channel. How is it possible, 50 years after Shannon's "philosophy of PCM" was first proclaimed, to argue, even hypothetically, *against* coding? And yet—"For many important classes of channels that arise in practice, Shannon's theorem does not apply and, in fact, performance *is* [emphasis in original] necessarily sacrificed using a digital approach" [6, p. 881].

---

6. I refer to transmitter-oriented strategies as "normative" in the sense that the classical Shannon papers focused on *coding*—by definition, a process implemented at the transmitter—as the "normal" solution for achieving control over the noisy channel.

Let us reconsider the problem of information transmission in a naturalistic context. In other words, if we return to natural information processing systems, such as human vision (as a point of reference for image-coding problems), we find that these systems do not operate for the most part on coded signals. The eye takes in data from a rich visual environment that is at once full of noise and yet offers no prepackaged error correction codes. The world does not transmit itself to us, as a signal. Somehow the vision system is capable of discriminating objects nonetheless, constructing an intelligible image from a noisy, complex, *uncoded* input signal [7].

How does the eye make sense of these data? One thing is clear: Vision is an active, *constructive* process. The eye has to put the data together, to compose its objects and events from raw physical signal elements. To do so, it relies first on strong statistical regularities inherent in the data.[7] These regularities provide the points of leverage for visual construction and encoding processes in the retina and the brain, which in total enable the vision system to synchronize itself, to detect objects, even in poor channel conditions (low light, fog, partial obstruction) and in the presence of noise (such as when a bird is discerned in the foliage of a tree or a fish in a river). The eye is an *active* receiver.

Other natural receivers function in a similar manner. Even language comprehension, which employs a coded signal, clearly involves a very active receiver process. The common experience of listening to a conversation in a noisy environment illustrates how much energy is expended by the receiver to discriminate and reconstruct the signal. In the classical Shannon transmitter-oriented model, the receiver should spend no more energy in a noisy channel than in a clear channel. The quality of the received signal may suffer, but the processing load on the receiver should be constant. Yet in natural communications systems, the receiver works harder in a noisy channel than in a clean one. That is why we like bright light for reading and a quiet table for conversation.

Communications theory begins to merge here with the broader study of pattern recognition and classification, in which the receiver's problems of the detection and discrimination are treated in a far more general manner, ramifying through diverse fields such as cognitive psychology, Bayesian statistics, and machine learning [8]. Figure 7.1 presents one fairly simple

---

7. "Most scenes ... display an enormous amount of similarity, at least in their statistical properties. By characterizing this regularity, investigators have gained important new insights about our visual environment—and about the human brain" [7, p. 238].

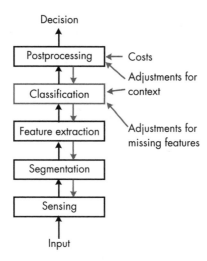

FIGURE 7.1 *Pattern recognition—an alternative receiver paradigm. (From: [8].* © 2001 *John Wiley & Sons Inc.)*

conceptual model of the pattern recognition process, with the following processing stages:

- *Sensing:* A sensor converts images or sounds or other physical inputs into signal data.
- *Segmentation:* A segmentor isolates sensed objects from the background or other objects.[8]
- *Feature extraction:* This process measures object properties that are useful for classification.
- *Classification:* This process assigns the sensed object to a category.
- *Postprocessing:* This process takes into account other considerations such as the effects of context and the cost of errors [8, p. 10].

A communications engineer will recognize familiar issues clothed in somewhat different terminology. What is striking is that the pattern recognition process is strongly receiver-centric. Although there is still a vague acceptance of the notion of preexisting objects in the world under

---

8. Duda et al. comment that "segmentation is one of the deepest problems in pattern recognition" [8, p. 10] and cite some very interesting examples, based on speech input data.

observation (which would correspond to the transmitter-designed signal elements), as a practical matter the study of pattern recognition and classification problems is philosophically committed to the primacy of active extraction of the patterns by the observer.[9]

The idea of an *active receiver*—that is, a receiver that does not depend strictly on the transmitter to prepackage the signal—has been gaining acceptance during the past decade, although second-generation systems do not yet make much use of active receiver technologies. Receiver-centric technologies are often referred to—bizarrely—as blind processes,[10] for example, blind equalization [10, 11], blind synchronization [12], blind carrier recovery [13], blind receiver beam forming [14], and so on, which typically involve cutting-edge signal processing techniques. The principle can also encompass rather simple data recovery procedures. Let us consider three postprocessing concepts that are being employed by active receivers at the logical level to improve performance without active assistance from the transmitter: (1) reliability analysis and erasure strategies, (2) error detection based on residual redundancy, and (3) error concealment.

7.1.1 Reliability Assessment and Erasure Strategies

Some of the simplest procedures involve allowing to receiver to monitor the reliability of its own decisions and to marshal the knowledge developed over time about the patterns of errors it sees:

> Knowledge that the channel errors are clustered and information about the location of bursts or deep fades on the channel should be passed on to the decoder. Indeed, monitoring the energy of each symbol enables

---

9. Duda et al. define *noise* (for example) as "any property of the sensed pattern which is not due to the true underlying model but instead to randomness in the world or the sensors" [8, p. 12]. This implies a classical Kantian-type metaphysic, but worn lightly.

10. The term *blind* is another piece of engineering poetry that seems less than explanatory in many contexts. For example: "The adjective 'blind' stresses the fact that 1) source signals are not observed and 2) no information is available about the mixture.... The so-called blindness should not be understood negatively" [9, p. 2009]. I would argue that in fact both criteria are incorrect, and so-called "blind" processes are in fact those which make a strong attempt to "see" what is going on in the channel, based on a close analysis of many dimensions of the received signal (beyond the decoded bits of message information per se).

us to distinguish between reliable and unreliable symbols, [and] unreliable symbols are said to be erased ... [15, p. 55][11]

Erasure is a well-established option in coding schemes, and has lost its novelty perhaps. It is worth reflecting on the fact that the creation of a third option for the receiver, effectively adding "neither" to the standard "either-or" of classical logic, is historic. It is arguably the first pragmatically driven departure from the "law of the excluded middle," which underlies most forms of symbolic logic, as well as traditional set theory and its derivatives. Beyond this, the idea of allowing the receiver to intentionally destroy a portion of the message that it cannot confidently decode should still shock us, I think. It is an important first step toward the design of an active receiver.

The strategy of erasure was initially suggested by development of coding schemes that allowed for the detection of more errors than could be corrected. A standard example (discussed in Chapter 5) is Reed–Solomon coding, which can detect that a given symbol is in error, but may not be able to supply the correct value [18]. It was found that in many cases performance could be improved by erasing these unreliable signal elements and replacing them with some sort of interpolated or dummy value, rather than leaving them in place, even though the correct value could not be determined.

This form of erasure is still strongly tied to the encoding technique used by the transmitter. The idea can be extended in ways that begin to be more independent of the transmitter and the original code structure. For example, in frequency-hopping systems where the message is broken up into a series of short packets, and each packet is transmitted on a different frequency out of a large set of available frequencies, a receiver can often determine that a particular hop may well have been exposed to interference from another transmitter. Indeed, in frequency-hopping systems, it is expected that some hops will be hit by jamming or interference from foreign transmitters. The receiver, if it decides that a particular hop has been hit, can erase the entire packet. Combined with a suitable signal structure, typically including *interleaving* of data payloads across multiple packets and appropriate error correction, this comprehensive erasure strategy can result in significantly improved performance [19].[12] A bad hop is erased, knowing that the signal can still be substantially reconstructed.

---

11. Among the ur-texts on the subject of reliability-based postprocessing techniques are those of Berlekamp [16] and Forney [17].

12. Pursley's work during the past decade has frequently focused on this application.

The trigger for the receiver to decide on erasure of a packet may be some measure of the rate of errors detected during the decoding process, which is similar to the elementary strategy of erasure based from certain code structures mentioned earlier and still somewhat transmitter dependent [19, p. 1224]. However, the receiver may also look at various global measures of the state of the channel, such as the overall power level, to determine whether an interferer may have been present on a given hop. Erasures based on such side information about the channel are truly receiver-centric.

The use of erasures has been developed into a foundation concept for one very common multiple access architecture: the *ALOHA channel* and its variants [20]. In this type of system, also called a *collision channel,* the transmitters access the channel whenever they need to transmit, in an uncoordinated manner, and it is often up to the receiver to decide when two transmitters have interfered with each other, erase the affected packets, and take one of several steps to reconstruct the signal.[13] In many such systems, the receiver is really the controlling entity in the communications process.

7.1.2 Residual Redundancy and Codeless Error Detection

Another simple postprocessing technique is *data interpolation.* A missing data point in a well-characterized series can often be reasonably estimated based on preceding and succeeding values. This may seem innocent enough, but how exactly do we interpolate? Even a seemingly simple step such as interpolation actually encompasses a subtle information theoretic assumption: that the received signal retains some redundancy, or patterning, even after the application of source coding compression techniques at the transmitter. This *residual redundancy* enables the receiver to detect errors independently of the error correction coding that is added by the transmitter. It is even possible to detect errors in signals without error correction coding. In one particular scheme designed for broadcast-quality speech transmission:

> [t]he receiver infers an error has occurred whenever an individual sample-to-sample difference at the output is greater than [an appropriate measure of the average difference]. Upon detecting an error the received sample is replaced by the output of a smoothing circuit. [22, pp. 838–839]

---

13. This subject has a vast literature and is beyond the scope of this volume. One short, recent article on the wireless collision channel, with good references, is [21].

Leaving aside the remedy (smoothing), which we shall address in the next section, what is noteworthy here is that the error has been detected not by using error correction coding bits supplied by the transmitter, but by analyzing the patterns in the received signal and flagging an event that departs from the expected sample-to-sample transition probabilities. Inherently, the receiver is assuming that the signal will not change too rapidly. If too abrupt a change occurs between adjacent samples, it is treated as an indicator that an error has occurred. This is an entirely receiver-centric process.

This phenomenon has been called residual redundancy because, in principle, the existence of such patterns or transition constraints shows that there is still redundant information in the signal, which has not been completely compressed by the source coder. Some have suggested that, in effect, the redundancy that remains in an incompletely compressed signal is functionally a kind of natural channel coding: "Methods based on residual redundancy ... use this redundancy as a form of implicit channel coding, to perhaps remove the need for explicit channel coding [23, p. 1738]."[14]

An advantage of this implicit coding is that it is free—it does not require extra bandwidth, or expand the transmitted signal:

> The benefit of error detection techniques at the video decoder that rely on the smoothness property of video signals is that they do not add any bits ... [24, p. 978]

> The absence of an explicit channel coder means that there is no overhead incurred due to channel coding; the structure being utilized by the decoder is due to the residual redundancy in the source coder output. [22, p. 841]

One might argue that there is still a bandwidth penalty, whether we leave extra redundancy in the output of the source coder, and skimp on explicit channel coding, or aggressively compress the source and then add back in error correction overhead, but this is not really true. The bandwidth cost of adding error correction (explicit channel coding) is easy to calculate. The bandwidth cost of retaining an incompletely compressed signal is less definite. Arguably, there is *always* some residual redundancy, at least for any analog source. The source itself will always have certain inherent regularities, and by definition no quantization or source coding compression technique can ever be thorough enough to completely eliminate them:

---

14. See also [6].

> We start from the premise that source coding does not remove all the redundancy from the data stream.... This in turn implies that there will be certain codeword-to-codeword transitions which are more likely than others....
>
> The source coder is now acting as a joint source/channel coder. The "channel addition" is not because of any redesign of the source coder. Rather it is simply a recognition of the fact that the output of the source coder contains redundancy which can be used to perform forward error correction [!?]. [22, p. 841]

The confusion here should be noted. The use of residual redundancy is most certainly *not* forward error correction. It is a completely receiver-centric process, which operates on the residual left over after the transmitter has done its best. The residual is perhaps better understood as—well, as a *residue*—the leftover *analog-ness* after the digital coding process has been carried out. No analog signal can ever be completely or perfectly compressed. It can only be compressed up to a point—that point being set by the quality criterion that governs the quantization process. It suggests that there is an intriguing complexity underlying the all-too-simple characterization of information signals as *either* digital or analog. It would appear that any digital signal that began life as an analog source is never completely relieved of its underlying analog character, including its inherent redundancy and its contextuality.

7.1.3 Error Concealment

Interpolation can be extended and subsumed by a more general strategy: *error concealment*. Just as residual redundancy is analogous to forward error *detection* based on explicit channel coding, error concealment is analogous to forward error *correction*.

The principle is again a simple one: If the receiver knows that a particular data element is corrupted, that element can be thrown out and replaced by a new value that is derived in some way from the receiver's knowledge about the statistical regularities in the signal it has been receiving. The most important a priori assumption that the receiver can normally make is that the typical rate of change of parameters in the original signal is slow. The source signals are *smooth*:

> It is well known that images of natural scenes have predominantly low-frequency components, i.e., the color values of spatially and temporally

adjacent pixels vary smoothly, except in regions with sharp edges. In addition, the human eye can tolerate more distortion to the high frequency components [the edges] than to the low frequency components. [24, p. 985][15]

This smoothness has already been noted in the discussion of receiver-centric techniques for error detection (Section 7.1.2). It also lends itself readily to error concealment techniques based on interpolation. These techniques can become rather sophisticated. Interpolation—or indeed, as it is often called, *smoothing*—can be applied in the time domain of the image (where an error is covered up by comparing previous and succeeding images), or in the spatial domain (where the error is concealed by a value generated by comparing adjacent pixels or regions of the image), or by using two-dimensional techniques involving spatial/temporal interpolation or spatial/frequency interpolation [24, pp. 985ff].

The notion of creating signal elements at the receiver—*ex nihilo*, as it were—is a little unsettling to some researchers, who may vaguely feel that they are somehow pulling a fast one.

Successful concealment of errors [uses] the received data at the decoder, which is now perhaps more appropriately called a *reconstructor* [emphasis in original]. The reconstructed image will not be pixel-for-pixel equivalent to the original, but visually equivalent ... [4, p. 120]

Visually equivalent—that is, just as good. This sounds suspicious, certainly, and yet interpolation [25, p. 1711], extrapolation [25, p. 1711], smoothing [26], shaping [27],[16] concealment [28, 29], substitution [30],[17] and masking [31, 32] techniques are quietly becoming a standard part of the repertoire for the active receiver. Progress toward a theoretical underpinning for these techniques has been confined for the most part to terminological innovations, which nevertheless point suggestively *away* from classical Shannon theory:

---

15. Analogous statements can be made about audio signals, such as speech.

16. "In the perceptual audio codec, the quantization noise is usually shaped according to the perceptual model" [27, p. 467].

17. "Once the error detector has identified a frame as unreliable, simple waveform estimation techniques like silence substitution ... may be employed in place of the unreliable frame of speech" [30, p. 211].

> *Forward error concealment* [emphasis in original] refers to those tech-
> niques in which the encoder plays a primary role.... On the other hand,
> *error concealment by postprocessing* [emphasis in original] includes
> techniques in which the decoder fulfills the task of error concealment
> ... without relying on additional information from the encoder. [24, pp.
> 976–977]

The reference to perceptual models—visually equivalent rather than
mathematically equivalent—underscores another aspect of the transition
to a post-Shannon framework. In effect, the receiver can be said to include
a semantic model—a model of the intelligent user at the output end of the
entire communications chain. The receiver is capable of manipulating the
received signal based on its knowledge of this end-use model. As we bring
this idea into focus, the apparently simple notions that gird up classical
Shannon theory—for example, *redundancy*—start to refract a deeper complexity [22].

A full critique of coding per se will not be attempted here. But the surprising robustness of the uncoded signal is another conundrum for students
of the Shannon paradigm. The classical view (presented in Chapter 6) is that
redundancy in the source must be thoroughly wrung out for efficient communications and capacity maximization:

> In traditional source-coder design, the goal is to eliminate both the sta-
> tistical and [perceptual] redundancy of the source signal as much as
> possible to achieve the best compression gain. [24, p. 982]

The idea that this might not be quite true can sometimes provoke a
puzzled optimism, and suggestions in the subjunctive mood:

> It could be advantageous in mobile channels to leave part of the redun-
> dancy in the source and use it at the channel decoder for better decod-
> ing. From the channel decoding literature, it is well known that the use
> of soft decisions gains 2–3 dB in transmission radio power. [33, p. 1765]

Or again:

> One approach to solve this problem is by intentionally keeping some
> redundancy in the source-coding stage such that better error conceal-
> ment can be performed at the decoder when transmission errors occur.
> [24, p. 982]

Further questions arise at this point. Which part of the redundancy do we leave in, and which do we extract? How is the soft decision concept, and the use of *context*, generally speaking, related to redundancy in the source? Indeed, why not take the next obvious step? The idea of reinserting redundancy into the source (prior to and apart from any classical channel coding) has begun to ferment:

> Signal reconstruction and error-concealment techniques ... strive to obtain a close approximation of the original signal.... In general, to help error detection and concealment at the decoder, a certain amount of redundancy needs to be added.... We refer to such added redundancy as *concealment redundancy* ... [24, p. 976]

There is a sense here that our bag of engineering tricks is currently larger than our theoretical framework. Nevertheless, this catalog of third-generation techniques must now address the wide variety of receiver-oriented strategies that are finding their way into the 3G wireless cookbooks. Some of the most interesting new ideas are based on making use of the information created by the physical channel itself.

7.2 The Transfer Function: Modeling the Channel

The idea of interpolating—or more broadly, *interpreting*—the received signal to try to fill in the gaps and creatively reconstruct a damaged message, is not hard to grasp. As human receivers, we all engage in signal reconstruction fairly frequently: deciphering a garbled message on an answering machine or parsing through a poor-quality photocopy. What *is* much more surprising perhaps is that we can also interpret the received signal to learn something about the external physical environment. In fact, the received signal contains an *enormous* amount of information about the physical characteristics of the channel through which it has just passed. *None* of this channel information is directly available to the transmitter[18]; it is created by the physical process of passing through the channel. The information added to the signal is so rich and complex that in many cases it constitutes a signature or fingerprint that can be used to uniquely identify a particular signal, and to

---

18. In some cases, the transmitter can make its own estimate of the channel conditions, based on feedback from the receiver, for example, but these indirect measures are inherently quite constrained.

discriminate that signal from other interfering signals. More than that, it can allow signals to be packed more densely in the physical communications channel. Indeed, this channel information is so important that it arguably enables the communications process to redefine the barriers on system capacity imposed by classical Shannon theory (as we shall see in Chapter 8).

This is not the normal view of the channel's role in the communications process. The idea that channel-induced distortion could somehow be a good thing cuts against the grain of a century of communications engineering theory and practice.

For example, in Chapter 5 we discussed the value of various diversity techniques that can break up structured fading patterns introduced by the channel. By removing this channel memory,[19] and rendering the fading more noise-like, more random, it was argued that the channel coding and error management techniques can become more effective. The signal is hardened, and capacity is increased. One very powerful structure-removing technique is the time-domain process called *interleaving*, whereby data samples drawn over an extended period of time are buffered prior to transmission and then shuffled—interleaved—so that the effects of a burst of errors caused by some form of structured interference (especially multipath-generated self-interference or fading) are spread out through the signal more evenly, in a more noise-like manner (Section 5.4.2). By adding time diversity in this way, the signal is strengthened and system performance is improved. The principal trade-off is delay, but up to a point it seems a good trade-off to make. This logic seems well-founded, and interleaving is widely used.

What then are we to make of the following remarks?

> One of the key technical problem areas [in wireless communications] ... is the effect of fading on network performance.... Many protocols, coding schemes ... and techniques were developed to eliminate channel memory (e.g., interleaving). An alternative approach may be *to take advantage of channel memory* ... [emphasis added].
>
> In fact, turning a channel with memory into a memoryless one by interleaving is not necessarily an efficient way of using it, since the interleaving operation may substantially *reduce* [emphasis added] the channel capacity. [34, p. 1468]

---

19. Channels are said to have "memory" when there are correlations between different signal elements passing through them. A memoryless channel is one in which the error probabilities (for example) of each received symbol are independent of the preceding and following symbols.

Or again, more simply: "Channel capacity increases if the channel memory is exploited" [35, p. 54].

We will see below how the *standard* view of the communications process, adopted by Shannon and others almost without exception until quite recently, is becoming obsolete. That view holds that there is a human-made object (called the signal) that is exposed to natural forces (the channel, noise), and thus degraded (errors introduced). The signal begins in a pristine state, containing a perfect version of the message it has been designed to carry.[20] It can only deteriorate from then on. The challenge for the communications engineer is how to mitigate this deterioration, and the good tidings of Claude Shannon are that it is possible to control this deterioration within a known limit as long as we do not try to transmit too rapidly.[21]

The *new* view, which is only just beginning to coalesce, is that the channel *enriches the signal*, by adding new information over and above the information that was preinstalled by the transmitter. We can learn things from the received signal that the transmitter does not know and could not logically have included in the message. The channel also stamps each transmission with a unique code of its own—over and above the code that we build into the transmitted signal. We can use the channel's new code to distinguish between different users who may not be otherwise distinguishable. Moreover, these new bounties of information are *free*, in terms of information theory.[22] They do not cost bandwidth, or use capacity. Properly exploited, they can *increase* capacity.

---

20. We ignore for the moment—as the first slippery step—the issue of the information destroyed in order to create the signal in the first place, through quantization.

21. A putatively more sophisticated interpretation of the classical model of the communications process is the "hypothesis testing" construct. As expressed somewhat casually by Verdú: "A certain observed random quantity [i.e., the received signal] has a distribution known to belong to a finite set. Each of the possible distributions constitutes a different hypothesis. We must make a guess as which is the 'true' distribution (hypothesis) on the basis of the value taken by the observed quantity. Data demodulation is a hypothesis-testing problem: the observed quantity is a noise-corrupted version of the transmitted signal, and there are as many hypotheses as different values for the transmitted data" [36, p. 85]. This widely disseminated metaphor has a cuteness to it that misleads us, I think, into accepting the inherent (unstated) premise—namely, that the transmitted signal contains all the true "information" and the other components of the received signal are merely in the way of the detection process.

22. They may have an economic cost, in the form of a more intensive processing requirement at the receiver.

Lest this seem all too exuberant, consider this example: In the late 1990s, the FCC began a regulatory process designed eventually to mandate an enhanced 911-type service for users of wireless phones. The enhanced requirement called for the wireless system to be able to determine with reasonable accuracy the *location* of a caller placing a 911 call for emergency assistance. This requirement spurred a number of competing technical proposals. For example, one straightforward way to attack this problem is to install a GPS receiver[23] in the mobile handset, which can capture the location coordinates and then transmit them to the base station as part of the call setup process. This approach would have its limitations; it would not work for receivers inside buildings or in areas where the GPS signals might be blocked for some reason.

A variety of other solutions have been developed, including some based on classical Pythagorean triangulation—reception of the signal at two different base stations, which can then compare *time-of-arrival* (TOA) and *angle-of-arrival* (AOA) information to come up with a fix on the position of the mobile. These solutions also had their shortcomings, however; for one thing, they require new network infrastructure to collect and centralize the TOA and AOA data, and they might tend to underperform in severe multipath environments as the delay spreads and angular spreads tend to increase, reducing the resolution of the system.

Another approach turns multipath to advantage by creating multipath signatures that can be used to locate the transmitting mobile without multiple receivers (as in triangulation-based techniques) and without using transmission bandwidth (as in GPS-based techniques). The Radiocamera technology[24] uses the multipath signature as a unique fingerprint for each geographical location in a given coverage area (Figure 7.2). The mobile signal received at the base station carries with it this fingerprint, which is then matched with the corresponding pattern stored in a central database, and the mobile's location is determined. Surprisingly, the multipath fingerprint contains enough information to map a typical large cell (which may cover hundreds of square miles) down to a hundred yards or less. If we assume that the system can achieve a resolution of approximately 100 yards over a 100-square-mile area (i.e., a cell with a radius of about 5.5 miles), this implies that the channel is adding approximately 15 bits of efficiently

---

23. GPS, of course, stands for Global Positioning System, a network of satellites circling the globe that provide a continuous signal that ground-based receivers can use to locate themselves in terms of latitude–longitude coordinates.

24. Developed originally by U.S. Wireless Corp. of San Ramon, California.

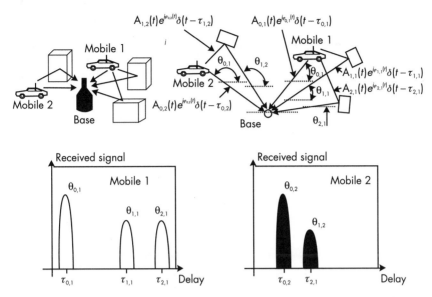

FIGURE 7.2 *Radio location by means of multipath signatures. (From: [37]. © 1998 IEEE. Reprinted with permission.)*

encoded location information for every message information bit that is transmitted![25]

The Radiocamera is a clever solution to a very specialized problem. The channel information is arguably much more robust, and could provide a great deal of information beyond just the transmitter's physical location. The crux of this illustrative example is to show how, by designing the base station receiver to decode this channel information (instead of merely discarding it), the receiver can acquire information that the transmitter does not have and could not logically have transmitted. The ratio of message information to

---

25. One hundred square miles mapped onto a grid of 100 yards of resolution generates nearly 31,000 unique locations, which would require 215 bits to encode uniquely. Of course the rate of change of the location is quite slow compared to the message bits, although it is not inherently limited. I should note that the anonymous reviewer of this manuscript raised a subtle question: How many message bits of the signal would be required by the receiver to be able to resolve the "extra" bits associated with the location information? In theoretical terms, I believe that a single chip should be enough, but in practical terms there is certainly an averaging requirement, which would mitigate to some extent the quantitative disproportionality of the message-information-to-channel-information ratio.

channel information gives us some idea of the enormous accretion of information—the information expansion of the signal—that takes place as the
signal passes through the channel.

To recapitulate, the communications process is not just about wrap-
ping the signal in protective armor and trying to fight through the noise and
interference in the physical channel, with the hope that enough of the message will arrive intact to allow the receiver to make sense of it. The emerging
view—which I think still lacks a coherent high-level formulation—is of a
process in which the transmitted message and the physical channel are
allowed to interact, to produce a new composite entity (the received signal)
that possesses all of the rich dimensionality of the analog physical world.
The receiver's task is to analyze this received signal, to reconstruct the transmitted message using all of the richness of the channel information as an
important resource. The *joint* task of the transmitter and the receiver is to
establish a shared logical framework to *control* the results of the receiver's
message-reconstruction process.

This control is ultimately indirect control, similar to the concept of *statistical* control in a large-scale clinical experiment, in which many potentially
confounding variables that cannot be directly manipulated are nevertheless
distinguished and controlled for through a strong experimental design. The
message is not really received, any more than we would say that the data in a
clinical drug trial are received. It is better to say that it is *constructed*, which
underscores the active, creative aspect of the receiver processes. Another
way to apply this analogy would be to view the received signal as a *dependent
variable*, and the transmitted signal, the physical channel, and perhaps other
interfering signals as *independent variables*.

The communications system mediates the relationship between the
transmitter and the receiver by means of a *model*—a model that encompasses two submodels: (1) a submodel of the transmitted signal—really, a
source model—and (2) a submodel of the physical channel. The stronger
and more detailed these models, the more information can be extracted from
the dependent variable regarding the relative contributions of different independent variables.

In the Section 7.1, we touched on the importance of having a source
model available to the receiver to allow it to actively interpret and amplify the
data it gleans from the received signal. The model of the channel—sometimes called the *channel transfer function* [38, p. 458][26]—contains, in

---

26. A closely related concept is the impulse response of the channel, which is the way in
    which the channel characteristically distorts a single transmitted pulse.

principle, the formula for reconstructing a detailed physical picture of the transmission environment. As we have noted, in some wireless systems (e.g., radar systems) the design objective is not to communicate a message, but to generate information about the environment. In such systems, the source model is very simple and the transfer function is paramount [39]. In a communications system, on the other hand, our objective is not per se to visualize the physical environment, but to accurately reconstruct the original signal. By analyzing the channel transfer function, however, we can, in principle, determine exactly what happened to the signal between the time that it left the transmitter and the time it arrived at the receiver. We can trace where it went and how much it was distorted by interacting with objects in the transmission path. We can use this information to powerfully assist the receiver in signal reconstruction, and even (as we have seen) to create entirely new information that was not part of the transmitted message at all.

7.2.1 Transmitter-Assisted Acquisition of the Transfer Function

How do we acquire knowledge of the transfer function? One way is to transmit a specific probe, designed expressly to learn the channel characteristics. Training sequences or pilot signals are built into the transmitted signal, usually at the beginning of a communications session. These are known data sequences that are sent through the channel, and they carry no user information. They are specifically designed to absorb information from the channel and deliver it to the receiver. The receiver examines the received sequence, which it knows corresponds to the training sequence. From this distorted image of a known sequence, the receiver is able to determine the distortion function that has been added by the channel.

7.2.2 Blind Acquisition of the Transfer Function

However, the transfer function is actually available for "free" in information theoretic terms. The receiver should be able to infer the channel transfer function from the received signal alone, without any overt help from the transmitter. Training sequences and pilot signals add overhead that take away from usable channel capacity. Increasingly, therefore, the research emphasis in the wireless field has focused on so-called "blind" receiver processes, which are designed to extract the transfer function and to diagnose and respond to the presence of interference without any overt help from the transmitter.

How can a blind receiver see? How does the receiver extract the channel transfer function on its own, without the assistance of the transmitter? The answer is twofold.

7.2.3 The Source Model as a Basis for Blind Acquisition

First, as we noted above, the receiver may possess a model of the transmitting source. If so, it already knows something about what the patterns and statistics of the source component of the received signal ought to be. In detecting and decoding a single symbol, the contributions of the source and the channel may be indeterminate. However, over time (called the learning period or the acquisition period), the receiver can use its a priori knowledge of what it *should* be seeing in order to help segregate the transmitter's contributions to the received signal from the channel's contributions (i.e., the transfer function). In other words, the source and the channel have *different* statistical characteristics, and if the source statistics are known, they can be used to isolate the effects of the channel.

This is similar to the process of deciphering an unknown secret code. The original message is the source; the enciphering system is like the channel transfer function. The coded message is a combination of both. The intended recipient of course has a decoder device or code table, and can quickly recover the original message. For her there is no ambiguity. But what about an eavesdropper or anyone else who does not know the ciphering system? History has shown that even strong codes can be broken. The lever for code cracking is almost always a knowledge of the source model. If we know that the coded message is originally an English text, we can use the known frequency distribution of English alphabetical symbols to help figure out the cipher.[27]

Some researchers would claim that this is the entire story: "Blind channel identification is based solely on the measurable channel output signal and some a priori knowledge of the statistics of the channel input signal" [42]. This is incorrect, however. There is another, far more powerful tool for cracking the natural codes that the channel creates and embeds in any wireless signal: our a priori knowledge of the physical model of that channel.

---

27. Sherlock Holmes knew this and used it in the "Case of the Dancing Men." Claude Shannon famously explored the statistics of English text sources in [40] as well as in the first part of [41].

7.2.4 A Priori Knowledge of the Physical Channel as a Basis for Blind Acquisition

The effects of the physical channel are various and complex—as complex as the physical universe itself, for that is what the wireless channel is. But if we adopt a high-level perspective suggested by our simple model of the communications signal as an electromagnetic entity in three dimensions—space, time, and frequency—we can simplify our analysis of the transfer function to focus on three basic effects, actually three manifestations of a single phenomenon: the *spreading* or *expansion* of the signal as it physically traverses the channel. This expansion in turn derives essentially from a single physical mechanism: *multipath propagation*.

By now, the multipath phenomenon should seem familiar. The physical transmission process in a wireless channel involves not one but many different paths between the transmitter and the receiver. The classical view of the multipath channel is the view presented in Chapter 3: Multipath reception involves the *noncoherent* mixing of multiple images of the transmitted signal, which creates what is in effect a form of self-interference. That is, the different multipath components of the received signal degrade one another. The principal manifestation of this self-interference, for a receiver that is in motion, is a severe and rapid fluctuation of the amplitude of the received signal—*multipath fading*.[28] This fast fading tears holes in the transmitted signal, and creates bursty error patterns in the bit stream decoded by the receiver. It also disrupts many link-level signal processing functions (such as convolutional error coding or power control), effectively breaking the coordination between transmitter and receiver. This view of multipath fading as an inherent threat to the integrity of the signal and even the communications process itself motivates a whole range of countermeasures designed to mitigate the effects of multipath fading, which is often seen as the overarching problem for mobile communications in general.[29]

7.2.5 "Multipath Is Your Friend"

The emerging view of the multipath channel is quite different.[30] Physically, of course, the description is the same: The transmitted signal arrives at the

---

28. There are phase and frequency effects as well.

29. "Rayleigh fading ... can in some respects be regarded as the most serious disturbance in a wireless communications system. Therefore, trying to find methods of improving system performance on this channel has been a major research area over the last 40 years" [43].

receiver over many paths, of different length, from different directions, depending on the positions of scatterers or reflectors relative to the receiver. The critical difference is that we no longer need assume that the different multipath components are incoherently combined and irretrievably blended into a single entity. We no longer need assume that the receiver cannot discriminate between the *different* multipath components. In fact, the modern understanding of the channel transfer function really begins with the exploration of the possibilities for distinguishing these signal components and either selecting among them, or combining them *coherently*.

Now, from this new standpoint, the relevant effect of multipath is the *expansion* of the signal as it passes through the channel. The signal expands physically, because of the multipath geometry, as a direct result of the multiplication of transmission paths between the transmitter and the receiver. Because it expands physically, we might say as well that it expands logically. It becomes more robust (!) and can carry more information (!!).

The expansion of the received signal, as noted above, is manifested in three ways [46] as discussed next.

Space-Domain Expansion

First, the signal expands in the spatial dimension. A relevant measure of this spatial expansion is the *angle of arrival* of the signal. In a hypothetical free-space channel without any reflectors, there would be only one path between the transmitter and the receiver, and the perceived angle of arrival of the received signal would be almost point-like (Figure 7.3). But in a real wireless channel there are many reflectors, and the signal images arrive from many directions [47]. In fact, the normal assumption for a mobile unit is that the receiver is *surrounded* by reflectors, and the spatial spreading is often held to be, nominally, a full 360° [37]. In other words, the received signal appears to be coming from many different directions, scattered at all points of the compass.[31] The channel geometry is not always symmetrical, however. The spreading seen by the cellular base station receiver is normally much narrower because the receiving antenna is elevated above most reflectors.[32]

---

30. One of the first articles to represent this new perspective is Turin's [44]: "Multipath reception is one form of diversity reception, in which information flows from transmitter to receiver via the natural diversity of multiple paths rather than the planned diversity of multiple frequency channels, multiple antennas, multiple time slots, etc. Thus instead of regarding the multipath phenomenon as a nuisance disturbance whose effects are to be suppressed, it should be regarded as an opportunity to improve system performance" [44, p. 332]. For a more recent declaration of the new model, see [45].

Top view

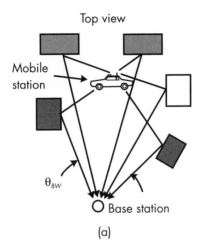

(a)

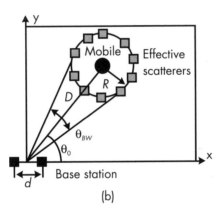

(b)

FIGURE 7.3 *Multipath-induced spatial spreading (AOA) for the mobile and the base station: (a) macrocell—base station perspective, and (b) Lee's model. (From: [37]. © 1998 IEEE. Reprinted with permission.)*

31. Indeed, the spatial dimension is three-dimensional; azimuthal variations are complemented by up-tilted and down-tilted signal components. Indoor signals may literally show full three-dimensional spreading in spatial terms.

32. "It is usually assumed that the scatterers surrounding the mobile station are about the same height as or higher than the mobile. This implies that the received signal at the mobile antenna arrives from all directions after bouncing from the surrounding scatterers.... However, the AOA [angle of arrival] of the received signal at the base station is quite different. In a macrocell environment, typically, the base station is deployed higher than the surrounding scatterers. Hence, the received signals at the base station result from the scattering process in the vicinity of the mobile station.... The multipath components are restricted to a smaller angular region" [37, pp. 11–12].

Time-Domain Expansion

Second, the signal expands in the temporal dimension. The direct path is the shortest path. Reflected paths vary in length considerably, and the *time of arrival* for different components of the received signal in a typical cellular system can be delayed by several microseconds up to tens of microseconds[33] (Figure 7.4).

Frequency-Domain Expansion

Third, if the receiver is in motion, it will be moving *toward* some of the signal components and *away from* others. Each path also has a different Doppler delay factor. Some paths are effectively blue shifted (i.e., the relative motion of the receiver is toward the direction from which that path is arriving), others are red shifted. This creates a Doppler spread—a spreading in the frequency domain[34] (Figure 7.5).

Channel Signatures

The classical view of the multipath channel assumed that these signal components all blur into one another, which is a fair approximation of what happens when the signal elements (transmitted symbols) are of long duration. The individual multipath components are assumed to be unresolvable, and each pulse or information-bearing symbol is subject to spreading that can only really be characterized statistically, based on the distribution of the components. In the classical view, the envelope of this statistical distribution becomes an important measure of the amount of distortion or degradation of the signal. In the time domain, this is referred to as the *delay spread*. In the frequency domain, we speak of the *Doppler spread* [50–54]. In the space domain, the variable is sometimes referred to as the *angular spread*[35] (Figure 7.6).

---

33. Clark et al. [48] cites figure of 4 µs for a typical urban delay spread, and 8 µs for "a typical hilly area." In unusual rural or mountainous regions, reported delay spreads are much larger.

34. Doppler spreads of 100–200 Hz are cited in the literature [e.g., 49].

35. Keep in mind that these are not really three different phenomena, but three views of the same thing. The focus of a given researcher is often determined by which of the three signal dimensions has been chosen for aggressive optimization. If we are trying to control the time domain of the signal (e.g., if we are studying synchronization or intersymbol interference, which are time-domain issues), then a frequency-domain effect will be seen in this light as a contributor to timing difficulties. This can lead to some misleading terminological distinctions. For example: "Acquisition is the pro-

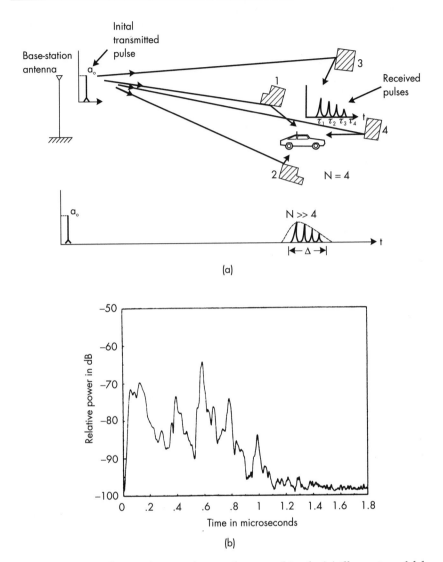

FIGURE 7.4   *Spreading in the time-domain due to multipath: (a) illustration of delay spread (From: [55]. © 1982 McGraw-Hill Inc.); and (b) a measured profile of average receiver power versus time delay at 850 MHz (From: [56]. © 1987 IEEE. Reprinted with permission.).*

Any two signals, at least if they originate in the same system, will have roughly similar statistics regarding delay, Doppler, and angular spread. That is the virtue of statistics—to average out the fine detail. However, if we

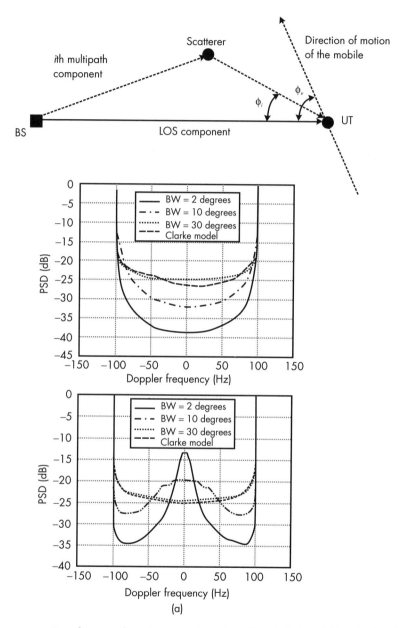

FIGURE 7.5 *Doppler spreading due to multipath, with red shift and blue shift: (a) a multipath component arriving at the mobile from the base station and Doppler spectra when using a directional antenna at the base station is compared with the Clarke model (From: [50]. © 2002 IEEE. Reprinted with permission.); and (b) averaged measured Doppler power spectrum (From: [51]. © 2002 IEEE. Reprinted with permission.).*

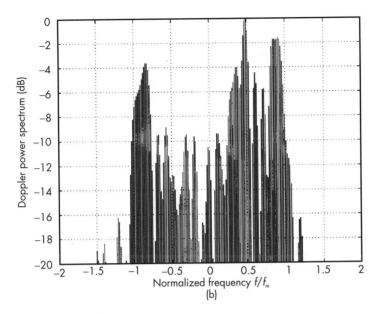

FIGURE 7.5 *Continued.*

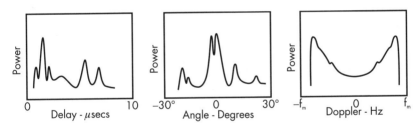

FIGURE 7.6 *Delay spread, Doppler spread, and angular spread as envelopes of statistical distributions. (From: [47]. © 1997 IEEE. Reprinted with permission.)*

cess by which a spread-spectrum receiver estimates the time delay of the received signal.... [Note the presumptive emphasis on the time domain.] However, [it is difficult to] directly model serial search systems that acquire Doppler [i.e., that acquire an accurate *frequency* reference] in addition to timing. Specifically, it is difficult to apply the closed form techniques when [the Doppler shift] is sufficiently large that [it] affects the timing. This situation is also known as *code Doppler*" [57, p. 2870]. In short, "code Doppler," which might sound as though it is something distinct from the basic spreading phenomenon, merely refers to a situation in which the Doppler spreading is large enough to create code timing offsets or overlaps.

assume that the transmitted signal consists of a series of discrete, extremely *short* pulses, then the expansion of each pulse in the time domain, for example, will manifest itself not as a blur of blended multipath components that can only be described as a statistical distribution, but as a series of discrete images of the transmitted pulse. In the time domain, for example, we should see a series of pulses delayed by short intervals (Figure 7.7). The pulse ramifies into a series of separate pulses spread out in time, each one an echo of the original pulse. The precise pattern of these time-staggered echoes depends on the location of reflectors in the channel, or, to put it another way, the pattern of echoes contains new information about the geometry of the channel. Because each receiver is located in a different physical position, no two receivers should see the exact same pattern. The pattern, in other words, is a *signature* (Figure 7.8).

The same thing happens in the space dimension. A single short pulse from the transmitter arrives at the receiver as a series of images with different angles of arrival. Each unit has its own spatial signature (Figure 7.9).

Finally, the same is true for the frequency dimension. Doppler mapping of individual pulses will show a Doppler signature for any given transmitter–receiver pair (Figure 7.9).

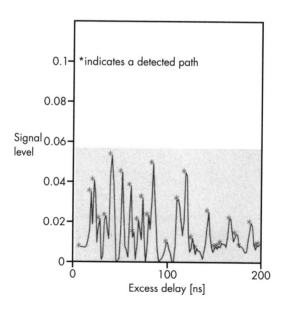

FIGURE 7.7 *A pulse broken into echoes in the time domain. (From: [58]. © 1991 IEEE. Reprinted with permission.)*

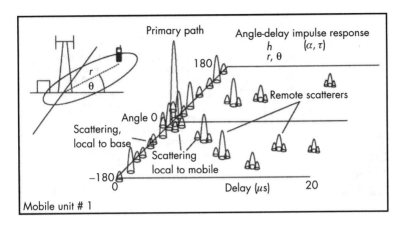

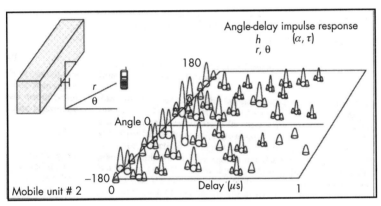

FIGURE 7.8 *Time-domain signature from two different mobiles. (From: [59]. © 1988 IEEE. Reprinted with permission.)*

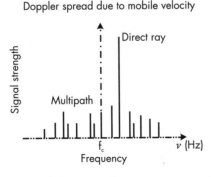

FIGURE 7.9 *Space-domain and frequency-domain signatures. (From: [60]. © 2001 IEEE. Reprinted with permission.)*

Two-dimensional signatures provide even better discrimination. For example, we can map the received signal in both space (angle of arrival) and time (delay), to produce a joint *space–time signature* (Figure 7.10) [61, 62]. It is claimed that "spatial and temporal (i.e., two-dimensional) signal processing ... will become a breakthrough technique for the third generation of wireless personal communications" [63, p. 28].

Another important two-dimensional signature is the *time–frequency signature,* in which multipath delay and Doppler shifts are mapped against each other[36] (Figure 7.11). T-F signatures are especially important for receivers operating in acoustic channels (e.g., for SONAR systems) [66]. It has been argued that human acoustic signal processing is based fundamentally on T-F signatures:

> There is a common belief that humans recognize signals (especially sounds) as objects in the time–frequency plane. Thus, we should try to represent these signals in terms of functions that are most concentrated in that plane. [67, p. 784]

The existence of multipath spreading in space, time, and frequency can, in principle, allow the receiver to discriminate the signature of the desired signal from other signatures generated by interferers. (Because those interferers will be located in different places, the multipath geometry will be different.) In fact, in some applications the relative *absence* of multipath can be a problem, and rich multipath environments with lots of reflectors (like cities) may allow for *higher* capacity than less dense scattering environments. (We will return to this point later.)

Furthermore, and perhaps profoundly, if the pulse is short enough, and our receiver has a well-defined structure and sufficient powers of resolution, the channel becomes "discrete-ized"; that is, we can determine ahead of time that the signal images should only occur at specific locations in the STF space.[37] Instead of a continuous blur, the so-called canonical representation becomes a three-dimensional lattice of specific three-dimensional coordinates (Figure 7.12). The individual multipath components of the signal will only occur at one or another of these lattice points, and not in between:

---

36. "A *joint time-frequency representation* (TFR) ... is a canonical time-frequency-based decomposition of the mobile wireless channel into a series of independent fading channels" [64, p. 123]. See also [65].

37. The best discussion of discretization I have found (although still somewhat superficial) is in [43]. See also [67].

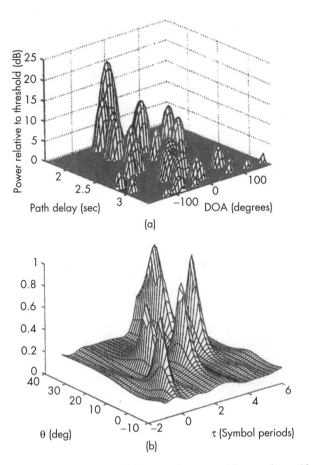

FIGURE 7.10 *Space–time signature: (a) a sample power-delay-angle profile generated using the GSGB model (From: [68]. © 1996 IEEE. Reprinted with permission.); and (b) two-dimensional spectrum of multipath parameters (From: [46]. © 1997 IEEE. Reprinted with permission.).*

A canonical representation of the received signal ... captures the essential degrees of freedom in the channel that are observable at the receiver, and corresponds to certain *discrete* multipath delays, Doppler shifts, and directions of arrival of the signaling waveform....

[This] is motivated by the fact that the signaling waveform has a finite duration T and an essentially finite bandwidth B. Hence, the signal exhibits only a finite number of *temporal* degrees of freedom that are captured by a set of uniformly spaced discrete multipath delays and Doppler shifts. Furthermore, assuming the antennas are spaced to avoid spatial aliasing, [the signal] possesses [a finite number of] spatial

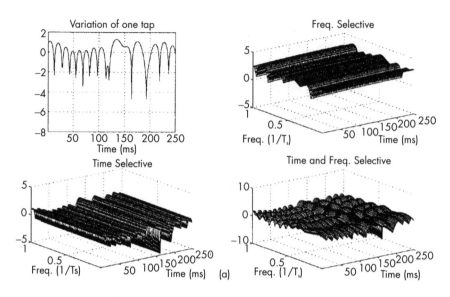

FIGURE 7.11 *Time–frequency signature: (a) fading channels generated by the basis expansion model (From: [10]. © 1998 IEEE. Reprinted with permission.); and (b) sampling of the time-frequency plane to create multipath-Doppler diversity channels and sampled time-frequency correlation function of a spread spectrum signaling waveform (From: [64]. © 1999 IEEE. Reprinted with permission.).*

degrees of freedom that can be captured by certain discrete DOAs even if the DOA distribution is continuous....

The representation exploits the fact that the underlying signal space possesses finite degrees of freedom due to the finite duration and essentially finite bandwidth of the signaling waveform and finite aperture of sensor array. [69, pp. 1669–1670, 1676][38]

The channel lattice (Figure 7.12) suddenly bears some similarity to the code spaces we have been dealing with in Chapters 5 and 6. This suggests

---

38. It strikes me that this article [69] has identified a potentially profound point that may have important practical implications. The physical matrix of multipath components (in S-T-F) may well be continuous, if we assume that the real world is filled with a huge number of reflectors. Be that as it may, the insight of these authors is that the receiver can only discriminate at certain fixed points in that continuous signal space: "the essential spatio-temporal degrees of freedom in the channel that are *observable* [emphasis in original] at the receiver" [69, p. 1670], and need only search at those

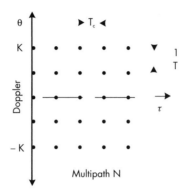

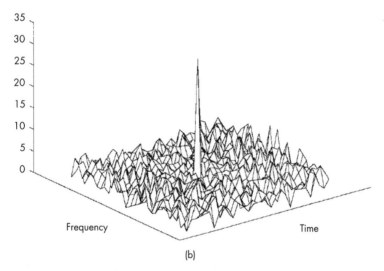

(b)

FIGURE 7.11 *Continued.*

discrete points "with virtually no loss of information" [p. 1669]. Similar comments are offered by Hansson and Aulin [43]: "The received signal ... is first discretized to a finite dimensional vector of N observables ... This would not in general lead to an optimal decision given the time continuous signal, but the purpose is to obtain close to optimal performance when the number of observables is large.... Suboptimal discretizations ... give performances close to optimal except at very high SNR" [43, p. 875]. The practical significance of this approach is that we could employ discrete mathematics to resolve the channel transfer function, which may open the door to more powerful techniques than those currently implementable within reasonable processing parameters—in other words, it may lead to more practical detectors.

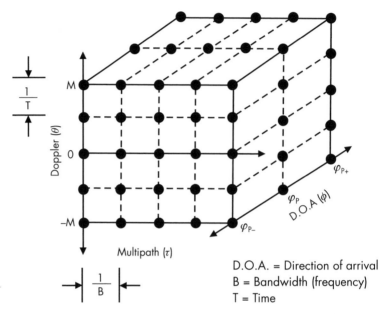

FIGURE 7.12 *Discretized lattice of signal coordinates. (From: [69]. © 2000 IEEE. Reprinted with permission.)*

that it may indeed be fruitful to view the channel transfer function as a *natural encoding process.* Like human-designed channel coding, the natural channel coding expands the information in the signal, effectively adding redundancy. The more reflectors, the more multipath echoes are generated. The more echoes, the more points in the channel lattice are occupied for each symbol.[39] The more points corresponding to each transmitted symbol, the stronger the natural channel encoding that is created, and the more robust the signal.[40]

In fact, we find that in the proper framework we can achieve *better* performance for *larger delay spreads* or *faster fading rates* [70]. In the classical perspective on the multipath channel, more intense fading should produce more errors, but we see the exact opposite under the proper conceptual lens (Figure 7.13): "Surprisingly ... fast fading is not destructive, rather it improves performance" [43, p. 883].[41]

---

39. "Large performance gains could be achieved by increasing the number of observables" [43, p. 883].

40. Recall that Shannon makes the point that longer codes give better performance.

This result is based on the implicit diversity added or created by the multipath channel.[42] Even more startling, the impact of implicit diversity appears to be greater when the signal is *not* intentionally strengthened by the addition of explicit diversity:

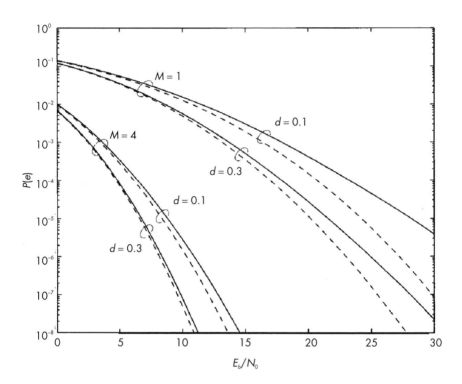

FIGURE 7.13 *Fading channel performance (M: Number of antenna elements; d: delay spread; solid lines: no fading; dashed lines: significant fading). Overall, improved performance "occurs for higher delay spreads or fade rates. This effect is due to implicit delay or Doppler diversity" [70, p. 1527].*

---

41. Also, in the introduction to a recent comprehensive survey of information theoretic aspects of fading channels, Biglieri, Proakis, and Shamai cite "interesting, unexpected outcomes—as, for example, the beneficial effect of fading in certain simple cellular models" [71, p. 576].

42. For a recent treatment of implicit diversity, and a thoughtful review of the early literature on the subject extending back to the 1960s, see [43, pp. 877ff].

> Implicit delay or Doppler diversity ... is most effective when there is no explicit diversity, that is, a receiver with one thread.... The performance of a one-thread receiver given a large delay spread or fade rate is potentially better than a two-thread receiver with a small delay spread or fade rate....
>
> There are greater performance gains as a result of Doppler diversity if the pulse shape is more spread out in time ... [70, pp. 1527–1528][43]

The reader should keep in mind that the term *Doppler diversity* refers here to the inherent spreading of the signal by the channel (viewed in the frequency domain) [51]. It is not part of the transmitted signal; it is part of the natural channel coding that is added by the beneficent universe. Again, though traditionally seen as a problem in signal processing, like fast fading, Doppler spreads may present another "free" resource: "The relatively modest Doppler spreads (100–200 Hz) encountered in practical fast fading scenarios can be transformed into significant ... gains ..." [51, p. 1691]. As we shall see below, new receivers are being developed that outperform existing receivers by accessing new dimensions of the channel transfer function, including the Doppler spread:

> Unlike existing receivers that treat Doppler as a nuisance, our framework demonstrates that Doppler is in fact another dimension for diversity that should be exploited to combat fading.... Even the relatively modest Doppler spread encountered in practice can be leveraged into substantial gains using our approach. [65, p. 130]

One study has even suggested marshalling Doppler signatures to support a multiple access architecture for satellites, dubbed *Doppler-based multiple access* (DBMA) [72]. Users would be discriminated and controlled based on the channel information embedded in their Doppler patterns.

Properly exploited, multipath propagation is an asset, not a liability. *This* is the second key factor, besides the source model, that can assist a smart receiver in blindly acquiring the signal. If that signal has been formatted properly (yes, the transmitter does have a role to play) to enable discrimination based on the spreading functions in space, time, and/or frequency, the receiver can develop tremendous amounts of new, free information to apply to its main task of accurately detecting the original message. There are

---

43. See also [43, p. 883]: "Rather surprisingly ... using pulses with longer duration may give a better performance on the Rayleigh channel."

even steps we can take to enhance the creation of this free information (see Chapter 8). Indeed, some information theoretic work suggests that the gains are rooted in the physics of the channel and may be independent of the signal format. Ozarow et al. [73] studied the impact of various forms of diversity on the capacity of a fading channel, including time diversity (a simple repetition code) and space diversity (multiple antennas). They also examined, however, the simple consequence of either a single-ray propagation path (i.e., something like line of sight with no reflectors) and double-ray propagation (the simplest multipath case):

> As expected [!], multipath has a beneficial effect on outage probability ... demonstrating the substantial improvement due to multi-path propagation. The double-ray propagation provides sort of a diversity mechanism ... [73, p. 366][44]

It should *not* be surprising that multipath can enhance performance. A single path can be blocked, and multiple reflections provide inherent redundancy and a better guarantee that the receiver will see the signal even if it does not have true line of sight.

Finally, if indeed multipath is beneficial to system performance, what happens when the channel does not provide enough multipath? The answer is obvious, and nonetheless startling: We can create *artificial* multipath:

> Narrowband CDMA systems *require* [emphasis added] artificial multipath generation on the downlink in order to achieve reasonable capacity. Artificial multipath is generated by using multiple transmit antennas ... with delays ... inserted between them. [74, p. 952]

The intentional creation of multipath interference should remind us of the other ways in which the purposeful addition of interference to the signal, such as partial response coding, can paradoxically strengthen the wireless link. Yet this is not merely an exercise in counterintuitive thinking. The study concludes that "artificial multipath ... is needed in order to achieve *practically useful capacities* [emphasis added]" [74, p. 960]. Indeed, under a different name, or several names (e.g., space–time coding, transmitter diversity, multiple-input/multiple-output communications), the idea of *purposefully* enriching the multipath channel by adding extra transmitters and extra

---

44. This study was based on a TDMA (narrowband) architecture, and made no design assumptions about the receiver architecture per se (e.g., RAKE was not assumed).

delay in the signal replicas is becoming one of the hottest topics for wireless research and development. Adding multipath interference can actually increase system capacity. As the authors of one recent paper argue:

> Multipath multiplies the reliable information rate that is possible in wireless communication.... This fact calls for a shift from the present multipath mitigation paradigm to a new one of multipath exploitation. [75, p. 851]

Perhaps nothing else illustrates so dramatically the shift in thinking that has been taking place in the wireless world.

In the following sections, we examine briefly and at a high level some of the receiver-centric strategies that are being developed for exploiting this channel information. (Signal construction strategies based on the same principles will be examined in Chapter 8.) These receiver designs have emerged in a haphazard fashion during the past four or five decades, and initially assumed the classical Shannon model of communications as an adversarial process: *Signal v. Noise.* They were motivated by the objective of characterizing and eliminating an undesired component of the received signal (i.e., distortion). They are only now beginning to be understood in light of the newer view of the channel we have been discussing here.

Receiver-centric strategies can be grouped or considered in two interesting ways: by the *number* of signal dimensions they address (and which ones), and/or by what they *do* with the received signal components they are able to discriminate.

Historically, the first applications involved discriminating the signal components in only a single dimension, such as time. A number of one-dimensional techniques have made their way into practice and fairly wide deployment. The newest concepts, on the other hand, focus on combining channel information derived from *two* dimensions, with space–time oriented approaches emerging as an early favorite.

The other way of categorizing signal-recovery techniques is defined by what the receiver does with the components it has managed to separate from the received signal. Broadly speaking, the receiver can either subtract/select, or add/combine. This distinction has been drawn clearly by Pursley:

> There are two fundamentally different approaches to the demodulation of the received signal. The first approach is to attempt to combine the multipath components in some way. For convenience, we refer to a receiver that is based on this approach as a multipath-combining

receiver. In a multipath-combining receiver, the energy of all the multipath components is used to extract the data. In the second approach, the receiver attempts to isolate a single component and extract the data from that one component only. Such a receiver is just a correlation receiver matched to the single component, so it does not take advantage of the energy in the other multipath components. However, it does discriminate against these components in order to eliminate (or at least significantly reduce) both intrasymbol and intersymbol interference. [76, p. 123][45]

To understand this distinction, we can go back to good old-fashioned *space diversity* (the fancy term for multiple receiving antennas, described briefly in Chapter 6), to derive a very simple analogy. Two antennas at the receiver take in two distinct images of the signal. In a fading environment, these two signal inputs should fade independently if the antennas have sufficient physical separation. This suggests one very basic strategy: selection. If we have two copies of the same signal, from two different antennas, we can look at both of them and select the better one [77, p. 1117]. By analogy, a number of signal-recovery strategies follow this approach: The receiver attempts to discriminate between the desired signal and the distortion introduced by the channel, and then to select the former and suppress the latter. Usually, this involves a process that is roughly equivalent to subtracting a copy of the distortion component from the entire received signal, leaving only the desired signal. Another term commonly used to describe this is *cancellation*.

The other way to implement space diversity in a two-antenna receiver is to find a method of combining the two signals. The combination should ideally be coherent, so that the result is always linear and additive. (Combination of signals with different phases—noncoherent combination—would sometimes produce cancellation, or fading, just like natural multipath.[46]) If coherence can be assured, then combination of the multiple signal images is attractive because each signal replica adds energy to the S part of the SNR [46]. Combining strategies will always outperform selection strategies[47]; hence, this philosophy of constructive combination of signal components is

---

45. See also [77].

46. There are a few species of noncoherent combining, such as equal gain combining "whereby all available branches are equally weighted and then added incoherently ... [However] equal gain combining is not optimal, due to the phenomenon often called 'noncoherent combining loss,' which means that combining more signals does not necessarily enhance performance" [77, p. 1117].

beginning to make its way into the communications architectures for third-generation wireless systems.

7.3 One-Dimensional Signal Recovery Strategies: Equalization and RAKE Receivers

The history of active receiver architectures has been dominated by two related concepts—the adaptive equalizer and the so-called RAKE receiver—which were first explored in the 1950s and 1960s.[48] Both techniques were proposed as stand-alone solutions to specific problems, rather than as elements of a comprehensive methodology of communications engineering. Both bear the stamp of sheer cleverness that is often associated with intuitive breakthroughs, and both have lacked (until recently) transparency as to the true intellectual foundations of the underlying signal processing theory. Both were long confined to *one-dimensional, time-domain implementations,* and it has only become clear quite recently how these solutions can be extended to other signal dimensions (such as frequency), and to more than one dimension at a time.

Roughly speaking, equalization is a subtractive strategy. It aims at discriminating between the component of the received signal contributed by the transmitter, and the distortion contributed by the channel transfer function. Once the distortion is known, it can be subtracted or canceled to recover a better image of the transmitted signal.

The RAKE is a combining strategy. Its goal is to separate the different components of the received signal that ordinarily would interfere with each other in a corrupted composite and then combine them coherently to realize a much stronger replica of the transmitted signal.

---

47. See [70, pp. 1120ff]. See also Clark et al. [78] where, roughly speaking, combining diversity outperforms selection diversity two to one, as a function of the number of branches.

48. "Two bodies of work in the literature are concerned with multipath receiver design. The older concentrates on the explicit diversity structure of resolvable paths; its thrust is to take advantage of this structure by optimally combining the contributions of different paths.... More recently, equalization techniques that were developed for data transmission over telephone lines have been applied to the radio multipath problem. Here receiver design concentrates on reduction of the effects of intersymbol interference, and the diversity-combining properties of the receiver are only implicit" [44, p. 332]. This last comment is off-target, I believe. Equalization is not a combining technique, as a rule, but a subtractive or cancellation strategy.

But another factor has strongly influenced the historical development of these two somewhat complementary approaches. Equalization works well for narrowband channels:

> [If] the bandwidth [of the signal] is much smaller that the coherence bandwidth of the channel ... then all the frequency components undergo the same attenuation and phase shift in transmission through the channel.... The channel impulse response can be easily measured and that measurement can be used at the receiver in the demodulation of the received signal [i.e., equalization]....
>
> Measurement of the channel impulse response [for wideband channels, on the other hand] is extremely difficult and unreliable, if not impossible. [71, p. 577]

Perhaps for this reason, equalizers have been used extensively in the narrowband era of mobile radio, especially second-generation TDMA architectures.[49]

The RAKE receiver, on the other hand, seems to be especially well suited to the needs of true wideband transmission. With the development of second-generation CDMA, RAKE designs, which had been almost completely ignored in commercial wireless systems for decades, suddenly assumed prominence. The wideband channel lends itself to the sort of constructive strategy that the RAKE embodies. Arguably this is the basis of whatever real advantage wideband systems may enjoy over narrowband systems: Wideband signals are more amenable to the use of constructive combining receiver techniques. (We examine this contentious question in Chapter 8.)

Both techniques are now finally being formalized and extended systematically, and we are realizing that there is an underlying common framework.[50] It seems likely that some of the distinctions I am relying on here may be superseded in the next few years by a more general understanding of multidimensional signal processing applied to the problems of the wireless receiver's physical manipulation of the received signal.

---

49. "The system designer may avoid the need for channel equalization by selecting Ts [the symbol duration] to satisfy the condition Ts Tm [where Tm is the delay spread induced by the channel].... Adaptive equalization is particularly applicable to reducing the effects of ISI in underspread wireless communications channels" [71, p. 627].

50. "The equalizing receiver ... and the RAKE receiver commonly used to combat frequency-selective fading can both be viewed as special cases of the same general structure" [79, p. 1493].

7.3.1 Subtractive Techniques: Equalization

Equalization was arguably the first truly modern, sophisticated signal processing technique to be deployed commercially.[51] Invented by R. W. Lucky in 1964, it was initially applied to the problem of intersymbol interference in wireline data communications [80]. ISI in this application arises from the fact that different frequency components of a voiceband modem signal traveling over a wireline circuit arrive at the receiver with different delays. The effect is to blur the signal in time, causing successive symbols to overlap if spaced too closely. To avoid this overlap, the bit rate has to be reduced. Unfortunately, different channels show different patterns of delay distortion, depending on the length of the wireline circuit and other factors affecting the general condition of the circuit (e.g., age and temperature). Because modems are used on preexisting circuits and may connect over different circuit paths for each new session, the problem of unpredictable time distortion of the wireline channel was a serious impediment to voiceband data communications. According to the official history of the Bell System, this was one of the three crucial signal processing problems identified by AT&T in the early 1960s and targeted for special attention.[52]

Lucky proposed a delightfully clever solution. First, the transmission is initiated with a known sequence of data bits, called a *training sequence*. Because the receiver knows what to expect, it can measure the actual received sequence and compare it to the expected sequence, to obtain the error or distortion characteristic (i.e., the transfer function) of that particular channel. Based on this knowledge, the receiver can adjust a specialized filter to correct for the now-known delay characteristics of the channel. For a wireline channel with very slowly changing characteristics, such an initial training at the start of a data transmission session is often sufficient.

---

51. This claim comprises a strong myopic bias of our "all-digital" era. Many of the signal processing innovations of the first 75 years or so of telephony are equally impressive, viewed in their own appropriate historical context. Even some of the tricks of automated or multiplexed telegraphy are still impressive, and carrier-based telephony as it was developed in the 1920s and 1930s was enormously important in enabling a national network to be built. Equalization, however, was the first to really focus on the signal elements (rather than the circuit elements), and although it can be implemented without digital technology, it is still inherently logical in character, in a way that points toward the DSP revolution that has now swept us all up.

52. "The other two goals were determining the most efficient modulation and deciding on an efficient error-correction technique" [80, p. 422].

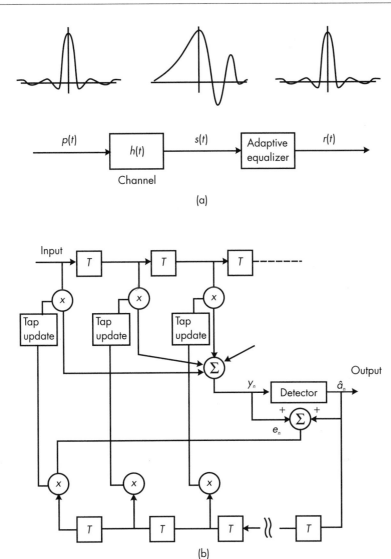

FIGURE 7.14 *Adaptive equalization: (a) adaptive equalizer (From: [38]. © 1990 Addison Wesley Inc.); and (b) adaptive zero equalizer (From: [81]. © 1992 Plenum Press.).*

The second half of the idea is the *tour de force*. Lucky extended the idea to allow the equalizer to become continuously adaptive, by using the ongoing stream of corrected data to provide a continuously updated measure of the transfer function, enabling the settings on the transversal filter to be updated on the fly (Figure 7.14). This implementation of a truly *adaptive* algorithm was probably the first commercial implementation of a digital feedback principle, a breakthrough in communications engineering, with broad significance:

> The feedback control information was obtained by comparing decided bits against the receiver's incoming, distorted analog pulse train in order to ascertain an error component. This technique of using prior decisions to determine an ideal on which to base adaptation became known as decision direction; it subsequently became the basis not only for equalization but for the recovery of timing and other necessary receiver parameters. [80, p. 424][53]

Adaptive equalization is made to order for digital cellular radio. In mobile communications, the time dispersion created by the wireless channel is far more severe than in wireline circuits. First, the delay spread of the channel is inherently large (up to tens of microseconds). Second, the channel transfer function fluctuates rapidly due to multipath phenomena as the mobile unit changes position relative to the reflectors in its environment. To control intersymbol interference in the time domain, equalization has become an essential element of signal processing architectures for second-generation wireless networks. The technology of equalization has developed tremendously in support of this challenging application, although in many respects the fundamental concept of decision feedback equalization discovered by Lucky is still the foundation of this field.[54] One focus of recent development work has been directed at blind or self-recovering equalization algorithms in which the initial setting of the filter coefficients at the receiver is accomplished rapidly and without the overhead cost of a training sequence [84, pp. 587ff].

From our perspective, equalization can be described in a more generalized way as process in which the received signal is analyzed into a message component (i.e., the component corresponding to, or contributed by, the

---

53. For a more recent treatment of feedback equalization, focused specifically on cellular applications, see [82].

54. There are good presentations of basic equalization concepts in most communications textbooks. For example, Blahut [38, pp. 458ff], Rappaport [83, Chap. 6], Proakis [84, pp. 554ff], Gibson [85, pp. 210ff], and Frerking [86, pp. 464ff]. One of the better recent surveys of equalization techniques and the supporting literature is to be found in Biglieri et al. [71, pp. 626–636].

transmitted signaling waveform) and a channel component, and the channel component is then subtracted[55] from the received signal (i.e., canceled) to leave a purified version of the original transmitted message.[56]

The classical implementation of equalization is in the time dimension, and often it is assumed that equalization is strictly a time-domain technique. However, it is becoming apparent that the underlying concept is more general. For example, in a new type of signal architecture called *orthogonal frequency-division multiplexing*, which we will examine in more depth in the next chapter, the problem of ISI is eliminated architecturally by spreading the transmission over many closely spaced frequency channels. This allows the symbol duration on each individual frequency channel to become quite long (i.e., the signaling rate on each subchannel is quite slow), and the problem of the time-delay spread becomes much less significant. The receiver can thus avoid the costs of equalization. However, this free lunch is paid for by the appearance of a new problem: *interchannel interference* (ICI[57]), which is a phenomenon that is strictly analogous to ISI, but manifest in the frequency domain [54]. Just as the time-delay spread creates ISI, the expansion of the signal in the frequency domain leads to ICI: "When the normalized Doppler frequency [i.e., the frequency-domain manifestation of multipath spreading] is high, the power of ICI cannot be ignored" [88, p. 1377].

ICI solutions are various, and somewhat overlapping, conceptually [89]. One promising approach is sometimes referred to as *frequency-domain equalization*.[58] The analysis of frequency characteristics of individual carriers

---

55. "Decision feedback equalization is a technique that uses previously detected symbols to synthesize the ... discrete impulse response that is subsequently subtracted from the received samples through a feedback loop" [82, p. 889].

56. It has been argued that equalization can be viewed as a constructive technique, not just a subtractive or cancellation technique. "An equalizer accomplishes more than distortion mitigation; it also provides diversity. Since distortion mitigation is achieved by gathering the dispersed symbol's energy back into the symbol's original time interval so that it doesn't hamper the detection of other symbols, the equalizer is simultaneously providing each received symbol with energy that would otherwise be lost" [87, p. 105]. I am not sure if this argument is entirely true, or always true. The general spirit of equalization is not constructive combination of separate parts of the transmitted symbol, but detection and cancellation of the transfer function of the channel. I can see the point of Sklar's argument that this may be viewed as redistributing some of the dispersed energy of the spread symbol back to its original time window, but I wonder whether his is really an accurate description of what is happening.

57. ICI is also sometimes used to refer to interchip interference in a direct sequence spread spectrum system. Given the confusion, this usage should be suppressed.

can lend itself to an analogous feedback-driven reshaping process in the frequency domain, to cancel frequency distortion introduced by the channel.

Finally, we may be ready to ask whether the signal can be equalized in the spatial domain? Indeed! Hayashi and Hara [93] advance a *spatial* equalization technique based on a multielement antenna array, with a specialized pilot signal to allow the receiver to learn the channel characteristics, and to accurately weight the inputs from different elements of an adaptive antenna array (Figure 7.15):

> We have proposed a spatial equalization method, where the beamformer calculates and adjusts the weights of adaptive array elements only using the estimated channel impulse response, and we have clearly shown that the weights of adaptive antenna array elements can be successfully controlled by the estimated channel impulse response. [93, p. 1250]

Raleigh and Paulraj [11] provide a more explicit characterization of the notion of equalization—normally thought of as a time-domain process—applied to the spatial domain in an antenna array. It is worth quoting them at length, because it helps make clear the more general principles of equalization, independent of the physical dimension to which they are applied:

> The techniques proposed in this paper do not consider conventional antenna concepts such as beam width, gain, and side lobe rejection. Instead, the antenna array is viewed from a statistical signal processing perspective. Each antenna array output represents a phase- and amplitude-coherent sensor which contains a weighted sum of the desired signal, the undesired signals, and noise. By applying an appropriate complex weight to each sensor signal and then summing the outputs, it is possible to cancel undesired interference and enhance the desired signal above the noise. [11, p. 219/458]

Note how the concept has been translated from the sensual aspects of the particular physical dimension (here, space) into a more abstract language of pure signal processing.

Thus, the idea of equalization may be generalized to any of the three physical signal dimensions. We can recast our understanding of the common process. T-equalization, F-equalization, and S-equalization are all based on the same handful of coordinated ideas:

---

58. See [88, p. 1376]. See also [90, 91]. For a somewhat different notion of frequency-domain equalization derived from work on DSL modems, see [92].

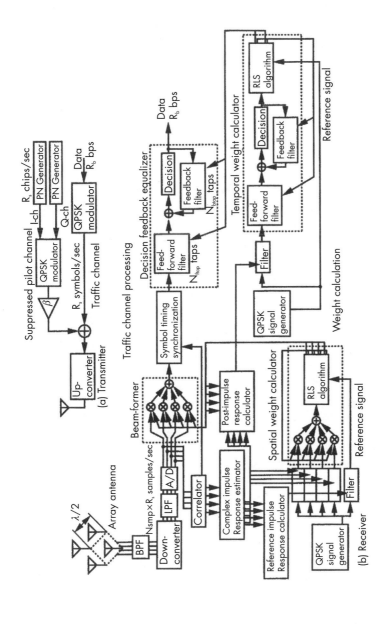

FIGURE 7.15 *Spatial equalization. (From: [93]. © 2001 IEEE. Reprinted with permission.)*

- *Multiple sampling of the signal:* Each signal element is assumed to be sampled repeatedly across the relevant dimension (e.g., in time, at various delays; or, in space, from various array elements).[59] The receiver has multiple "looks" at each received signal element.

- *Channel impulse response measurement:* The receiver estimates the channel transfer function in terms of its effects on a transmitted symbol (or sequence). This estimation may be either transmitter assisted, by means of a training sequence or a pilot, or it may be blind—that is, recovered by the receiver from the received signal without any help from the transmitter.

- *Weighting of multiple signal components:* The impulse response is used to assign the proper weights to the different components of the received signal. Because equalization is construed as a one-dimensional process (until we reach the following section), the signal components are all one dimensional. The relevant signal components for T-equalization are time samples. For F-equalization, they are inputs from different frequency subchannels. For S-equalization, they are inputs from different antenna array elements.

- *Cancellation:* The weighted information about the channel is used to cancel the contribution attributed to the channel impulse response, to leave a pure replica of the originally transmitted signal.

- *Feedback:* Once the equalizer is trained, some form of feedback based on the decided symbols (usually within the receiver itself) is used to continuously adjust and update the weights of the equalizer.[60]

It is interesting to note that, from an information theoretic perspective, equalization is a physical-level signal processing alternative to error correction coding on the logical level, and may allow performance gains on an

---

59. This need not imply any sort of oversampling or increase in the sampling rate. It does imply, however, that the receiver is searching for the information content or impulse response of a given target symbol across more than one sample. Conceptually speaking, it is a mirror image of the convolutional process or partial response process that may be implemented by the transmitter.

60. As may be imagined, there are an enormous number of variations on these themes, and some may not fit strictly within this framework, while still bearing the name equalization. There are also, arguably, many approaches to equalizer design that may not embed the notion of feedback literally.

uncoded signal that are similar to those that can be realized by forward error correction alone. Just as we saw that logical signal recovery techniques, such as error concealment, can allow good performance even with uncoded or lightly coded signals (Section 7.1), so may physical signal recovery techniques like equalization be traded off against the use of transmitter-centric techniques such as forward error correction. Cioffi et al. [94] suggest that the equalization (bolstered by precoding[61]) can achieve as much in a fading channel as the error correction coding alone can achieve in a Gaussian (non-fading) channel:

> [It is] a general principle that if a coding method can achieve a certain coding gain at a certain target error probability on an ideal channel, then an adaptation of that method can achieve the same gain over *uncoded* [emphasis added] transmission with the [decision-feedback equalizer] on an arbitrary ISI channel. [94, p. 2595][62]

7.3.2 Constructive Techniques: RAKE Architectures

Subtractive or cancellation techniques work by separating signal compenents and discarding or canceling out some of them. Constructive techniques work by separating signal components and then recombining them in an additive (usually coherent) fashion. As noted earlier, combining strategies always outperform selection strategies, in principle [77].[63] The premier example of a constructive combining technique is the RAKE receiver.

The RAKE is an old idea [44]. In 1957, even before the invention of the equalizer, Price and Green put forward an extremely innovative concept for a receiver, which turns out to be a generalization or extension of the equalization principle to the receiver architecture as a whole: the so-called RAKE receiver [96].[64] The signal processing framework is similar to the equalizer: Like the tapped delay line used in the T-equalizer, the T-RAKE uses a delay line with a series of taps called *fingers* to sample the received signal several

---

61. That is, any technique implemented at the transmitter to effectively predistort or precondition the signal in anticipation of the changes that will be effected by the channel transfer function.

62. See also [95].

63. For a more fundamental information theoretic treatment and verification of the same principle, see [73].

64. A brief account of the historical circumstances of the invention is given in [97, pp. 92ff].

times at slightly shifted sampling instants.[65] The diagram of the circuit suggested to someone the image of a garden rake, and the metaphor took hold (Figure 7.16).

Instead of trying to separate and remove the effects of multipath delays, as the equalizer does, the RAKE receiver is designed to separate and then recombine the individual rays of the multipath signal:

> The ideal tapped delay line receiver ... attempts to collect coherently the signal energy from all the received signal paths that fall within the span of the delay line and carry the same information. [97, p. 446][66]

In principle, this can be a *coherent* or constructive combination, in which the different signal images generated by the multipath channel are resynchronized with each other and added to produce an even stronger signal and an improved SNR [98]. In effect, the RAKE reassembles the fragmented multiple images of the original transmission back into a single coherent whole, undoing the effects of multipath. Instead of losing signal energy by subtracting it as distortion, the RAKE receiver recovers signal energy and uses it to enhance the performance of the receiver.

The original implementations of the RAKE design for military spread spectrum systems were costly and hardware intensive, and for decades the RAKE concept remained beyond the limits of commercial feasibility [97]. Following that original paper by Price and Green [96], nearly 40 years elapsed before integrated-circuit technologies evolved to the stage where RAKE designs could be implemented in a commercial system. Yet with the emergence of CDMA spread spectrum architectures for second-generation wireless systems in the early 1990s, the RAKE receiver became a crucial link in the new technology, to enable CDMA signals to cope with multipath interference [99]. As third-generation CDMA systems near commercial deployment, interest in sophisticated RAKE designs is growing.

The RAKE has come into its own because of demands placed on the receiver by the signaling strategies implemented in wideband CDMA architectures (see Chapter 8). Wideband signals are generally created by recasting a low-rate data signal into a higher rate transmission format. (We examine how this is done in Chapter 8). As a consequence, the individual

---

65. "The equalizing receiver ... and the RAKE receiver commonly used to combat frequency-selective fading can both be viewed as special cases of the same general structure" [79].

66. Proakis uses virtually identical language in [84, p. 732].

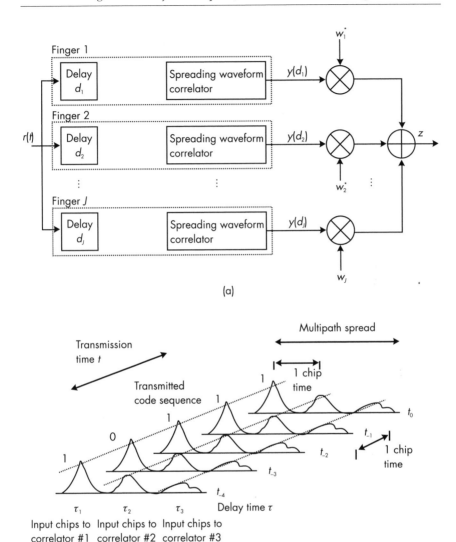

FIGURE 7.16 *The RAKE receiver: (a) RAKE receiver structure (From: [100]. ©
2000 IEEE. Reprinted with permission.); and (b) example of received
chips by a three-finger RAKE receiver (From: [87]. © 1997 IEEE.
Reprinted with permission.).*

signal elements— pulses or chips—are of very short duration. With very
short pulses, the multipath components of the received signal no longer
overlap or blur together, but tend to separate out into distinct images of the
original pulse.

> When the transmitted signal has a bandwidth greater than the coherence bandwidth of the channel, the frequency components [of the signal] are subjected to different gains and phase shifts ... [and] the multipath components in the channel ... are resolvable. [71, p. 577]

The RAKE is well suited to detect these different multipath images and combine them.

The RAKE principle has undergone intensive study in recent years, and has been extended to a variety of idealized and semi-idealized designs that explore the limits of potential performance improvements [98]. One important design dimension is the number of fingers, which fixes the number of multipath components that can be resolved. Another dimension is the way in which the fingers are positioned and configured, to maximize the combining gains [101]. At one end, practical RAKE implementations have a small number of fingers that are usually positioned by the delays built into the chip rate.[67] At the other end, idealized designs have been studied in which it is assumed that there are an infinite number of fingers(!!), or at least "more fingers than the number of multipaths" and "the finger positions and weights should change instantaneously with the fading" [100, p. 1538].[68] Not surprisingly, these idealized receivers provide excellent simulated results.

The RAKE idea has also been extended to multiuser detection (see Section 7.5). The standard RAKE is designed to detect only the multipath images from one transmitter, but the architecture can be extended to allow the RAKE to detect and separate signals and their multipath components, originating with many different transmitters. This so-called *decorrelating RAKE* would seem to imply a combined additive/subtractive strategy [104]. Bottomley et al. [100] indeed describe a generalized RAKE architecture where some fingers are deployed to capture and combine multipath components of the desired signal, while others are positioned on top of interferers to help cancel them:

> The window [of potential finger positions, in time] spans from several chip periods *before* the earliest arriving multipath component to several

---

67. "In the special case of the traditional RAKE receiver, the finger delays would equal the channel delays" [100, p. 1538].

68. See also Win and colleagues [102]: "We introduce the term *all Rake (A-rake)* receiver to describe the receiver with unlimited resources (taps or correlators) and instant adaptability, so that it can, in principle, combine all of the resolved multipath components" [15, 16], and see Win and Kostic [103].

chip periods after the latest arriving multipath component ... *L* fingers are positioned on the multipath components (to collect energy) [i.e., these are *combining* fingers] and the remaining *J–L* fingers are placed ... to suppress interference [these are the *subtractive* fingers]. [100, p. 1540]

The RAKE that combines both strategies (additive and subtractive) is capable of significantly better performance and higher capacity (Figure 7.17). This also illustrates the common conceptual core underlying the gamut of receiver-centric strategies. RAKEs are usually implemented in order to perform constructive combination of multipath components, but they can equally be used to support an equalizing strategy of targeting and canceling interfering signals.

Like the equalizer, the RAKE receiver was first implemented in the time domain. More recently, it has been realized that the RAKE principle can also be extended to the frequency domain:

A new receiver structure for exploiting Doppler diversity, which we refer to as the "Doppler RAKE receiver." ... The Doppler RAKE receiver is the dual of the conventional RAKE receiver in that it samples the Doppler axis to achieve ... diversity. [64, pp. 127–128]

F-domain RAKE receivers are also being explored as alternatives to conventional T-domain RAKEs in order to simplify the processing requirements for mobile receivers [105].

RAKE techniques also suggest similarities with multiple-antenna systems, where different receiver antenna elements capture and in some cases combine different images of the transmitted signal, which is a reasonable segue perhaps to a brief discussion of the vast field of array processing.

7.4 Spatial Techniques: Array Processing

Despite its antecedents (mainly the use of equalizers on certain wireline channels beginning in the 1960s), the exploration of the one-dimensional time signature of the received signal is a relatively new field of wireless technology. It is only within the last 10 years or so that digital wireless systems have emerged that can really make good use of this information. As well, receiver architectures based on handling multiple time images of the signal depend on fairly intensive signal processing techniques that have only recently become commercially feasible.[69] The frequency-domain versions of both the adaptive equalizer and the RAKE receiver are still in their infancy,

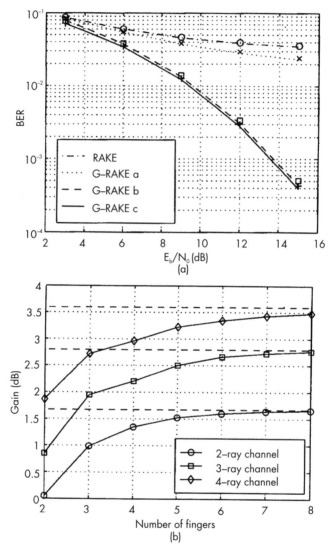

FIGURE 7.17 *RAKE receiver performance: (a) comparison of link simulation (mark-ers) and analysis (lines) of the conventional RAKE receiver and general-ized RAKE (G-RAKE) receiver; and (b) gain in the SNR for the G-RAKE receiver compared to the L-finger RAKE receiver for L = two-, three-, and four-ray channels. (From: [100]. © 2000 IEEE. Reprinted with permission.)*

commercially speaking. Moreover, detailed time and frequency information is still of limited use in second-generation wireless systems. It is fair to say that other than the outstanding success of T-equalization in countering ISI for appropriate types of digital transmission systems, the value of receiver architectures based on processing the time and frequency signatures of the received signal has not been firmly established (in *commercial* terms). Even today's T-RAKEs, while in use for IS-95 CDMA, are still rather rudimentary realizations of the general RAKE concept.

The story is quite different for the realm of spatial processing techniques. This is indeed a vast field, based on several important military applications, including direction finding, EW, and radar systems, which have driven decades of intensive research and development. The value of spatial information embedded in the received signal is much more obvious. If we can learn the direction from which a signal has originated, we can use that information in a number of militarily obvious ways. They are not typically *communications* applications and support an different cost model. Moreover, unlike T-domain and F-domain processes, which lead to complex receiver hardware (and software) implementations, many spatial signal processing techniques can be implemented with arrangements called arrays, in which a number of more-or-less conventional receivers are simply ganged in parallel. The outputs of these multiple receiver chains may require some special handling, but the cost of the processing is embedded in the infrastructure end of the system (rather than the cost- and size-challenged mobile unit).

The importation of these techniques into the field of wireless communications has really just started. The use of smart antennas and other space-domain techniques in *commercial* wireless systems is nascent, and the first applications are also quite rudimentary. But behind this somewhat tentative trend there is a huge inventory of military and related noncommunications technology, such that even a high-level survey of space-domain signal processing techniques would require a book-length presentation in its own right.[70]

---

69. As recently as the early 1990s, the task of equalizing a TDMA signal was viewed as a serious challenge in the implementation of the IS-54 standard (even though the equalization technique itself, as a receiver-centric process, is not specified by the standard). The RAKE receiver, brought to commercial attention by the CDMA IS–95 architecture during the same time frame, was viewed as an exotic and slightly doubtful solution at first. It is really only now, at the beginning of the new decade, that even these basic techniques are achieving a routine status.

70. A good, recent collection of articles on the subject, targeted specifically to commercial wireless applications for the most part, can be found in [106]. Most of the articles referred to in this section are reprinted in this book.

Following our framework, we will focus here on receiver-centric spatial processing techniques. For the most part, these are subtractive techniques, which (like equalization) attempt to discriminate between a desired signal component and an undesired component (either an interferer, or an undesired multipath image of the desired signal) and to suppress the latter. The term spatial filtering is sometimes used, by analogy with other sorts of filtering processes [107]. But increasingly the more common generic term for space-domain signal processing systems is smart antenna.

All smart antennas are based on an antenna array—a grid or other configuration of regularly spaced antennas, each connected to various receiver processing elements, providing thus a number of spatially independent samples of the received signal. The samples have different phase and timing information, which are capable of providing information about the spatial location of the transmitter or reflector.

Let us consider a generic version of an antenna array (Figure 7.18). Such an array can implement a number of functions:

1. *Direction finding:* The signals can be processed to determine the direction from which they have originated.

2. *Spatial filtering:* Signals arriving from a given direction can be enhanced, and those from other directions can be suppressed.

3. *Coverage shaping (beam forming):* The receiver can be directed to pay attention, so to speak, to specific zones or quadrants. Signals inside these beams will be enhanced relative to signals located outside of them.

4. *Spatial equalization:* As noted above in Section 7.3.1, the inputs from an antenna array can also be handled in a more abstract signal processing manner to drive a spatial equalizer modeled on the generalized equalization concepts developed above.

These features can be implemented in a fixed manner or adaptively. An adaptive array can, in principle, detect the angle of arrival of a desired signal and steer a receiver beam directly toward that transmitter. It can also simultaneously identify other, interfering transmitters coming from different directions and can fashion a null in the antenna pattern to suppress them. In the most elaborate implementations, called *fully adaptive arrays,* the system can track both desired users and interfering transmitters in real time, moving the beam by adjusting the weighting and mixing parameters of the different array elements [108]. In principle, a smart antenna of this sort could track

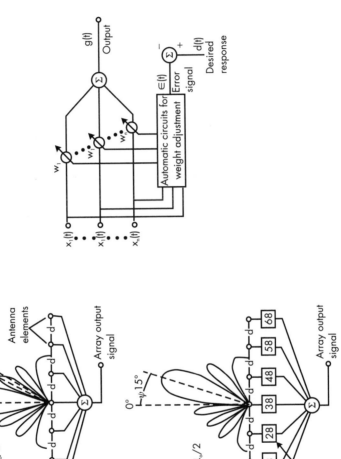

FIGURE 7.18 *The antenna array concept. (From: [107]. © 1967 IEEE. Reprinted with permission.)*

and handle all interferers and all multipath signal components, although the number of antenna elements would become very large.

Actually tracking individual users and/or interferers is quite challenging. Early commercial uses of so-called "smart" antenna technology have focused on coverage shaping or beam forming.[71] Fixed beams are quite easy to form and may be viewed as an extension of older nonsmart antenna systems (where directional selectivity was achieved by the design of propagation patterns from single antenna elements). The fixed beams can be viewed as more tailored versions of cells where the gain comes from better frequency reuse in the spatial domain. Adaptive techniques are still largely in development, as of this writing, but we may expect to see antennas becoming even smarter during the next few years. This technology is well established in the military field, and commercialization is mainly a function of cost-reduced implementations.

The array is possibly the most straightforward example of the way in which the receiver can generate information on its own for "free" (in information theoretic terms). With a proper array (enough elements to provide good resolution), the receiver can develop a detailed picture of its physical environment; indeed, this is precisely what systems like radar and sonar are designed to do. The transmitter, of course, does not possess this information and cannot really operate with it, although it can assist in generating it. The communications receiver may not have a need to visualize its environment per se, but it stands in a unique position to apply these potentially enormous quantities of surplus information to its basic task of detecting and decoding the transmitted signal.

7.5 Multidimensional Signal Recovery Strategies

The recognition that many signal recovery strategies can be viewed in a pure signal processing framework, abstracted to a degree from the physics of the underlying signal dimension (e.g., time or frequency), has begun to suggest to some researchers the possibility of extending these processes to more than one physical dimension simultaneously. Although few of these multidimensional techniques have yet reached commercial reality, it seems likely that multidimensional receiver architectures will emerge as one of the cornerstones of next-generation high-performance wireless systems.

71. Godara's survey [109] is fairly comprehensive. See also [110].

7.5.1 Multidimensional Equalization

One-dimensional approaches to signal recovery all begin in the same way, by measuring or estimating the channel impulse response. Because the impulse response can be characterized in more than one dimension simultaneously, it should be possible to drive more than one equalization dimension. Hayashi and Hara [93] note that:

> A spatial equalization method ... calculates and adjusts the weights of adaptive array elements only using an estimated channel impulse response.... Taking into account that the estimated channel impulse response can be also used for temporal equalization, it is quite natural that we can easily extend the idea of "impulse response estimation-based control" in the spatial equalization to a basic strategy in *spatio-temporal equalization* [emphasis added]. [93, p. 1250]

Space–time equalization [111–114] is one realization of joint space–time processing, which is one of the fastest developing areas in wireless technology, as we shall see further on (and in Chapter 8).

Another two-dimensional equalization concept is offered by Sung and Brady:

> We introduce a *frequency-space domain equalization* [emphasis added] (FSDE) using an antenna array.... An FSDE on the received OFDM [orthogonal frequency division multiplexing] signal [i.e., a signal constructed of multiple overlapping frequency sub-channels—see Chapter 8] is performed by space–time Fourier transform (STFT) and frequency-domain equalization (FDE). While a time-domain Fourier transform in STFT demodulates the OFDM signals, a space-domain Fourier transform attenuates the OFDM signals arriving at the antenna array through unwanted paths.... A frequency-domain equalizer is applied to the output of the STFT to compensate for the remaining interchannel interference. [115, from the Abstract]

It seems safe to predict that there will be further exploration of multidimensional signal cancellation as third-generation systems are deployed.

7.5.2 Multidimensional RAKE Receivers

The RAKE architecture can also be extended to more than one dimension. Mun et al. describe a two-dimensional RAKE that "exploits the space- and time-domain structure of the received multipath signal" [116, p. 1312]. The hardware for this receiver employs an antenna array with multiple time-domain fingers on each branch [46] (Figure 7.19). They claim that the space–time RAKE outperforms the conventional time-only RAKE by a factor "proportional to the number of [antenna] array elements" [46, p. 1315]. Other space–time RAKE structures are described by Naguid and Paulraj [117], Onggasanusi et al. [118], and Liu and Zoltowski [119] (Figure 7.20).

A different two-dimensional RAKE architecture—"a time-frequency generalization of the RAKE receiver" [64, p. 123]—combines the signal images discriminated in both time and frequency: "The Time-Frequency RAKE receiver, which correlates the received waveform with multipath-Doppler-shifted copies of the transmitted symbol waveform ..." [120, p. 83]. Simulations indicate a potential improvement of 3–6 dB in system performance compared to the one-dimensional time-domain RAKE [64, p. 129] (Figure 7.21).

7.6 Multiuser Detection and Interference Cancellation

The ultimate receiver would be something like the interference demon we have met in Chapter 4. Stationed at the front end to the receiver, the demon

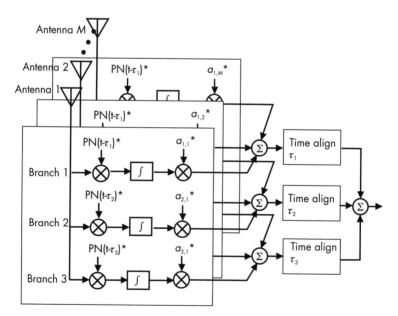

FIGURE 7.19 *A two-dimensional RAKE receiver. (From: [116]. © 2001 IEEE. Reprinted with permission.)*

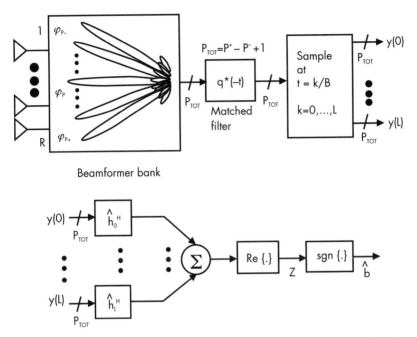

FIGURE 7.20   *Space–time RAKE. (From: [118]. © 2000 IEEE. Reprinted with permission.)*

would be able to see each component of the received signal as it arrived. It would be capable of discriminating the desired signal from all of the undesired signals and perfectly subtracting or canceling the interferers—leaving the desired signal in nearly pristine isolation (marred only by the inevitable halo of purely random noise that not even the demon can eliminate).

Recently, we have begun to dream of such a receiver, which, it turns out, is based on a very simple principle: If the receiver is capable of detecting and demodulating the desired signal, it should be able to detect and demodulate just as well any other signal of the same format. In a multiuser system, where a number of users share the same physical channel, and where all share the same basic signal structure, a receiver could in principle demodulate not just the target user's signal but all the other signals as well. The nontarget signals, once known, could be subtracted precisely from the ensemble received signal. Once all the nontarget signals have been subtracted, the target signal remains in a virtually interference-free state (Figure 7.22).

To restate this proposition in the language in which it is customarily presented: We begin with the *single-user detector*. Operating in a multiuser channel, this detector could in principle detect and demodulate any one of

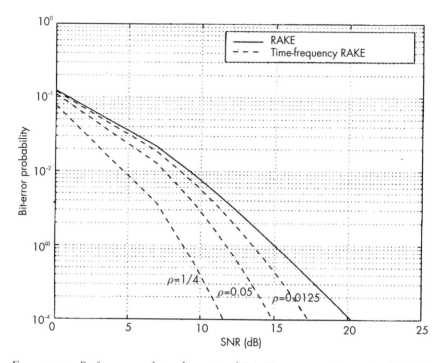

FIGURE 7.21 *Performance of two-dimensional RAKE versus one-dimensional RAKE.
(From: [64]. © 1999 IEEE. Reprinted with permission.)*

the users [121]. Normally, it detects only one and regards the rest as a noise-like background.[72] (This is the main principle of CDMA operation, which we shall touch on in Chapter 8.) However, we can extend this detector in a straightforward manner so that it can operate on multiple users of the same format at the same time. This may either mean creating parallel demodulation threads capable of literally operating simultaneously or it may mean that the processor is fast enough to perform the same operation (i.e., detection and demodulation of a signal) several times on the same sample of the received signal before it has to take in the next sample.[73] Whether executed in parallel, or sequentially, the process is—conceptually—simple.

---

72. "A single-user detector is defined as a receiver structure that requires no information regarding other (interfering) users present in the system and demodulates the data signal of one user only ... When a [single] user is detected, the interference due to cross-correlations with signals of different users is ignored; thus the system becomes interference-limited instead of noise-limited" [122, p. 24].

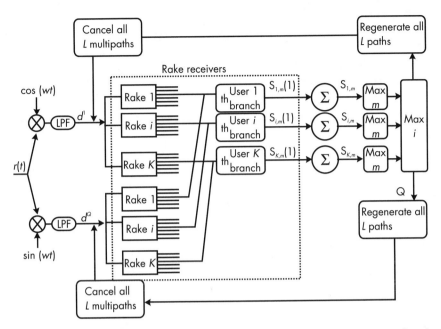

FIGURE 7.22 *The multiuser detector. (From: [123]. © 1995 IEEE. Reprinted with permission.)*

This approach is called multiuser detection, and the result is multiuser interference cancellation. It is most associated with the name of Sergio Verdú at Princeton University, and many believe that it is the next frontier in signal processing for wireless communications [2].[74] The simplicity in concept is, of course, bedeviled by the details of implementation.

The hard part is the detection of many different signals mixed together [125]. Even the detection of a single user in a CDMA spread spectrum architecture involves a large amount of signal processing. The more or less simultaneous detection of N intermixed signals is a task that increases in difficulty more or less *exponentially* as a function of N.[75] We can spend more

---

73. "The successive cancellation must operate fast enough to keep up with the bit-rate and not introduce intolerable delay. For this reason, it will presumably be necessary to limit the number of cancellations" [123, p. 53].

74. See also the excellent review article by Moshavi [124] and also [123].

75. Different algorithms tweak the processing burden to some degree, but the complexity/gain trade-off remains unfavorable for all multiuser detection schemes that really try to detect multiple signals. "Despite the fact that the performance of this [maximum likelihood sequence estimator] receiver is optimum, its computational com-

money on receiver hardware (the *parallel cancellation* architecture) or on processing horsepower (the *successive cancellation* architecture),[76] but even so, true multiuser detection is still considered impractical (i.e., impossible?) within the development window for 3G systems.[77] So far, most of the results are from simulations, where different schemes have promised considerable gains in performance. One reported field test found that mobile transmitting power could be reduced by 6 dB in a wideband CDMA system, through the use of linear parallel multistage interference cancellation [128].

This chapter has ranged widely across some of the newest and least coherent fields of wireless technology research. We have touched on a number of conundrums—the possible limits to the channel coding paradigm that has dominated communications engineering since Shannon, the beneficial effects of multipath fading, and the information theoretic implications of a more active approach to the use of channel information—which are probably still too fresh to sort out fully. The purely passive receiver is fast becoming an anachronism. New receiver architectures are being developed that will take a much more active role in reconstructing the transmitted signal and sorting it out from interfering signals, as well as exploiting the free information supplied by the channel itself.

Inevitably, changing ideas about receiver functionality lead back to a reconsideration of the transmitter architecture. Hand-in-glove with the trend toward active receivers, there is emerging a new approach—we might well say a new *philosophy*—toward signal construction. It is a philosophy that casts aside, sometimes implicitly, sometimes ostentatiously, the traditional assumptions regarding signal structure, orthogonality, and interference management. This is the subject of Chapter 8.

---

plexity, in terms of the number of operations $O(2^K)$, grows exponentially with the number of users (K), making the implementation of such a receiver extremely difficult" [123, p. 26]. See also [126].

76. Koulakiotis and Aghbami [123] present a reasonably comprehensive taxonomy of multiuser detection receiver structures. See also [124].

77. "One way to improve capacity is to employ sophisticated receiver structures. Performance close to the single-user receiver can be achieved through optimum multiuser detection. Unfortunately, this receiver is impractical as its complexity grows exponentially with the number of users" [127, p. 2294].

REFERENCES

[1] Shannon, C. E., "Channels with Side Information at the Transmitter," *IBM J. of Research and Development,* Vol. 2, 1958, pp. 289–293. Reprinted in Slepian, D., (Ed.), *Key Papers in the Development of Information Theory,* Piscataway, NJ: IEEE Press, 1974.

[2] Verdú, S., *Multiuser Detection,* New York: Cambridge University Press, 1998.

[3] Goldsmith, A. J., and P. P. Varaiya, "Capacity of Fading Channels with Channel Side Information," *IEEE Trans. on Information Theory,* Vol. 43, No. 6, November 1999, pp. 1986–1992.

[4] Hemami, S. S., "Robust Image Communication over Wireless Channels," *IEEE Communications Magazine,* Vol. 39, No. 11, November 2001, pp. 120–131.

[5] Girod, B., and N. Färber, "Feedback-Based Error Control for Mobile Video Transmission," *Proc. IEEE,* Vol. 87, No. 10, October 1999, pp. 1707–1723.

[6] Chen, B., and G. Wornell, "Analog Error-Correcting Codes Based on Chaotic Dynamical Systems," *IEEE Trans. on Communications,* Vol. 46, No. 7, July 1998, pp. 881–890.

[7] Olshausen, B. A., and D. J. Field, "Vision and the Coding of Natural Images," *American Scientist,* Vol. 88, No. 3, May–June 2000, pp. 238–245.

[8] Duda, R., P. E. Hart, and D. G. Stork, *Pattern Classification,* 2nd ed., New York: Wiley, 2001.

[9] Cardoso, J.-F., "Blind Signal Separation: Statistical Principles," *Proc. IEEE,* Vol. 86, No. 10, October 1998, pp. 2009–2025.

[10] Giannakis, G. B., and C. Tepedelenlioglu, "Basis Expansion Models and Diversity Techniques for Blind Identification and Equalization of Time-Varying Channels," *Proc. IEEE,* Vol. 86, No. 10, October 1998, pp. 1969–1986.

[11] Raleigh, G. G., and A. Paulraj, "Time Varying Vector Channel Estimation for Adaptive Spatial Equalization," *Proc. IEEE Global Telecommunications Conf.,* Singapore, November 1995, pp. 218–224. Reprinted in Rappaport, T., (Ed.), *Smart Antennas: Adaptive Arrays, Algorithms and Wireless Position Location,* Piscataway, NJ: IEEE Press, 1998.

[12] Ghogho, M., A. Swami, and T. Durrani, "Blind Synchronization and Doppler Spread Estimation for MSK Signals in Time-Selective Fading Channels," *Proc. IEEE Int. Conf. on Acoustics, Speech, and Signal Processing,* Istanbul, Turkey, June 2000, Vol. 5, pp. 2665–2668.

[13] Luise, M., and R. Reggiannini, "Carrier Frequency Recovery in All-Digital Modems for Burst-Mode Transmissions," *IEEE Trans. on Communications,* Vol. 43, Nos. 2–4, February–April 1995, pp. 1169–1178.

[14]   van der Veen, A.-J., and A. Paulraj, "An Analytical Constant Modulus Algorithm," *IEEE Trans. on Signal Processing*, Vol. 44, No. 5, May 1996, pp. 1136–1155. Reprinted in Rappaport, T., (Ed.), *Smart Antennas: Adaptive Arrays, Algorithms and Wireless Position Location*, Piscataway, NJ: IEEE Press, 1998.

[15]   Soliman, S., and K. Mokrani, "Performance of Coded Systems Over Fading Dispersive Channels," *IEEE Trans. on Communications*, Vol. 40, No. 1, January 1992, pp. 51–59.

[16]   Berlekamp, E. R., *Algebraic Coding Theory*, New York: McGraw-Hill, 1968.

[17]   Forney, G. D., "Generalized Minimum Distance Decoding," *IEEE Trans. on Information Theory*, Vol. 12, No. 2, April 1966, pp. 125–131.

[18]   Wicker, S. B., and V. Bhargava, (Eds.), *Reed–Solomon Codes and Their Applications*, Piscataway, NJ: IEEE Press, 1994.

[19]   Pursley, M. B., and C. S. Wilkins, "Adaptive-Rate Coding for Frequency-Hop Communications over Rayleigh Fading Channels," *IEEE J. on Selected Areas in Communications*, Vol. 17, No. 7, July 1999, pp. 1224–1232.

[20]   Abramson, N., "The Throughput of Packet Broadcasting Channels," *IEEE Trans. on Communications*, Vol. 25, No. 1, January 1977, pp. 117–128.

[21]   Thomas, G., "Capacity of the Wireless Packet Collision Channel Without Feedback," *IEEE Trans. on Information Theory*, Vol. 46, No. 3, May 2000, pp. 1141–1144.

[22]   Sayood, K., and J. C. Borkenhagen, "Use of Residual Redundancy in the Design of Joint Source/Channel Coders," *IEEE Trans. on Communications*, Vol. 39, No. 6, June 1991, pp. 838–839.

[23]   Van Dyck, R., and D. J. Miller, "Transport of Wireless Video Using Separate, Concatenated, and Joint Source-Channel Coding," *Proc. IEEE*, Vol. 87, No. 10, October 1999, pp. 1734–1750.

[24]   Wang, Y., and Q.-F. Zhu, "Error Control and Concealment for Video Communication: A Review," *Proc. IEEE*, Vol. 86, No. 5, May 1998, pp. 974–997.

[25]   Girod, B., and N. Färber, "Feedback-Based Error Control for Mobile Video Transmission," *Proc. IEEE*, Vol. 87, No. 10, October 1999, pp. 1707–1723.

[26]   Wang, Y., Q.-F. Zhu, and L. Shaw, "Maximally Smooth Image Recovery in Transform Coding," *IEEE Trans. on Communications*, Vol. 41, No. 10, October 1993, pp. 1544–1551.

[27]   Painter, T., and A. Spanias, "Perceptual Coding of Digital Audio," *Proc. IEEE*, Vol. 88, No. 4, April 2000, pp. 451–513.

[28]   Lam, W.-M., and A. R. Reibman, "An Error Concealment Algorithm for Images Subject to Channel Errors," *IEEE Trans. on Image Processing*, Vol. 4, No. 5, May 1995, pp. 533–542.

[29]   Salam, P., N. B. Shroff, and E. J. Delp, "Error Concealment in MPEG Video Streams over ATM Networks," *IEEE J. on Selected Areas in Communications*, Vol. 18, No. 6, June 2000, pp. 1129–1144.

[30]   Wong, W. C., R. Steele, and C.-E. Sundberg, *Source-Matched Mobile Communications*, Piscataway, NJ: IEEE Press, 1995.

[31]   Schroeder, M., B. S. Atal, and J. L. Hall, "Optimizing Digital Speech Coders by Exploiting Masking Properties of the Human Ear," *J. Acoustic Society of America*, December 1979, pp. 1647–1652.

[32]   Johnston, J., "Estimation of Perceptual Entropy Using Noise Masking Criteria," *Proc. ICASSP-88*, May 1988, pp. 2524–2527.

[33]   Hagenauer, J., and T. Stockhammer, "Channel Coding and Transmission Aspects for Wireless Multimedia," *Proc. IEEE*, Vol. 87, No. 19, October 1999, pp. 1764–1777.

[34]   Zorzi, M., R. Rao, and L. B. Milstein, "Error Statistics in Data Transmission over Fading Channels," *IEEE Trans. on Communications*, Vol. 46, No. 11, November 1998, pp. 1468–1477.

[35]   Soliman, S., and K. Mokrani, "Performance of Coded Systems over Fading Dispersive Channels," *IEEE Trans. on Communications*, Vol. 40, No. 1, January 1992, pp. 51–59.

[36]   Verdú, S., *Multiuser Detection*, New York: Cambridge University Press, 1998.

[37]   Ertel, R. B., et al., "Overview of Spatial Channel Models for Antenna Array Communication Systems," *IEEE Personal Communications*, Vol. 5, No. 1, February 1998, pp. 10–22.

[38]   Blahut, R., *Digital Transmission of Information*, Reading, MA: Addison-Wesley, 1990.

[39]   Sheinvald, J., M. Wax, and A. J. Weiss, "Localization of Multiple Sources with Moving Arrays," *IEEE Trans. on Signal Processing*, Vol. 46, No. 10, October 1998, pp. 2736–2743.

[40]   Shannon, C. E., "Prediction and Entropy of Printed English," *Bell System Technical J.*, Vol. 30, No. 1, January 1951, pp. 50–64.

[41]   Shannon, C. E., "A Mathematical Theory of Communication," *Bell System Technical J.*, Vol. 27, July 1948, pp. 379–423.

[42]   Ding, Z., "Multipath Channel Identification Based on Partial Systems Information," *IEEE Trans. on Signal Processing*, Vol. 45, No. 1, January 1997, pp. 235–240.

[43]   Hansson, U., and T. M. Aulin, "Aspects on Single Symbol Signaling on the Frequency Flat Rayleigh Fading Channel," *IEEE Trans. on Communications*, Vol. 47, No. 6, June 1999, pp. 874–883.

[44]   Turin, G. L., "Introduction to Spread Spectrum Antimultipath Techniques and Their Application to Urban Digital Radio," *Proc. IEEE*, Vol. 68, No. 3, March 1980, pp. 328–353:

[45]   Raleigh, G., and V. K. Jones, "Multivariate Modulation and Coding for Wireless Communication," *IEEE J. on Selected Areas in Communications*, Vol. 17, No. 5, May 1999, pp. 851–866.

[46]   Paulraj, A. J., and C. B. Papadias, "Space–Time Processing for Wireless Communications," *IEEE Signal Processing Magazine*, Vol. 14, No. 5, November 1997, pp. 49–83. Reprinted in Rappaport, T., (Ed.), *Smart Antennas: Adaptive Arrays, Algorithms and Wireless Position Location*, Piscataway, NJ: IEEE Press, 1998.

[47]   Ghogho, M., O. Besson, and A. Swami, "Estimation of Directions of Arrival of Multiple Scattered Sources," *IEEE Trans. on Signal Processing*, Vol. 49, No. 11, November 2001, pp. 2467–2480.

[48]   Clark, M. V., et al., "MMSE Diversity Combining for Wideband Digital Cellular Radio," *IEEE Trans. on Communications*, Vol. 40, No. 6, June 1992, pp. 1128–1135.

[49]   Sayeed, A. M., A. Sendonaris, and B. Aazhang, "Multiuser Detection in Fast-Fading Multipath Environments," *IEEE J. on Selected Areas in Communications*, Vol. 16, No. 9, December 1998, pp. 1691–1701.

[50]   Petrus, P., J. Reed, and T. Rappaport, "Geometrical-Based Statistical Macrocell Channel Model for Mobile Environments," *IEEE Trans. on Communications*, Vol. 50, No. 3, March 2002, pp. 495–502.

[51]   Kermoal, J., et al., "A Stochastic MIMO Radio Channel Model with Experimental Validation," *IEEE J. on Selected Areas in Communications*, Vol. 20, No. 6, August 2002, pp. 1211–1226.

[52]   Zhao, Y., and S.-G. Häggman, "Intercarrier Interference Self-Cancellation Scheme for OFDM Mobile Communications Systems," *IEEE Trans. on Communications*, Vol. 49, No. 7, July 2001, pp. 1185–1191.

[53]   Müller, T., and H. Rohling, "Channel Coding for Narrowband Rayleigh Fading with Robustness Against Changes in Doppler Spread," *IEEE Trans. on Communications*, Vol. 45, No. 2, February 1997, pp. 148–151.

[54]   Giannetti, F., M. Luise, and R. Reggiannini, "Simple Carrier Frequency Rate-of-Change Estimators," *IEEE Trans. on Communications*, Vol. 47, No. 9, September 1999, pp. 1310–1314.

[55]   Lee, W. C. Y., *Mobile Communications Engineering*, New York: McGraw-Hill, 1982, p. 40.

[56]   Cox, D., H. W. Arnold, and P. T. Porter., "Universal Digitial Portable Communications: A System Perspective," *IEEE J. on Selected Areas in Communications*, Vol. 5, No. 5, June 1987, pp. 764–773.

[57] Fuxjaeger, A. W., and R. A. Iltis, "Acquisition of Timing and Doppler-Shift in a Direct-Sequence Spread-Spectrum System," *IEEE Trans. on Communications*, Vol. 42, No. 10, October 1994, pp. 2870–2880.

[58] Yegani, P., and C. D. McGillem, "Model for the Factory Radio Channel," *IEEE Trans. on Communications*, Vol. 39, No. 10, October 1991, p. 1447.

[59] Paulraj, A., and B. C. Ng, "Space–Time Modems for Wireless Personal Communications," *IEEE Personal Communications*, February 1988, pp. 36–48.

[60] Fontan, P., et al., "Statistical Modeling of the LMS Channel," *IEEE Trans. on Vehicular Technology*, Vol. 50, No. 6, November 2001, p. 1557.

[61] Miller, S. Y., and S. C. Schwartz, "Integrated Spatial-Temporal Detectors for Asynchronous Gaussian Multiple-Access Channels," *IEEE Trans. on Communications*, Vol. 43, Nos. 2–4, February–April 1995, pp. 396–410.

[62] Vanderveen, M. C., A.-J. van der Veen, and A. Paulraj, "Estimation of Multipath Parameters in Wireless Communications," *IEEE Trans. on Signal Processing*, Vol. 46, No. 3, March 1998, pp. 682–690.

[63] Kohno, R., "Spatial and Temporal Communication Theory Using Adaptive Antenna Arrays," *IEEE Personal Communications*, Vol. 5, No. 1, February 1998, pp. 28–35.

[64] Sayeed, A., and B. Aazhang, "Joint Multipath-Doppler Diversity in Mobile Wireless Communications," *IEEE Trans. on Communications*, Vol. 47, No. 1, January 1999, pp. 123–132.

[65] O'Neill, J. C., and W. J. Williams, "A Function of Time, Frequency, Lag and Doppler," *IEEE Trans. on Signal Processing*, Vol. 47, No. 3, March 1999, pp. 789–799.

[66] Iltis, R., and A. Fuxjaeger, "A Digital DS Spread Spectrum Receiver with Joint Channel and Doppler Shift Estimation," *IEEE Trans. on Communications*, Vol. 39, No. 8, August 1991, pp. 1255–1267.

[67] DeBrunner, V., M. Ozaydin, and T. Przebinda, "Resolution in Time-Frequency," *IEEE Trans. on Signal Processing*, Vol. 47, No. 3, March 1999, pp. 783–788.

[68] Liberti, J., and T. Rappaport, "A Geometrically Based Model for Line of Sight Multipath Radio Channels," *IEEE Vehicular Technology Conf.*, Atlanta, GA, April 1996, pp. 844–848.

[69] Onggosanusi, E., A. M. Sayeed, and B. D. Van Veen, "Canonical Space-Time Processing for Wireless Communications," *IEEE Trans. on Communications*, Vol. 48, No. 10, October 2000, pp. 1669–1680.

[70] Baas, N. J., and D. P. Taylor, "Matched Filter Bounds for Wireless Communication over Rayleigh Fading Dispersive Channels," *IEEE Trans. on Communications*, Vol. 49, No. 9, September 2001, pp. 1525–1528.

[71]   Biglieri, E., J. Proakis, and S. Shamai, "Fading Channels: Information-Theoretic and Communications Aspects," in Verdú, S., and S. McLaughlin, (Eds.), *Information Theory: Fifty Years of Discovery*, Piscataway, NJ: IEEE Press, 2000, pp. 575–648.

[72]   Ali, I., et al., "Doppler Based Multiple Access (DBMA): A Novel Flow Control Mechanism for LEO Satellite Systems," *MILCOM 97 Proc.*, November 2–5, 1997, Vol. 1, pp. 103–108.

[73]   Ozarow, L. H., S. Shamai, and A. D. Wyner, "Information Theoretic Considerations for Cellular Mobile Radio," *IEEE Trans. on Vehicular Technology*, Vol. 43, No.2, May 1994, pp. 359–378.

[74]   Jalali, A., and P. Mermelstein, "Effects of Diversity, Power Control, and Bandwidth on the Capacity of Microcellular CDMA Systems," *IEEE J. on Selected Areas in Communications*, Vol. 12, No. 5, June 1994, pp. 952–961.

[75]   Raleigh, G., and V. K. Jones, "Multivariate Modulation and Coding for Wireless Communication," *IEEE J. on Selected Areas in Communications*, Vol. 17, No. 5, May 1999, pp. 851–866.

[76]   Pursley, M. "The Role of Spread Spectrum in Packet Radio Networks," *Proc. IEEE*, Vol. 75, No. 1, January 1987, p. 123.

[77]   Eng, T., N. Kong, and L. B. Milstein, "Comparison of Diversity Fading Combining Techniques for Rayleigh Fading Channels," *IEEE Trans. on Communications*, Vol. 44, No. 9, September 1996, pp. 1117–1129.

[78]   Clark, M. V., et al., "MMSE Diversity Combining for Wideband Digital Cellular Radio," *IEEE Trans. on Communications*, Vol. 40, No. 6, June 1992, pp. 1128–1135.

[79]   Davis, M., A. Monk, and L. B. Milstein, "A Noise Whitening Approach to Multiple-Access Noise Rejection—Part II: Implementation Issues," *IEEE J. on Selected Areas in Communications*, Vol. 14, No. 8, October 1996, pp. 1488–1499.

[80]   Millman, S., (Ed.), *A History of Engineering and Science in the Bell System: Communications Sciences (1925–1980)*, Holmdel, NJ: AT&T Bell Laboratories Press, 1984, pp. 422–425.

[81]   Gitlin, R., J. Hayes, and S. B. Weinstein, *Data Communications Principles*, New York: Plenum, 1992, p. 572.

[82]   Balaban, P., and J. Salz, "Optimum Diversity Combining and Equalization in Digital Data Transmission with Applications to Cellular Mobile Radio—Part I: Theoretical Considerations and Part II: Numerical Results," *IEEE Trans. on Communications*, Vol. 40, No. 5, May 1992, pp. 885–907.

[83]   Rappaport, T. S., *Wireless Communications: Principles and Practices*, Upper Saddle River, NJ: Prentice-Hall, 1996.

[84]   Proakis, J. G., *Digital Communications*, 2nd ed., New York: McGraw-Hill, 1989.

[85]   Gibson, J. D., *Principles of Analog and Digital Communications*, 2nd ed., New York: Macmillan, 1993.

[86]   Frerking, M., *Digital Signal Processing in Communication Systems*, New York: Van Nostrand Reinhold, 1994.

[87]   Sklar, B., "Rayleigh Fading Channels in Mobile Communication Systems, Part II: Mitigation," *IEEE Communications Magazine*, July 1997, pp. 102–109.

[88]   Choi, Y.-S., P. J. Voltz, and F. Casara, "On Channel Estimation and Detection for Multicarrier Signals in Fast and Selective Rayleigh Fading Channels," *IEEE Trans. on Communications*, Vol. 49, No. 8, August 2001, pp. 1375–1387.

[89]   Armstrong, J., "Analysis of New and Existing Methods of Reducing Intercarrier Interference Due to Carrier Frequency Offset in OFDM," *IEEE Trans. on Communications*, Vol. 47, No. 3, March 1999, pp. 365–369.

[90]   Jeon, W. G., K. H. Chang, and Y. S. Cho, "An Equalization Technique for Orthogonal Multiplexing Systems in Time-Variant Multipath Channels," *IEEE Trans. on Communications*, Vol. 47, No. 1, January 1999, pp. 27–32.

[91]   Ahn, J., and H. S. Lee, "Frequency Domain Equalization of OFDM Signals over Frequency Nonselective Rayleigh Fading Channels," *Electronic Letters*, Vol. 29, No. 16, August 5, 1993, pp. 1476–1477.

[92]   Pollet, T., et al., "Equalization for DMT-Based Broadband Modems," *IEEE Communications Magazine*, May 2000, pp. 106–113.

[93]   Hayashi, K., and S. Hara, "A New Spatio-Temporal Equalization Method Based on Estimated Channel Response," *IEEE Trans. on Vehicular Technology*, Vol. 50, No. 5, September 2001, pp. 1250–1259.

[94]   Cioffi, J. M., et al., "MMSE Decision-Feedback Equalizers and Coding—Part II: Coding Results," *IEEE Trans. on Communications*, Vol. 43, No. 10, October 1995, pp. 2595–2604.

[95]   Eyuboglu, M. V., and G. D. Forney, "Trellis Precoding: Combined Coding, Precoding and Shaping for Intersymbol Interference Channels," *IEEE Trans. on Information Theory*, Vol. 38, No. 2, March 1992, pp. 301–314.

[96]   Price, R., and P. E. Green, "A Communication Technique for Multipath Channels," *Proc. IRE*, Vol. 46, No. 3, March 1958, pp. 555–570.

[97]   Simon, M. K., et al., *Spread Spectrum Communications Handbook*, rev. ed. New York: McGraw-Hill, 1994.

[98]   Viterbi, A. J., "The Orthogonal-Random Waveform Dichotomy for Digital Mobile Personal Communication," *IEEE Personal Communications*, First Quarter 1994, pp. 18–24.

[99]   Grob, U., et al., "Microcellular Direct-Sequence Spread-Spectrum Radio System Using N-Path RAKE Receiver," *IEEE J. on Selected Areas in Communications*, Vol. 8, No. 5, June 1990, pp. 772–780.

[100] Bottomley, G. E., T. Ottosson, and Y.-P. E. Wang, "A Generalized RAKE Receiver for Interference Suppression," *IEEE J. on Selected Areas in Communications*, Vol. 18, No. 8, August 2000, pp. 1536–1545.

[101] Glisic, S., and M. D. Katz, "Modeling of the Code Acquisition Process for Rake Receivers in CDMA Wireless Networks with Multipath and Transmitter Diversity," *IEEE J. on Selected Areas in Communications*, Vol. 19, No. 1, January 2001, pp. 21–32.

[102] Win, M. Z., G. Chrisikos, and N. Sollenberger, "Performance of Rake Reception in Dense Multipath Channels: Implications of Spreading Bandwidth and Selection Diversity Order," *IEEE J. on Selected Areas in Communications*, Vol. 18, No. 8, August 2000, pp. 1516–1525.

[103] Win, M. Z., and Z. A. Kostic, "Impact of Spreading Bandwidth on Rake Reception in Dense Multipath Channels," *IEEE J. on Selected Areas in Communications*, Vol. 17, No. 10, October 1999, pp. 1794–1806.

[104] Liu, H., and K. Li, "A Decorrelating RAKE Receiver for CDMA Communications over Frequency-Selective Fading Channels," *IEEE Trans. on Communications*, Vol. 47, No. 7, July 1999, pp. 1036–1045.

[105] Wang, S.-Y., and C.-C. Huang, "On the Architecture and Performance of an FFT-Based Spread-Spectrum Downlink RAKE Receiver," *IEEE Trans. on Vehicular Technology*, Vol. 50, No. 1, January 2001, pp. 234–243.

[106] Rappaport, T. S., (Ed.), *Smart Antennas: Adaptive Arrays, Algorithms, and Wireless Position Location*, Piscataway, NJ: IEEE Press, 1998.

[107] Widrow, B., et al., "Adaptive Antenna Systems," *Proc. IEEE*, Vol. 55, No. 12, December 1967, pp. 2143–2159. Reprinted in Rappaport, T. S., (Ed.), *Smart Antennas: Adaptive Arrays, Algorithms, and Wireless Position Location*, Piscataway, NJ: IEEE Press, 1998.

[108] Gabriel, W. F., "Adaptive Processing Array Systems," *Proc. IEEE*, Vol. 80, No. 1, January 1992, pp. 152–162. Reprinted in Rappaport, T. S., (Ed.), *Smart Antennas: Adaptive Arrays, Algorithms, and Wireless Position Location*, Piscataway, NJ: IEEE Press, 1998.

[109] Godara, L., "Application of Antenna Arrays to Mobile Communications, Part II: Beam-Forming and Direction-of-Arrival Considerations," *Proc. IEEE*, Vol. 85, No. 8, August 1997, pp. 1195–1245. Reprinted in Rappaport, T. S., (Ed.), *Smart Antennas: Adaptive Arrays, Algorithms, and Wireless Position Location*, Piscataway, NJ: IEEE Press, 1998.

[110] Van Veen, B. D., and K. M. Buckley, "Beamforming: A Versatile Approach to Spatial Filtering," *IEEE ASSP Magazine*, April 1988, pp. 4–24. Reprinted in Rappaport, T. S., (Ed.), *Smart Antennas: Adaptive Arrays, Algorithms, and Wireless Position Location*, Piscataway, NJ: IEEE Press, 1998.

[111] Chiani, M., and A. Zanella, "Spatial and Temporal Equalization for Broadband Wireless Indoor Networks at Millimeter Waves," *IEEE J. on Selected Areas in Communications*, Vol. 17, No. 10, October 1999, pp. 1725–1734.

[112] Kohno, R., "Spatial and Temporal Communication Theory Using Adaptive Antenna Arrays," *IEEE Personal Communications*, February 1998, pp. 28–35.

[113] van der Veen, A.-J., S. Talwar, and A. Paulraj, "A Subspace Approach to Blind Space-Time Signal Processing for Wireless Communications Systems," *IEEE Trans. on Signal Processing*, Vol. 45, No. 1, January 1997, pp. 173–190.

[114] Wax, M., and A. Leshem, "Joint Estimation of Time Delays and Directions of Arrival of Multiple Reflections of a Known Signal," *IEEE Trans. on Signal Processing*, Vol. 45, No. 10, October 1997, pp. 2477–2484.

[115] Sung, S., and D. Brady, "Spectral Spatial Equalization for OFDM in Time-Varying Frequency-Selective Multipath Channels," *Proc. 2000 IEEE Sensor Array and Multichannel Signal Processing Workshop*, March 16–17, 2000, Cambridge, MA, pp. 434–438.

[116] Mun, C., M.-S. Choi, and H.-K. Park, "Performance of 2-D RAKE Receiver in a Correlated Frequency-Selective Nakagami-Fading," *IEEE Trans. on Vehicular Technology*, Vol. 50, No. 5, September 2001, pp. 1312–1317.

[117] Naguid, A. F., and A. Paulraj, "A Base-Station Antenna Array Receiver for Cellular DS/CDMA with M-ary Orthogonal Modulation," *Proc. of 28th Asilomar Conf. on Signals, Syst., Comput.*, Pacific Grove, CA, November 1994, pp. 858–852.

[118] Onggosanusi, E. N., A. Sayeed, and B. D. Van Veen, "Canonical Space-Time Processing for Wireless Communications," *IEEE Trans. on Communications*, Vol. 48, No. 10, October 2000, pp. 1669–1680.

[119] Liu, H., and M. D. Zoltowski, "Blind Equalization in Antenna Array CDMA Systems," *IEEE Trans. on Signal Processing*, Vol. 45, No. 1, January 1997, pp. 161–172.

[120] Bhashyam, S., A. M. Sayeed, and B. Aazhang, "Time-Selective Signaling and Reception for Communication over Multipath Fading Channels," *IEEE Trans. on Communications*, Vol. 48, No. 1, January 2000, pp. 83–94.

[121] Poor, H. V., and S. Verdú, "Single-User Detectors for Multiuser Channels," *IEEE Trans. on Communications*, Vol. 36, No. 1, January 1988, pp. 50–60.

[122] Koulakiotis, D., and A. H. Aghbami, "Data Detection Techniques for DS/CDMA Mobile Systems: A Review," *IEEE Personal Communications*, Vol. 7, No. 3, June 2000, pp. 24–34.

[123] Duel-Hallen, A., J. Holtzman, and Z. Zvonar, "Multiuser Detection for CDMA Systems," *IEEE Personal Communications*, April 1995, pp. 46–58.

[124] Moshavi, S., "Multi-User Detection for DS-CDMA Communications," *IEEE Communications Magazine*, October 1996, pp. 124–136.

[125] Honig, M., U. Madhow, and S. Verdú, "Blind Adaptive Multiuser Detection," *IEEE Trans. on Information Theory*, Vol. 41, No. 4, July 1995, pp. 944–960.

[126] Calderbank, A. R., G. Pottie, and N. Seshadri, "Cochannel Interference Suppression Through Time/Space Diversity," *IEEE Trans. on Information Theory*, Vol. 46, No. 3, May 2000, pp. 922–932.

[127] Luise, M., et al., "Guest Editorial: Signal Synchronization in Digital Transmission Systems," *IEEE J. on Selected Areas in Communications*, Vol. 19, No. 12, December 2001, pp. 2293–2295.

[128] Suzuki, T., et al., "Field Test Performance and Analysis of a Base Station to Cancel Wideband CDMA Interference," *IEICE Trans. on Communications*, Vol. E84-B, No. 3, March 2001, pp. 383–391.

8

Signal Expansion Strategies: Beyond Orthogonality

We are halfway through the looking glass. The traditional valences of wireless channel phenomena such as multipath and Doppler spread seem to have been reversed. Interference is no longer simply toxic. Like a powerful drug with serious but controllable side effects, it may, used in the right dosage, promote good communications health. The idea of strict physical orthogonality—that is, of strictly separating information-bearing signals from different users into distinct space–time–frequency bins with physical buffers between them—has been called into question. In the last chapter we saw how passing through the physical channel causes the signal to *expand* in all three physical dimensions and, more surprisingly, we learned that many researchers are beginning to view this spreading, which is normally seen as a problem, to be instead a *benefit* that can be exploited to improve the robustness of the transmission, and—what is even more astonishing—to *increase* the information-carrying capacity of the signal. We have seen how, viewed from this perspective, a human-made signal seems somehow to be enriched during transmission by new information created by the channel itself, which the receiver can apply powerfully to its own purposes. Indeed, by even the roughest calculations, it would appear that the channel enriches the signal by an enormous factor—the added information is many times greater than the original content of the transmitted message.

In this chapter, we extend these lessons toward their logical asymptote: If signal spreading is beneficial, why not spread *intentionally*? Why not break down the orthogonal signal boundaries explicitly and thoroughly and create a new type of signal structure that is in some sense fully expanded, or at least as fully expanded as we can possibly manage within the limits of our receiver technology? If natural signal spreading is really a boon for communications, then shouldn't it be possible to *design* the expansion in a controlled way to achieve even better results?

Let us consider, then, how to expand the signal intentionally. Indeed, at least *three* very different strategies can be used to accomplish this. Each strategy approaches the problem from a different historical starting point, and the fundamental similarities among these approaches have become clear only relatively recently. As we might guess from the recurrence of the magic number three, each strategy attacks the challenge of signal spreading from a different signal dimension: space, time, or frequency.

The oldest form of designed signal spreading is known as *spread spectrum,* with its pedigree going back to classified military communications systems first developed more than 50 years ago.[1] It is a familiar subject; in fact, in some respects it is almost too familiar; some of its important attributes are neglected even by its adherents. Spread spectrum has become the basis for the first commercial quasi-post-Shannon wireless architecture (IS-95 CDMA), and has been selected as the basic architecture for true 3G standards soon to emerge [*wideband CDMA* (WCDMA) and CDMA2000]. Very roughly speaking, spread spectrum in its common form[2] involves a kind of forced multiplication of the signal in the *time* domain, the effect of which (per Fourier analysis) is to expand the signal in the frequency domain.

The second form of intentional signal spreading is based on a very different set of ideas emanating from several seminal articles published in the early 1970s. It involves a forced multiplication or spreading of the signal in the *frequency* domain (again, speaking roughly), with a resultant expansion effect in time (at the symbol level, as we shall see). This type of signal structure is called by several names. In wireless applications now emerging, it is often but not always called *orthogonal frequency-division multiplexing,* or OFDM (although this is a compound misnomer, because arguably OFDM is neither orthogonal in the normal sense at all, nor is it really much like what is normally referred to as frequency-division multiplexing). In the field of

---

1. See [1] and historical references cited therein.
2. The common form is *direct-sequence* spread spectrum.

high-speed digital subscriber lines, the same technique is known as discrete multitone modulation, and has become the most popular of several competing DSL schemes. Other terminological and architectural variants include multicarrier modulation and multicarrier CDMA.

The third form of spread signal architecture is so new that it is still only half glimpsed, as it were, even by its proponents. It is based on a forced spreading in *space*, by transmitting the same signal many times over many different antennas, with various extra signal processing wrinkles thrown in. The technology is still in its infancy and, characteristically, there is a plethora of similar concepts veiled in very different terminologies. The closest thing to a generic label is perhaps what is referred to as the *multiple-input/multiple-output* architecture, although many of these systems also employ some sort of time spreading, so *space–time coding* is another common search phrase. One of the splashier proposals has emerged recently from work at Bell Labs and goes by the name of *BLAST* (styled as Bell Labs Layered Space–Time), which itself comes in several colors. More prosaic versions of the same strategy are sometimes labeled simply *transmitter diversity.*

In this chapter, then, we review the ways in which the new view of the wireless channel, which was the main subject of Chapter 7, has begun to actively influence the design of wireless signal construction strategies. Common to all of these techniques is the more or less explicit abandonment of orthogonality, in the normal sense, to design signals that do not require physical buffering in space, time, or frequency from other similar signals. The end result and indeed the motivating goal in all of these cases is often a *new multiple access architecture* that does not require *physical* separation of signals from different users. Strange as it seems to engineers raised in the classical pre-Shannon world of interference avoidance and physical buffering, these spread signal architectures allow multiple users to occupy the same S-T-F space simultaneously and completely. Receivers are designed to discriminate on the basis of coded signatures that do not depend on these seemingly basic physical dimensions in order to allow separation of different users.

The combination of multiple spread signals into full-blown multiple access architectures is beyond the scope of this chapter and this volume.[3] That said, the reader is assumed to be generally familiar with the principles of code division, and this knowledge will be helpful in following the presentation here.[4] The focus of this chapter is on the physical construction of a

---

3. Multiple access implementations are intended to be the main subject of Volume 2 of this three-part work.

4. My earlier books give an overview of CDMA [1, 2].

spread signal. From the standpoint of the physical layer technologies, the hallmark of spread signal design is that the information-carrying capacity of the channel is enhanced. Whether that capacity is allocated to a single user or divided among a large number of lower rate users is largely, though not entirely, irrelevant to the spreading technique.

8.1 An Analogy: Stereo and Beyond

The perverse, provocative, paradoxical view of the wireless signal construction process that we explore in this chapter can be approached by way of an analogy that exposes how oddly natural and straightforward the new paradigm really is. The analogy I wish to draw is with a natural information processing system that is part of the foundation of human experience. This system incorporates an unusual receiver architecture that is designed (if the idea of design is still permitted here) to process acoustic information—the human *ear*.[5]

The study of the ear—the physiology and psychology of hearing—is one of the intellectual sources of communications science. Alexander Graham Bell began, of course, as a teacher of the deaf, a practical pathologist of the ear, and he spent long hours dissecting cadavers to study the structure of the tympanum and the cochlea. One of his first experimental communications devices involved connecting a dismembered ear to a mechanical stylus in an attempt to replicate the mechano-physiology of sound reception. Somewhere in the course of his obsession, the vital idea took hold that the ear could be viewed as a model of a simple system to convert sound into some other type of signal. The diaphragm of the telephone that he invented is a kind of prosthesis for the eardrum, converting sound into a small electrical current. Indeed, in the original versions of the Bell telephone it was voice power that directly drove the microphone—the idea of modulating an electrical carrier came later.

Research on hearing processes continued to provide important insights into early speech compression systems, and we still can find today, buried at the core of any *voice* transmission system, an idealized ear—that is, a model of the signal as it is received and processed by the human ear. For example, the important technical parameters of PCM such as sampling rate and code

---

5. Much of this information was drawn from material available on a wonderful Web site created by Northwestern University's Music School: http://www.northwestern.edu/musicschool/classes/3D/index.html.

structure are based on highly simplified forms of the response patterns of the ear. Shannon's idea of rate distortion as a criterion for determining acceptable quantization losses to create a digital representation of an analog signal is implicitly calibrated to perceptual models of what can and what cannot be heard, that is, discriminated by means of the ear. In other words, the very definition of information itself, in Shannon's own terms, is obscurely bound up with the design limits of the ear.[6]

How does the ear really do its job? Consider, say, a performance of the "Ode to Joy" chorus of Beethoven's Ninth Symphony, from a communications perspective. Assume, for simplicity, that it is already in the form of a recording. Pop the CD into the CD player, pull up a chair, and the transmission begins. Music and voices fill the air. People settle into themselves, their heart rates stabilize, and blood pressure drops. What is going on here?

Approaching this arrangement—transmitter, signal, channel, and receiver—from the standpoint of a wireless communications engineer, several things strike us right away. First, we take note of the unusual receiver configuration. The receiver has two channels, or, more accurately, there are two separate and independent receivers—two human ears, each of which constitutes a complete receiver capable of demodulating the signal and converting it into neural code, forwarded for further processing downstream in the cerebral cortex. Somewhere deep in the brain the experience of a unified sound experience is synthesized from these two independent channels. That part is still a deep mystery.

Why two receivers? It is not immediately clear. The answer we might be tempted to offer, redundancy—which is a corollary of space diversity (multiple antennas) in a wireless system designed to average out the effects of multipath fading—would be wrong. We do not have two ears just in case we go deaf in one of them. Nor would it seem that multiple receivers are intended to cope with destructive acoustic fading or delay spread; our two ears do not seem to help much in situations where acoustic echoes (gross multipath) dominate the channel. (Consider how hard it is to hear voice messages transmitted over loudspeakers in large train stations, for example.)

We have two ears because two ears can collect *more information* than just one ear. Part of that information relates to the *spatial location* of the

---

6. I will not pursue this admittedly provocative observation here—but I stand by it. The definition of information in Shannon's writing is clearly dependent on conversion of the original analog phenomenon into digital form—there is no such thing as analog information!—and this step in turn is based on the semantic criterion, which is based on the ear and its capabilities for discrimination.

transmitter. Our two ears constitute a spatial sensor array, and we can local-ize sounds with remarkable accuracy: We can resolve a minimum audible angle as narrow as 2° or less depending on the exact task.[7] This is much bet-ter than any commercial RF antenna array, even those with many more elements.

Indeed, the ear performs with too much accuracy it would seem, or at least more than we can immediately account for. As wireless engineers, we know that a two-element array should be *limited to two dimensions* in its abil-ity to discriminate angle of arrival.[8] Given a horizontal placement of the two sensor elements, we should expect to see good discrimination of left, right, and center, but not in front or behind, up or down.[9] Three sensors ought to be needed in order to distinguish three spatial dimensions. In other words, a sound coming from a source directly over your head ought to be indistin-guishable from a sound originating from a point directly behind your head. Yet humans can easily distinguish the two. Apparently there is more to the array structure of the ear than just two sensor elements.

More detailed study of the physiology of the hearing process has dis-closed that the human acoustic information processing system actually develops additional auditory pathways within the human body. That is, the ear actually creates additional, very fine-grained multipath information dur-ing the last segment of the acoustic channel:

> The sound waves that reach the listener's two eardrums are affected by the interaction of the original sound wave with the listener's torso, head, pinnae (outer ears), and ear canals. The composite of these properties can be measured and captured as a head-related transfer function,

7. "The highest resolution is evident in the horizontal dimension, especially in front of the listener where the minimum audible angle is 2-degrees or less.... That angle in-creases to near 10-degrees at the sides and narrows to near 6-degrees in the rear. By comparison, the resolution in the vertical dimension is quite low. The vertical mini-mum audible angle in front is near 9-degrees and steadily increases until overhead when it reaches 22-degrees" [3]. See also [4].

8. Interesting enough, this puzzle was first studied by Lord Rayleigh, in 1907 [5].

9. "Classical psychoacoustics focused on the separation of the two ears and proposed the duplex theory of sound localization.... Interaural intensity difference (IID) and interaural time difference (ITD) each make a significant impact on perceptual judg-ments.... These observations do not, however, provide sufficient explanation for hu-man localization. In fact, IID and ITD only affect ... perceived position along a left–right axis between the ears. With only ITD and IID, a person cannot judge whether an acoustic event is in front, above, behind or below" [3, p. 8].

HRTF. *The complexity of the interaction of the sound wave with the acoustics of the listener's body makes the HRTF at each ear strongly dependent on the direction of the sound* [emphasis added]. [3, p. 3]

In a classical Shannon framework, we might have analyzed these various extra resonances as interference or noise, an inescapable degradation due to the physical limitations of human tissue. We would have focused on the ability of the ear to function *despite* these implementation losses. It is becoming clear from a variety of physiological studies that this is probably the wrong way to look at it. If the auditory system were based on the classical signal versus noise model, the ear might well have been much more isolated in various ways, acoustically, from the rest of the body. Instead, we find that the ear-receiver is embedded into the bone structure of the skull, which transmits sound well, and is coupled structurally with the complex acoustic cavities in the head in ways that suggest that the extra resonances are intended to be enhanced, not suppressed. The HRTF is a designed feature of the system, intended to expand the multipath signature and improve our localization capabilities without adding extra receiver elements.

There are numerous acoustic factors which add complexity and richness to HRTFs. For example, there is a clear magnitude peak in the region around 3000 Hz that is caused by the resonance of the ear canal. There are notches and other fine details in the magnitude response caused by *the constructive and destructive interference of the direct wave with sound reflected off the body* [emphasis added]. Reflected sound below 2000 Hz is mainly from the torso, and above 4000 Hz it is mainly from the pinnae; in between, there is a region of overlapping influence. [3, p. 5][10]

In short, we find that the receiver actively develops and uses additional paths in the acoustic channel, to create important additional information about the signal.

There is more. The HRTFs are in fact much richer than needed strictly for angular spatial localization.[11] The rich multipath signature contains other

---

10. See also [6].

11. "Even though HRTFs are very rich in acoustic detail, perceptual research suggests that the auditory system is selective in the acoustic information that it utilizes in making judgments of sound direction. [For example] Evidence reveals that monaural phase information is irrelevant to spatial perception, and that interaural phase information is extremely important" [3, p. 8].

types of information as well, such as *distance*. We can normally tell approximately how far away the source of a sound is, as well as its direction. If we consider the evolutionary determinants of the hearing system, it would seem obvious that both direction and distance of a sound source are required to enable an animal to effectively respond to it (e.g., as a prey item or as a threat). Yet even HRTFs cannot be the total answer to the development of accurate range information. It has been found experimentally that HRTFs change very little when the distance of the sound source is more than about 6 feet from the head [3, p. 6]. Larger scale distance information is extracted primarily from what the acoustic engineer calls the reverberation pattern, and what the RF engineer might refer to as the time-delay signature [3].[12]

Beyond fully localizing the source in space (direction and range), the HRTF and the auditory processing system are capable of developing other types of information about the channel. In listening to a recorded music performance, the listener is able to sense many things about the room within which the sound transmission is occurring, such as its size and aspects of its shape. It may be possible to tell whether the room is carpeted, or even from what sort of materials the surfaces are constructed.[13] Even the temperature and humidity of the air can leave a discernible imprint on the acoustic signal.

We do not normally analyze our knowledge of sounds into these components; instead, the channel information is often attached to a particular task or embedded as a part of our rich overall experience of a particular situation, but people are capable of construing huge amounts of information from seemingly limited sound signals. I have noticed that most people can tell not only the likely location of a coin dropped on a hard surface (and where to look for it), but even the denomination of the coin and sometimes the type of surface. Certainly, in rooms designed specifically for acoustic events (e.g., symphony halls), the physical characteristics of the environment are designed quite carefully, precisely because the listener is able to discern differences in the environment that may be construed positively or negatively for the listening experience. I remember the controversy that developed several years ago after renovations at Carnegie Hall in New York, when musicians and others began to claim that they could hear a layer of concrete that had been added underneath the stage—even though the contractor and

---

12. See also [7].

13. "The listener may also perceive the perceptual qualities associated with concert halls such as definition and spaciousness. Maybe the listener also hears certain properties of the environment such as the size and shape of a room or the reflective properties of the walls and furnishings" [3, p. 2].

Carnegie authorities vehemently denied it. After several years of continuing complaints, the floor of the stage was excavated and indeed it was found that a concrete layer had been installed. I am not sure that I would have heard the difference or known how to interpret it, but this shows that a trained ear can extract very specific information from the channel transfer function of an acoustic signal.

We will get to the transmitters in a moment, but we may say that in some ways the transmitter can be seen as the mirror image of the receiver. Figure 8.1 shows a design for a sophisticated stereo system, whereby a single input is divided into two channels (for the right and left speakers), and each channel is shaped by a filter (transfer function) that is in turn weighted by an appropriate HRTF corresponding specifically to the right or left ear of a particular listener. The channels are also adjusted for gain (loudness) and delay, based on the presumed orientation of the listener's head to the sound source. This system could also be viewed in reverse: The left and right filters are the listener's two ears, shaped by unique HRTFs and characterized by different gain and delay patterns in the received signals. These two channels

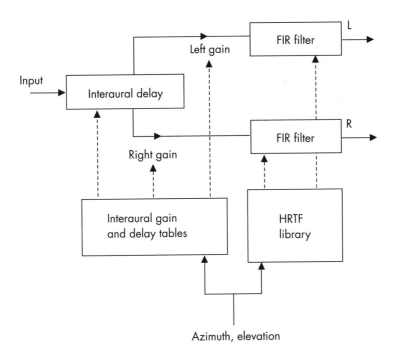

FIGURE 8.1 *The complex acoustic channel. (From: [3]. © 1995 Computer Music Journal.)*

are eventually combined by the brain into a single sound experience, and parameters such as azimuth and elevation are generated as outputs.

Beyond the spatial characteristics of the signal, we can examine the acoustic signal in terms familiar to us from our discussions in the previous chapters, such as the time domain and the frequency domain. We find a familiar distortion process that also yields important information to the post-Shannon receiver (Figure 8.2). Researchers have studied the impulse response of the auditory system, with findings that should sound familiar:

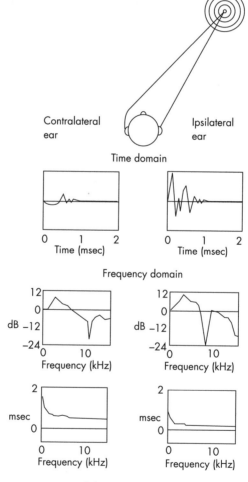

FIGURE 8.2 *Distortion in time and frequency domains of the acoustic signal. (From: [3]. © 1995 Computer Music Journal.)*

From the standpoint of the time domain, the signals that reach the two ears are no longer impulsive. The energy has been spread over 1–3 msecs by the acoustic interaction with the listener's body. Comparing the two ears, the sound arriving at the [near] ear is generally more intense and arrives earlier than that at the [far] ear. These differences between the two ears are called the interaural intensity difference (IID) and the interaural time difference (ITD) respectively. When a sound source is completely to the side ... the ITD reaches a maximum near .7 to .8 msec. [3, p. 4]

In other words, as the channel spreads the signal, each receiver (each ear) sees a different signature in the spreading function. The detailed translation of the impulse response into a time–energy curve, plotted against the angle of arrival, shows a much more complex pattern (Figure 8.3).

To fully synthesize these perspectives on the auditory system may still be a bit out of reach, pending studies of the neurophysiology of sound perception that are slowly tracing the channel back up into the brain. The experience of sound is an integrated experience. We do not really hear two distinct signals, one from each ear, in the way that we do receive two distinct signals from the tactile sensors on our right and left hands. We hear unified sound events, and attached to them are many layers of important information such as location and other characteristics of the environment that constitute the acoustic channel. This complex combination of two signals by the brain is still enveloped in mystery and viewed by those in the field with a sense of awe:

Each ear has a different transformation, and the transformation changes as the head and/or the source moves. The auditory system performs the

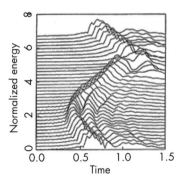

FIGURE 8.3 *Time–energy curve versus AOA in the acoustic channel. (From: [3]. ©*
1995 Computer Music Journal.)

phenomenal task of integrating the information arriving at the two ears into a single, fused perceptual image of the acoustic event in space: the auditory system extracts out the directional information and reconstructs an estimate of the original source spectrum. This is accomplished in spite of the fact that there is no direct structural representation of spatial information in the auditory system as there is in the visual system ... No wonder that research into three-dimensional sound has lagged behind research into three-dimensional vision! [3, p. 8]

This last point—the complexity of the auditory translation—has just begun to be explored by neurophysiologists. It is indeed clear that hearing involves a more subtle combination of multiple signals than vision. Tracing the neurons in the brain that map the inputs from individual sensors (such as retinal cells or cochlear hair cells in the ear), physiological researchers are finding that the auditory map interacts with the visual map, and is to some degree calibrated by it, but the information from the auditory system is independently derived. The following comments are from a recent study of the interaction between auditory maps and visual maps in barn owls. (Barn owls? I have intercalated what I hope is an appropriate commentary based on the main topic at hand—wireless communications.)

Most brain maps are projection maps, in which neurons in the retina, say, project to the relevant brain area in such a way that the spatial arrangements of the neurons is the same. [8, p. 30][14]

In other words, the visual map constructed in the brain is, we would say, linear, or a bit map representation of the signal. At least at the stage of creating the visual map in the visual cortex, the system is not heavily focused on the structure of the signal.[15]

In contrast, the auditory space map... does not reflect the [physical or spatial] organization of the auditory nerves, but is actively computed when information from the two ears is combined in the brain. The trans-

---

14. The article quoted here is a summary of a more technical presentation of the underlying research, which is found in the same issue [9].

15. Obviously, the vision system encompasses a vast amount of further postprocessing of the visual map, to distinguish edges, shapes and objects, and colors, and many other complex features that require nonlinear transformations. It does seem, however, that the initial visual map is something like a bit map of the retinal impulses.

> lation of auditory cues into a topological representation of space is clearly
> a complex transformation, even for a static auditory scene. [8, p. 30]

That is, the auditory system works immediately and directly on the *structure* of the signal, as it is decomposed in the time and frequency domains, and the space map appears thus to be constructed from these non-linear computations. The ear is truly an even more active receiver than the eye. In broad terms, the form of this transformation has been plotted: "Sounds are first transmitted to the brain's central nucleus ... where they are grouped by various criteria such as frequency" [8, p. 30]. That is, the structure of the signal components is distilled and clarified.

> The resulting neural signals are then ... organized into an arrangement
> that represents the location of sounds in space. The auditory map is
> modified by visual signals ... and aligned with a visual map, producing a
> multimodal map of space. [8, p. 30]

The receiver uses side information that is generated from an independent channel (the visual channel) as a kind of pilot to help calibrate the auditory map.

In humans, and perhaps barn owls, which are both strongly visual species, the visual space map may occupy primary position in creating the animal's sense of its environment, and the auditory map probably operates for most of us at a fairly subliminal level most of the time. We use vision as the primary system for locating stimulus events in space, with audio supplying important but subsidiary cues in most cases. However, other species have evolved that use the auditory map as the primary or even the only basis for creating the spatial model of their environments. Bats and whales are important mammalian examples. These species have learned to enhance the auditory map by generating their own sound events to create reflections on demand, rather than waiting for the environment to randomly supply these events. The ability of bats to navigate through very fine obstacles in rapid flight using only their sonar system is a well-known and still amazing demonstration that multipath processing of sound signals can be used to create an enormous amount of information supporting a highly detailed spatial map.

We may have delved deeply enough into a foreign discipline for our purposes. To recapitulate:

1. The acoustic information processing system is not intended merely to receive and decode a narrowly construed information-bearing signal, based on the classical Shannon model. It is designed to create

an accurate map of the entire physical environment using acoustic signals (with help, apparently, from the visual side channel).

2. To do so, the auditory system must exploit the multipath information created by the physical channel. Hence, there are two receivers whose outputs are later combined in a complex nonlinear way that relies heavily on detecting the structure of all the signal components, rather than simply separating the desired signal from the background noise.

3. To further enhance the power of the auditory system, the receiver is designed to create additional multipath components—to add even more structure to the signal—by using the various pathways through the body and the structure of the ear. Head, torso, outer ear, and ear canal all add extra resonances to the received signal, distorting the signal in ways that create richer signatures containing more information. This enables the auditory system to discriminate spatial information, for example, with much greater accuracy than should normally be possible with just two sensor elements (two ears).

4. The ear apparently analyzes the spreading of the signal in the time and frequency domains to obtain its read on the structure of the environment, in a way that is suggestive of the use of time and frequency signatures in a wireless receiver to discriminate signal components.

This picture of the acoustic receiver should give us some comfort in viewing the wireless receiver in a different light, as we have tried to show in Chapter 7: as an active, constructive system that can, in principle, take advantage of structured channel information to develop a much stronger model of signal detection, with various useful outputs of extra information (such as location) that were not a part of the original transmitted signal at all.

What about the transmitter, and the generation of the acoustic signal in the first place? Signal construction is the real focus of this chapter.

The classical Shannon communications model, applied to the transmission of Beethoven's "Ode to Joy," would suggest that the optimum signaling strategy would be to position a single transmitter or speaker at a reasonable short distance directly in front of the listener, in the middle of an open environment well away from walls or other large objects that could create multipath interference. We would also want to instruct the listener to remain rigidly still, so as not to change the angle of incidence of the sound source on his two ears. Indeed, we might even instruct him to cover one ear so as to eliminate the redundant and potentially interfering signal. This conjures a Monty Pythonesque image of an intensely concentrating audiophile (perhaps in a bowler hat) sitting on a small chair in the middle of an open field, staring intently at a small speaker a few feet in front of him, and imagining that he has obtained a near-optimum experience of Beethoven's music.

Contrast this with what today's state-of-the-art acoustic transmission systems actually look like. For example, Bose has recently designed a high-performance audio system for the Porsche Carrera [10]. The system employs no less than seven separate transmitters (speakers) within the confines of the two-seater's interior. The design of this system is reported to have taken 3 years and required a painstaking analysis of the acoustic environment of the cabin. The goal was to create a sonic illusion of a wide field of sound. Cost? About $3,000.

Indeed, we all realize that stereophonic sound generated from at least two speakers is superior to monophonic audio. Serious students of sound reproduction are intensely critical, however, of the limitations of a two-speaker system:

> Traditional stereo reproduction provides you with some spatial information, but not enough to recreate the full dimensionality of being in a room with a live musical performance. Rather than sensing that you are within a 3-D space, loudspeaker reproduction creates the impression that you are in front of the sound space.... You are relegated to the role of an immobile observer with *impoverished sensory information* [emphasis added]. [3, p. 2]

To this end, multichannel systems have been developed, and different techniques applied to split the normal left/right pair of recorded signals into four or more artificially differentiated signal components transmitted through multiple speaker elements. Many systems now routinely provide for separate transmission of different signal components based on frequency (e.g., bass speakers designed to be placed separately from high-frequency tweeters). More advanced techniques pick apart the signals in the electronic domain and create artificial signal images that are intended to enrich the audio. Work on three-dimensional sound is progressing, but the mainstay of sound enrichment is the multiplication of transmitters ("the best 3-D solution for large listening spaces is to use an array of loudspeakers" [3, p. 16]).

Similar considerations enter into the design of concert halls and even audio rooms in smaller settings, where far from avoiding multipath, the practical solutions all involve designing a signal that is well and evenly spread.

> When a concert hall is being designed there are many factors ... that must be considered to create good sound. It is important to avoid the use of large flat walls because reflections will be unevenly distributed. Walls that are curved inside [i.e., concave] create more problems because sound is focused to certain hot spots that are very intense, while other places lack reverberation. [11, p. 1]

In other words, good design here is not about suppressing reflections or reverberations. To the contrary, it is largely about making sure that the signal is spread evenly so that all seats in the building enjoy good and rich multipath characteristics. The time-domain analysis of each potential receiver position—each seat—should show a similar, voluptuous delay pattern (Figure 8.4).

The analogy is complete. Multipath is the acoustic engineer's friend, and the very survival of bats and other species depends on creating and processing an extremely dense, rich received signal that is full of extra information components that the classical Shannon model would exclude or convert to noise. The use of multiple receivers and multiple transmitters is accepted as the way to create a superior signal and improve communications performance. The enriched signal carries an astonishing amount of information about the spaciousness, humidity, temperature, directionality, even the shape and volume of the room, the presence of carpeting, furniture, draperies, the unseen layer of concrete under the floor, and so forth.

Kendall summarizes the perspective of the acoustic engineer:

> When an acoustic event occurs in a natural environment, sound waves from that event propagate in all directions. The waves encounter objects in the environment with which they interact by reflection and diffraction. The constructive and destructive interference of all the resulting waves creates a rich acoustic admixture that is further enriched when there are multiple sound sources. [3, p. 3]

If there is one single way to best characterize what this extended analogy teaches, it is that signal spreading is a vital strategy in the construction of strong communications links. Let us now consider the most common techniques for spreading an RF signal.

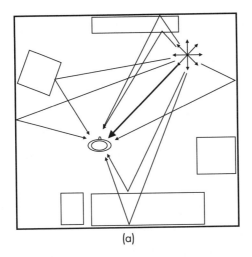

(a)

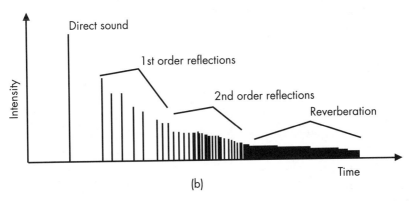

(b)

FIGURE 8.4 *Richness of the acoustic channel: (a) acoustic multipath; and (b) the time domain acoustic signature. (From: [3]. © 1995 Computer Music Journal.)*

8.2 Spreading Forced Through Multiplication in the Time Domain: Direct-Sequence Spread Spectrum

The typical data rate for a wireless voice signal is around 10 Kbps today. Using the general rule of thumb that it should take one frequency cycle to encode one information-bearing symbol, a binary signaling system carrying a 10-Kbps information-bearing message should occupy about 10 kHz of bandwidth measured in the frequency domain. If we ignore the implementation inefficiencies and rather moderate spreading tendencies inherent in most realistic modulation schemes (which imposes a guardband requirement, that is, F-buffering), we indeed find that this data-rate-to-frequency-bandwidth translation does roughly hold.

This is a very narrow signal—let us call it now the *unspread* signal—and it will be seriously affected by frequency-selective multipath fading. The coherence frequency bandwidth (F_{coh}) is typically at least 500 kHz at cellular frequencies, which means that—viewed from a frequency-domain perspective—the unspread signal runs the risk of falling into some very large holes. We can liken it perhaps to a bicyclist trying to ride a racing bike across a roadway that is pockmarked with large potholes a foot or more in diameter. Hitting one of those holes, the wheel of the bike will fall in and damage is inevitable (or at least a very bumpy ride). If, however, we could somehow expand the signal, so that it occupied a bandwidth much larger than F_{coh}, then it would be much less vulnerable. Just as a truck, with large wide tires, can easily roll over a 1-foot pothole in the road without even feeling it, a spread version of the signal would be much more robust in a multipath fading environment.

How can we spread this signal, assuming that we still have only 10 Kbps of actual information payload to carry? How can we force this narrow signal to expand? The answer is quite simple, although there are two quite different ways of looking at it.

One perspective is the *logical* or code-oriented perspective. A very straightforward strategy is to simply multiply each information symbol by, say, 100, and to transmit the multiplied data symbols in the same time interval, that is, to speed up the transmission rate. Each symbol is repeated 100 times, and the 10-Kbps signaling rate is speeded up to 1,000 Kbps (Figure 8.5). Why is this repeating code more robust? The receiver now has 100 copies of each data symbol to work with and, in principle, it can lose a great many of them to the frequency potholes and still successfully receive the intended message.

But the same phenomenon can be viewed from a *physical* perspective, where it is much more interesting. Physically speaking, it is the increased speed of the data rate and the shortening of each individual symbol that are important. The fundamental, albeit mysterious trade-off between time and frequency comes into play. As the time period of each transmitted pulse becomes shorter (holding other things constant, of course), the frequency spectrum of each pulse becomes wider. As the pulse duration becomes even shorter than the impulse response time of the channel—that is, as the pulse becomes short enough to sneak through the channel before it can be significantly distorted by the channel—the spectrum broadens out to create a new sort of physical signal:

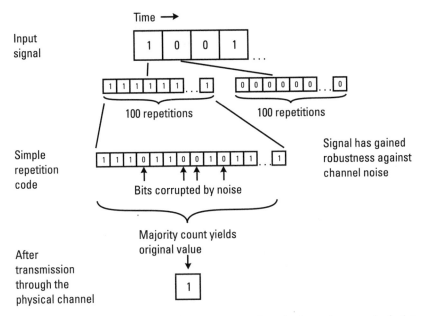

FIGURE 8.5 *Expansion by means of a repetition code—the brute force method of signal expansion.*

Sufficiently short pulses can be treated as though they possess a flat spectrum over an indefinitely wide range of frequencies. The expression *sufficiently short* means that the duration of the pulse is much shorter than the response time of any part of the system under consideration. Such a pulse can be completely characterized by its strength or area; its precise shape is immaterial. [12, p. 35]

This sort of signal is called a *spread spectrum signal,* and it possesses a number of interesting properties.

First, it is said to be characterized by flat fading. It is much wider than F_{coh} and is not nearly as vulnerable to frequency-selective multipath fading. This resistance to narrowband fading is inherent in the signal, physically, and has nothing to do with the coding applied at the logical level.

Second, viewed in the time domain, each pulse makes it through the channel before the channel has a chance to change. There is no meaningful distortion *within* each pulse. Different pulses, of course, see different channel configurations, but they are able to be discriminated as distinct entities by the receiver. To restate that, a properly designed receiver (e.g., a RAKE receiver) can discriminate each pulse clearly. In fact, the receiver can

discriminate different images of each pulse created by the multipath channel. Instead of blending together in a scrambled blur, the different multipath
components of the pulse can be resolved. As we have seen in Chapter 7,
they can be, in principle, combined coherently. Multipath fading is eliminated by resolving it into individual signal components; instead of interacting destructively (causing fading), they can be combined constructively.

Third, the signal represents or allows a favorable trade-off of bandwidth for power. The power–bandwidth trade-off was recognized by Shannon in his work on pulse code modulation. PCM is really the first practical
form of spread spectrum; the voice signal is encoded at a high data rate (64
Kbps) and the signal is thus spread considerably beyond the approximately
3-kHz frequency bandwidth of the original analog voice signal. The result is
that we gain in robustness; the PCM signal can function well at a SNR a
thousand times worse than that required for good analog transmission. To
say it differently, the power of the signal relative to the noise, which is
viewed as having a constant power, can be enormously reduced.

The power–bandwidth trade-off has been treated as a quasi-mystical
precept by some spread spectrum proponents, and it has been suggested that
Shannon himself somehow proved that spread spectrum is the best of all
possible signaling strategies. This is not true, but it is an embedded assumption in many partisan presentations. However, it is true that one practical
result of using a spread spectrum signal is that it can allow the transmitter
power to be reduced.[16] In wireless systems, this can mean improved battery
life for portables. In extremely spread systems—where the spread signal is
thousands of times wider than the data signal—it can become possible to
operate a communications link (with some additional measures in the coding
domain) at *negative* SNRs, which once seemed like a high paradox.[17] In short,
where there is a strong pragmatic motivation for reducing the power of the

16. A friend (Michael Hirsch) once commented on this in what I thought was one of the
only really true observations about the relative merits of CDMA versus TDMA, by
saying that TDMA systems must be designed to transmit "at least enough power,"
while CDMA systems are designed to transmit with "just enough power." That is,
TDMA systems are, practically speaking, required to maintain a certain margin over
the minimum required signal level, whereas CDMA systems are required to minimize the surplus energy transmitted above the absolute minimum required by the receiver. Of course, the price paid is that CDMA must use extremely tight power
control and must devote a significant portion of the signaling bandwidth to establish
this power control. TDMA, on the other hand, can use much looser power control
and need not devote nearly so much bandwidth to this function.

17. See [1, p. 352ff].

signal (and where enough spectrum is available to support the luxury of spreading), spread spectrum offers this advantage. One such application is for the construction of signals that are difficult for the unauthorized listener to detect, which is often a requirement for military communications and other types of secure or secret communications. The ability to bury a signal in the background noise is advantageous here.

Fourth, the spread signal possesses a new property called *processing gain,* which is roughly defined as the ratio between the bandwidth of the spread signal and the bandwidth of the unspread data signal. In the example cited above, a 10-Kbps data signal was spread (through a repeating code) to 1 Mbps; this spreading factor of 100 is said to translate into a processing gain of a factor of 100, or 20 dB. This 20 dB can be added to the S side of the SNR calculation. In some spread spectrum systems, processing gains of as much as 30 dB can be realized. In principle, this is a physical layer effect rather than a coding effect per se. The spread signal contains 100 times as much information as it needs, strictly speaking, to represent the information contained in the data message, and this inherent redundancy translates into a stronger signal (which has been paid for, extravagantly, with bandwidth). That is what the idea of processing gain attempts to capture.

Finally, the spread spectrum signal creates a new set of opportunities at the logical level to construct a coded signal capable of much higher performance. As we now understand, the spread spectrum signal possesses much greater carrying capacity, by definition, than it actually requires for the information-bearing message. The number of transmitted symbols is much greater than the number of data symbols. How should we invest this extra capacity in the construction of a stronger signal? In the example developed above, the forced spreading of the low-rate data signal was accomplished by simply repeating the data symbols at a much higher rate. It would seem that we should be able to use this extra carrying capacity more intelligently. Intuitively, we feel there must be more powerful coding strategies that would bring even greater gains. Moreover, the wastefulness of a simple repeating code looks like a gross misuse of precious spectrum, and counter to our objective at the next higher system level (multiple access) of allowing as many users as possible to share the available communications bandwidth. The signal processing arguments in favor of spreading are all well and good, but what about the capacity problem?

This brings us to the subject of CDMA.

In commercial wireless applications, spread spectrum is the foundation for a popular and controversial multiple access architecture: code-division multiple access. The construction of multiple access schemes like

CDMA is properly a topic for Volume 2, but we may briefly touch on some of the ways in which the spread spectrum signals are designed at the logical level in support of a CDMA architecture. The key concept is direct-sequence spreading.

Rather than simply replicating the data signal to achieve the physical spreading effect, the standard technique is to mix the data signal with a second coded signal, called a *spreading code*. The second signal runs at a much higher rate than the data signal, and the resulting sum of the two also runs at the higher rate; this means that the output of this mixing process can physically spread the transmitted signal to occupy a much larger frequency bandwidth, as we want.

At the logical level, this mixing can be thought of as addition of the code symbols of the two signals together (Figure 8.6). The information symbols in the data signal are now effectively spread over a larger number of new symbols that were created by adding each data symbol to a corresponding number of symbols in the spreading code.

What sort of spreading code should be used?

First, the prevailing wisdom in the development of spread spectrum CDMA is that we should choose a spreading code with noise-like properties. That is, viewed as a signal on its own, the spreading code should look like white noise: If it is a binary code, it should have a pseudorandom pattern of ones and zeroes, in equal numbers over the long term. The cycle or repetition time of the code sequence should be long. Because of their noise-like nature, such codes are sometimes called pseudonoise sequences or PN codes.

A noise-like code should also have the property of low autocorrelation with time-shifted images of itself. This is the basis for discriminating different multipath images, which should show good decorrelation properties to

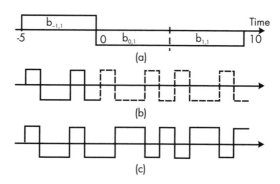

FIGURE 8.6 *Direct-sequence signal addition: (a) data sequence, (b) spreading sequence, and (c) composite.*

enable the receiver to separate different rays for subsequent processing (e.g., constructive combination in a RAKE receiver). It is essential for good performance in the wireless channel.

The other chief requirement on the spreading code is derived from the multiple access requirement. We must be able to discriminate different users' signals effectively. Any two given spreading codes should show good cross-correlation properties. Each should appear like noise to the other (Figure 8.7).

The design of appropriate spreading codes is a special subfield of coding theory. The argument that noise-like spreading codes are the most desirable direct sequences is rooted in the assumptions we have touched on previously that the best performance (within practical engineering constraints) can be achieved when the signal is most clearly distinguished against a flat, featureless background of evenly distributed energy lacking in any identifiable signal-like structure.[18] The truth, I think, is more subtle. The receiver should be able to view the undesired energy in the channel, whether it comes from multipath images of its own desired signal, or from other users, as noise-like *and* non-noise-like (random *and* structured) at the same time. If the spreading key is applied, the spread signal component should resolve itself clearly into a known pattern. If the spreading key is not applied, the spread signal component should look like random white noise. In practice, in today's limited RAKE receivers, the receiver will apply the spreading key to the first few fingers to detect the strongest handful of multipath images—and then it will leave the rest of the energy from the remaining multipath components unkeyed and noise-like. Similarly, the standard CDMA receiver strategy today is to treat all other users in the same band as noiselike, and not to apply spreading keys to extract the structure of those interferers. With the emergence of multiuser detection receivers (discussed in Chapter 7), the receiver may apply despreading keys to one or more of the strongest interference sources and use the structure inherent in them to remove their contributions to the received signal composite. Unless we assume that all coextant users in a shared multiple access channel are thus despread and removed, some of them will still be left in a noise-like state, from the standpoint of the target receiver.

There are many variations on the design of the direct-sequence spreading function. One illustrative suggestion, advanced by Wornell, is called *spread signature CDMA* [13, 14]. In normal direct sequence each data symbol is

---

18. See discussion on noise versus interference in Chapter 4.

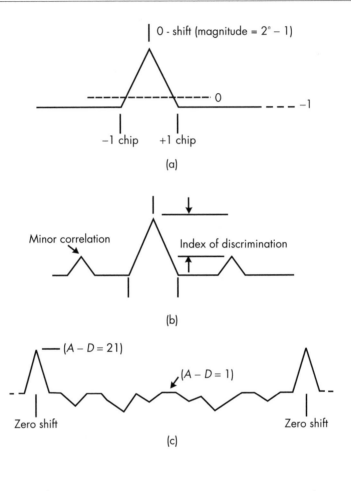

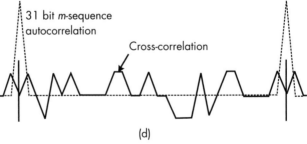

FIGURE 8.7 *Autocorrelation and cross-correlation: (a) m-sequence autocorrelation function; (b) typical nonmaximal code autocorrelation function; (c) autocorrelation for 21-chip maximal code; and (d) comparative autocorrelation and cross-correlation for 31 chip m-sequences. (From: [15]. © 1984 John Wiley & Sons Inc.)*

mapped onto a spreading sequence, but each loaded sequence (corresponding to a single underlying data symbol) is nonoverlapping (orthogonal) with respect to preceding and following sequences. Wornell proposes to expand the length of the spreading sequence such that consecutive sequences partially overlap one another.

> The key motivation for using longer signature sequence lengths ... is ... a greater temporal diversity benefit. In particular, the spread of the signature determines the temporal extent over which a symbol is transmitted (independent of the symbol rate). The longer this symbol duration the better the immunity to fades within the symbol interval. [14, p. 589]

If I understand this properly, Wornell's proposal is similar in spirit to convolutional coding or partial response signaling (Chapter 5), in which the information value of a source data symbol is spread over a number of channel coded symbols to achieve greater robustness. Here the spreading sequences are themselves convolved and spread over an even larger time interval, to achieve greater inherent time diversity and robustness against interference.[19] Once again, we see how signals are endowed with extra interference, purposefully inserted, to create overall resistance to channel degradations.

An even more ambitious proposal is made by Bhashyam et al. [16] for what they call time-selective signaling, but which might be better described as time–frequency spreading codes, a two-dimensional signaling strategy to construct signals that are processed by a time–frequency RAKE receiver (Figure 8.8) [16, 17]. Again, the secret ingredient is carefully controlled interference that is built into the signal as it is expanded in both time and frequency:

> The basic idea behind our signaling approach is to achieve time-spreading by increasing the symbol duration—*symbols may overlap in time* ... [emphasis added]. In effect, time-selective signaling facilitates maximal exploitation of Doppler diversity at the cost of *introducing controlled intersymbol interference* [emphasis added]. [16, p. 83]

---

19. "The canonical channel representation suggests that for maximal channel diversity, the symbol duration should be increased as much as possible.... Our approach to time-selective signaling [employs] spreading waveforms ... that typically last longer than the intersymbol duration, thereby introducing overlap between symbols" [16, p. 85].

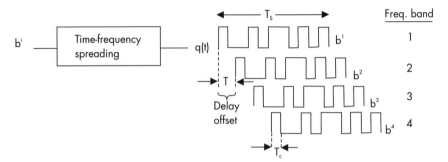

FIGURE 8.8 *Two-dimensional signal spreading. (From: [16]. © 2000 IEEE. Reprinted with permission.)*

The spreading code structure is sophisticated, to manage precisely the effects of the injected interference (Figure 8.9):

> The ISI introduced by overlapping codes diminishes with increasing length of the code. As the code length increases, the number of symbols in the overlap increases, but the correlation properties of the code also improve, thereby resulting in reduced ISI contribution due to each overlapping symbol. [16, p. 89]

To recap, direct sequence is the most common method of constructing a spread spectrum signal for commercial applications like cellular radio, although other quite different spreading techniques are available, such as frequency hopping, and are preferred for some applications.[20] The physical end product of the direct-sequence spreading and coding process is a transmitted signal that occupies a wide frequency bandwidth: Second-generation CDMA uses a bandwidth of 1.25 MHz, whereas third-generation architectures will use carriers of 5 MHz or more in bandwidth. The direct-sequence spread spectrum signal is wide enough that it can easily roll over the

---

20. Frequency hopping is another type of signal structure that can accomplish a similar spreading, and at least in some forms, a frquency-hopping spread signal falls under the same sort of analytical structure as direct-sequence spread signals. The hopping sequence must be designed according to principles similar to those used for direct-sequence spreading codes, for example, but in general, frequency hopping has not emerged as a preferred solution for commercial architectures for reasons related to the economics of the receiver hardware, although it is popular in many military applications because it is arguably more robust than direct sequence against certain types of jamming strategies.

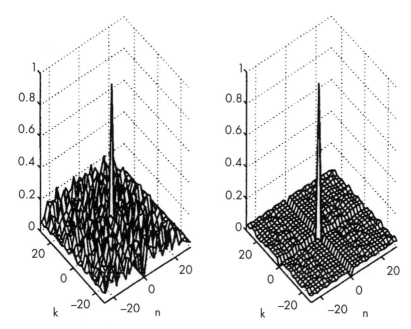

FIGURE 8.9 *Spreading code structure. (From: [16]. © 2000 IEEE. Reprinted with permission.)*

spectrum potholes created by sharp frequency-selective fades; or, to put it another way, the severity of the fading is averaged down across the band from as much as 40 dB to only a few decibels of variability due to multipath effects. Another way to say the same thing is that the widely spread signal possesses inherent frequency diversity. In the time domain, the symbol rate is so fast that the multipath phenomenon is resolved into individual rays or multipath echoes. A properly designed receiver can detect these individual signal components and combine them. Moreover, each underlying data symbol is spread over a much larger number of direct-sequence symbols, usually called chips, which means that the signal can also be said to possess inherent time diversity.

However, keep in mind that the spread signal, although more robust, imposes a large penalty in lost spectrum efficiency. A data signal that might have occupied a narrowband channel of 10 kHz to 30 kHz has been grossly inflated and now occupies perhaps 100 times as much spectrum. If this were the entire story, direct-sequence spread spectrum would hold little commercial interest. We need a multiple access architecture that can somehow recover this lost capacity.

The solution is to allow multiple spread signals to co-occupy the same physical communications space, the same physical channel (Figure 8.10). This is the basis of CDMA, and although it has become in some sense commonplace, it is worthwhile emphasizing its complete and radical departure from the pre-Shannon framework of physical orthogonality. In a direct-sequence/CDMA system, a number of signals simultaneously occupy the same space, time, and frequency coordinates. There is no way to discriminate between users based on any of the *physical* parameters of the signal. The classical notion of interference management by the use of physical buffers has been abandoned. All transmitted signals are allowed to mutually interfere with one another, and the received signal seen by any given user is a composite of all of them, and looks like wideband noise.

In practice, multiple access architectures based on direct-sequence spread spectrum signals, like CDMA, require very complex signal processing frameworks. The multiple access task is to create and maintain a precisely designed set of mutually interfering signals that can later be unscrambled and separated back out into their individual components. It is a bit like the proverbial house of cards: The structure can be set up, but it requires a lot of care, and the whole system must be properly isolated from unplanned bumps and jogs coming from outside sources. In other words, CDMA is typically more sensitive to interference in many respects than orthogonal

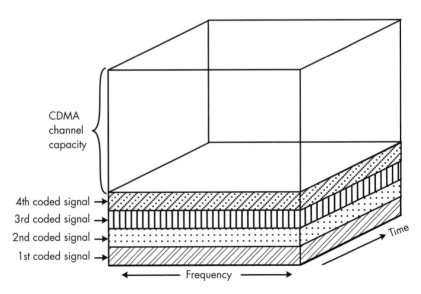

FIGURE 8.10   *CDMA signals overlaid on one another.*

multiple access architectures like TDMA, and special measures must be taken to keep the system stable.

The practical result is that CDMA has to give back much of its bandwidth to support embedded interference-mitigation overhead. For example, extremely accurate power control is required to prevent different user signals from jamming one another at the receiver (the so-called near/far problem). The amount of the total bandwidth required to implement this power control can be quite large (~20%), although it can also be shared to some degree with other pilot functions. A more serious overhead problem comes from the need to allow CDMA mobile units to communicate simultaneously with more than one base station. When base stations are operating with a frequency reuse of 1, mobile units near the boundaries of the cells may be linked with multiple base stations simultaneously, which reduces capacity. The CDMA industry has packaged this problem and resold it as a "benefit," called *soft handoff*. Soft sounds good, I suppose, and there may be some silver lining effects, but the real bottom line is that soft handoff means a huge loss in capacity. In real-world CDMA systems, the ratio between actual traffic on the system and billable traffic is typically about 2 to 1 in urban settings. That is, if we remove the double counting created by soft handoff when a mobile is linked with two or three different base stations at the same time, we find approximately 50% overhead that is part of the investment to manage interference within the CDMA logical framework. It is not that interference has disappeared; it has been translated into a different domain. So has the idea of orthogonality.

Nevertheless, the relinquishment of physical signal orthogonality is a huge conceptual leap, and has helped transform the understanding of wireless communications during the past 10 years. The new perspectives on communications, which we are lumping under the heading of post-Shannon architectures, have been given impetus by the development and commercial success of CDMA in the late 1990s. There is probably no overstating this. CDMA has embraced many ideas and seeded many controversies (and some nonsense). Unfortunately (for the logic of the presentation), we must leave the CDMA thread at this point, to be resumed in the second volume, where we will examine the construction of multiple access systems in depth. However, the development of communications architectures that did not depend on physical separation of users' signals challenged the field intellectually, probably more than anything since the development of digital communications in the 1940s. The realization that signal spreading can be beneficial, the understanding that the channel creates information, and the idea that interference can be controlled and purposefully manipulated rather than simply suppressed and avoided—all of these ideas were huge conceptual breakthroughs that have begun to inspire other ideas for signal construction ... to which we now turn.

8.3 Spreading Forced Through Multiplication in the Frequency Domain: OFDM and Multicarrier CDMA

Direct-sequence spreading multiplies the signal in the time domain—by mapping the data symbols onto a much larger number of spreading code symbols (chips)—in order to force the signal to spread in the frequency domain. The familiar parallelism between time- and frequency-domain processes suggests that a similar strategy can be developed in which the signal is multiplied in the frequency domain in order to force the signal to expand in the time domain. The generic term for this form of signal spreading is orthogonal frequency-division multiplexing. There are two ways to approach the OFDM concept: (1) as a signal-construction strategy ("How do we do it?"), and (2) as a channel-adaptation strategy ("Why do we do it?").

8.3.1 Constructing the Complex F-Domain Signal

Let us begin with the "how" question, and another analogy. A piano[21] is a frequency-domain engine for generating spectrally complex acoustic signals out of simple elementary tones. It will serve as a framework for exploring the second important strategy for signal expansion.

A piano can be played in many ways, of course. To connect with the flow of the previous narrative, we will contrive first to play it in a rather unnatural manner. We will lay a wooden beam across the keys, connected to a mechanical arm that can depress the beam and cause all of the keys to sound simultaneously. The arm is activated by an old-fashioned telegraph key, such that tapping on the key will depress the beam and cause the entire keyboard to sound. Assume this is our communications device, and we can use it to transmit a message in Morse code or to transmit a direct-sequence spread spectrum signal. (We can assume that the key-pressing rate can be very fast.) Perhaps we should modify the piano to be a bit more like an electronic organ or synthesizer, such that the keys sound when depressed, and

---

21. One recent textbook on Fourier techniques begins thus: "This book is about one big idea: You can synthesize a variety of complicated functions from pure sinusoids in much the same way that you produce a major chord by striking nearby C, E, G keys on a piano" [18, p. ix].

stop sounding when released, so that there is no sustain or decaying reverberation of the strings. The entire piano/organ is capable of producing sharp on–off binary signals.

The sound created by this special piano would be somewhat like a classic spread spectrum signal. Each pulse would occupy the entire available frequency bandwidth. The contraption could next be programmed to transmit direct-sequence signals with a unique spreading code and an embedded data signal. The signaling rate would be, of course, very fast. Now imagine that several such pianos are arranged in the same room to operate in the same way, the only difference being the spreading codes. The cacophony that would ensue is analogous to the multiuser CDMA channel, and may give us a renewed respect for the difficulties faced by the CDMA receiver that must try to discriminate a message from just one of those transmitters out of the blend of many similar signals.

Now, assume instead that we can construct 88 individual mechanical fingers, to press down independently on the 88 individual keys in the piano keyboard. Instead of sending the same data with all 88 keys (treated in effect as a single large key), we could divide the data we have to send into 88 individual data streams. These 88 data streams would be used to key 88 separate subchannels (one for each key). The entire transmitter is now channelized in the frequency domain, and the signal is composed of 88 separate components. Note also that the transmission rate can be modified. The rate for each individual key is 1/88th the original signaling rate (Figure 8.11). The symbol-keying speed of each subchannel is greatly reduced. (This will turn out to be an important point.) Viewed in gross physical terms, however, the total signal bears a physical similarity to the one we started with. It occupies

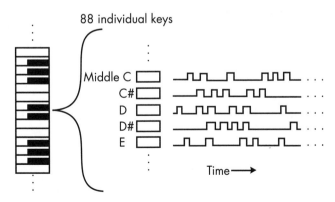

FIGURE 8.11 *Prototype OFDM with 88 data streams, and serial-to-parallel conversion at the transmitter.*

the entire frequency band, and indeed the entire STF space. The spreading effect and the lack of physical orthogonality are roughly similar to the direct-sequence spread spectrum signal.

The idea of subdividing a given large bandwidth into individual frequency-domain channels and transmitting a portion of the message over each channel should appear familiar. Strictly speaking, this is *frequency-division multiplexing* (FDM), which is the oldest form of multiple access for wireless communications.[22] The only immediate difference between this new model and conventional FDM is that instead of assigning each channel to an individual user (or pair of users), we are assigning all the channels (in a given set) to one user (or pair of users), and allowing that user to divide his total bit stream over a large number of channels. The division of the bit stream into substreams by the transmitter and the recombination of the received substreams into one output signal at the receiver is a standard multiplexing function. No problem there.

What *is* problematic is the transmitter–receiver hardware implementation. To build such a system using conventional FDM technology, we would need 88 individual transmitters and 88 separate receivers—a costly and cumbersome hardware ensemble. The key breakthrough for the practical implementation of a frequency-multiplied spread spectrum signal was the discovery of a new method of combining and separating frequency subchannels that does not require separate transmitters and receivers for each subchannel. The root concept is one of the oldest and most foundational principles of all signal processing: Fourier analysis.

Fourier analysis is eponymically associated with Joseph Fourier, a nineteenth-century French polymath who developed the technique in the process of trying to understand the flow of heat in solid bodies, although it has arguably been independently discovered several times in different contexts, and is even said to underlie the cycles and epicycles by which Ptolemaic astronomy explained the motions of the planets in a supposedly geocentric universe [18, pp. 12–14]. The core idea is simple: that any complex signal, viewed in the frequency domain, can be understood as a composite of a number of simple sinusoidal signals that have been added together [20, pp. 49ff]. The transform is reversible, or bidirectional. Complex signals can be decomposed into their simple elements (the Fourier transform proper), and simple elements can be combined to create complex signals (the inverse Fourier transform). The power of the technique is that it

22. "The earliest MCM [multicarrier modulation] modems borrowed from conventional FDM technology, and used filters to completely separate the bands" [19, p. 5].

allows the analysis, or synthesis, of highly complex signals out of very simple elements. It possesses a sort of *alphabetic* power—just as we can create an encyclopedia of information using just a small number of discrete elements (letters of the alphabet), Fourier techniques show how we can use simple elements to build almost any sort of signal we want.

Because of their power, Fourier techniques are said to be the most widely used of all signal processing techniques, with applications to astronomy, thermodynamics, the analysis of electrical signals from the heart, and many other fields including communications engineering. The main obstacle to their use was traditionally the problem of computation. The original Fourier transform required very complex hand calculations and for decades its use was limited to relatively simple (non-real-time) problems. What was needed was a way of accelerating and mechanizing the computation.

The second half of the breakthrough occurred in 1965, with the publication of a paper by Cooley and Tukey, which has been called the most widely cited technical paper ever written, or perhaps merely "the most important algorithm of the 20th century" [21].[23] James W. Cooley and John W. Tukey discovered (or rediscovered, it was later claimed[24]) an algorithm now called the *fast Fourier transform* (FFT), which increased the speed of computation at least a hundredfold [23]. This enabled the application of Fourier analysis to a vast set of problems that had previously been too hard for conventional Fourier computation to crack in a reasonable time. The FFT also happened to come along just as electronic computers and, not long after, microprocessors were becoming available on a wide scale. These two developments, the new algorithm and the emergence of computing engines driven by Moore's Law, exploded Fourier techniques into dozens of new commercial implementations: "In fact, it is no overstatement to say that the discovery of the FFT created the field of digital signal processing" [22].[25] The

---

23. See also [18, pp. 295ff]. In the folklore surrounding this algorithm there are all sorts of illustrative claims; for example, it has been estimated that in 1990, on the installed base of supercomputers made by Cray Research—some 200 machines—40% of all the CPU cycles were spent computing the FFT. See [22].

24. The FFT and IFFT are such important algorithms that they have risen to the status of "history of science" milestones and have attracted the attention of scholars with knowledge of and access to a vast historical literature outside the normal scope of a working mathematician. It has been argued that the FFT was actually anticipated as long ago as 1805 in a paper by the great mathematician Carl Friedrich Gauss, which was later published in his collected works in 1866 and then forgotten for a century. Proving that there is nothing new under the sun is often a significant motivation for historical scholarship [18, p. 14ff].

FFT enabled, as a practical matter, the expansion of the analytical framework from two-dimensional to three-dimensional: "It made working in the frequency domain equally computationally feasible as working in the temporal or spatial domain" [23]. Today the FFT and its twin, the *inverse FFT* (IFFT), are part of the standard library of signal processing techniques.[26]

The FFT changes the implementation strategy for the novel FDM approach we have sketched above, as the reader can probably already anticipate. Consider now again those 88 subchannels, each one of which is effectively a simple frequency carrier, modulated with on–off keying. To construct a single complex waveform from 88 individual simple waveforms is what the inverse FFT is designed to do. The result is a single new complex waveform that contains all the information that was originally carried by each of the subcarriers, but it is itself nevertheless a signal that can be handled by a single transmitter (instead of 88 separate transmitters). At the receiver, the same magic runs backward. A single receiver captures the physical signal and outputs it to an FFT engine, which decomposes it into all 88 original data streams. These are demodulated, the data recovered, and then reinterleaved in the correct order to create the original high-speed signal (Figure 8.12).

This FFT-based version of FDM was apparently first proposed in the early 1970s [25], although some authors point to conceptual work done in the 1960s [26–29] or even earlier [19].[27] But the practical enablement of this concept required both the FFT/IFFT algorithm and the availability of

---

25. The importance apparently cannot be overstated, and some have even expressed concern that the FFT distorted the normal course of science: "The prominence of the FFT may also have contributed to slow other areas of research. The FFT provided scientists with a big analytic hammer, and for many, the world suddenly looked as though it was full of nails—even if this wasn't always so. Problems which may have benefited from other more appropriate techniques were sometimes massaged into a DFT framework, simply because the FFT was so efficient" [23].

26. The expansion and generalization of frequency-domain analysis has continued to develop new nomenclatures and more powerful techniques, such as wavelet analysis, where the equivalents of FFT and IFFT are *fast wavelet transform* (FWT) and IFWT. Essentially, wavelets incorporate a time and frequency shift of the underlying basis function, while classical Fourier analysis addresses only a frequency shift. Undoubtedly this will be one of the fields to watch in the next decade. A very informative presentation can be found in [24].

27. Bingham [19] cites the first use of the concept dating from 1957. The source reference—which I have not checked—is Doelz et al. [30].

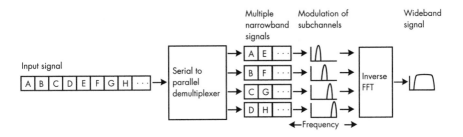

FIGURE 8.12 *OFDM system architecture.*

sufficiently powerful digital signal processors, and it was really not until the 1980s that it became feasible to build such a system for commercial use.

This new type of FDM signal has been called by various names. In the development of high-rate wireline communications systems known as digital subscriber lines (xDSL), the technique is known as *discrete multitone modulation* (DMT) [31]. In some studies it is referred to simply as *multicarrier modulation* (MCM) [19]. The channel has been called the multitone channel [32]. The term commonly used in wireless applications is orthogonal frequency-division multiplexing (or modulation) [14], which is the name we will use here. However, we must remind ourselves that in this case the word *orthogonal* does not refer to physical orthogonality, but to a form of logical orthogonality similar in spirit to the use of spreading codes with low cross-correlation functions in direct-sequence spread spectrum. To understand how this concept is developed, let us consider the practical problems involved in constructing the OFDM signal.

The first problem, which may already have occurred to the reader, is that any signal that is finite or limited in the time domain will, in practice, tend to spread somewhat beyond its nominal boundaries in the frequency domain. In other words, any sharply defined pulse, of whatever shape, will have some tendency to spread. In spread spectrum (direct-sequence type), we take advantage of this by shortening the pulses dramatically and forcing the signal to expand greatly in the frequency domain. In more traditional narrowband signals (like TDMA), the spreading is not purposefully accentuated, but there is a certain tendency to spread nevertheless. This is normally accepted as a slight penalty for operating with real-world hardware and time constraints, but it does mean that FDM systems normally need some kind of frequency-domain buffer between occupied carriers, or the signals on those carriers will tend to smear into one another and cause interference [Figure 8.13(a)]. How can we really pack the subcarriers in an OFDM-type signal any more closely than in a conventional FDM system?

There are two solutions. The less interesting solution is indeed simply to accept the penalty by spacing the individual tones so that their spectra do not significantly overlap. The payoff of OFDM, which we will get to shortly, is not actually dependent on the carrier spacing, but on the way in which the subchannels are adapted to the channel transfer function. So, this is a feasible solution in some uses. In DSL-type wireline applications, where the spectrum of the copper wire is large and fully available, discrete separate tones can be spaced to minimize physical overlap. It is sometimes argued that this is what distinguishes DMT (although other presentations use DMT and OFDM interchangeably [24, p. 112], and other presentations involve explicitly overlapping DMT signal structures [33]). Nonoverlapping spacing in a *wireless* channel is a wasteful strategy; indeed, it is simply FDM, from a channel perspective, with the same overhead or buffering penalties as conventional channelized access schemes.

The much more interesting solution is to dense pack the subcarriers more closely, taking full advantage of the fact that all carriers are constructed and coordinated by a single controller and passed through a single transmitter. That is, unlike conventional FDM where adjacent carriers are being modulated by different transmitters and the degree of coordination between those transmitters is assumed to be quite limited, in an OFDM signal all the subcarriers are produced by one transmitter. This means that we can indeed define and control the characteristics of each signal to ensure a much higher degree of coordination between signals in adjacent channels. The individual subcarriers can be spaced so that they produce overlapping spectra,[28] which means that the individual signal components will interfere with each other physically [Figure 8.13(b)]. However, we can *design the interference* in such a way that its effects can be largely self-canceling. This signal structure is best introduced by reference to a more well-known technique in the time domain, which is strictly similar in principle: Nyquist signaling.

A square pulse produced by the transmitter is seen at the receiver as a wavelike oscillation with a main peak preceded and followed by smaller peaks that spread far beyond the desired sampling interval. Two pulses close together will overlap and interfere with each other, creating ISI. To ensure

---

28. "The main difference between frequency division multiplexing (FDM) and OFDM is that in OFDM the spectrum of the individual carriers mutually overlap, giving therefore an optimum spectrum efficiency" [34, p. 2908].

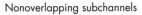

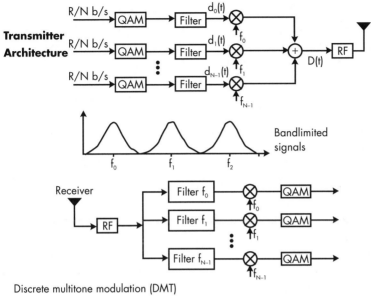

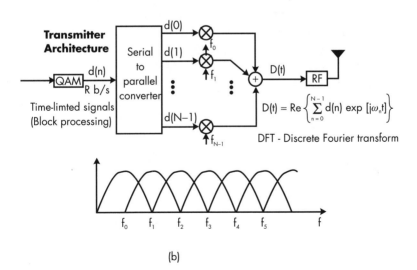

(b)

FIGURE 8.13 *OFDM subchannel spacing option (a) nonoverlapping and (b) Nyquist spacing.*

that successive pulses can be transmitted without ISI, the shape and timing of each pulse must be controlled, so that the waveform of the potentially interfering preceding pulse happens to be crossing the zero energy level at the precise instant that the following pulse is reaching its maximum.

Harry Nyquist elaborated the principles underlying this signaling strategy in his work on telegraph signaling in the 1920s, and showed that by careful coordination it is possible to signal at the Nyquist rate without creating ISI. The shape of the *transmitted* pulse is also important; different transmitted pulse shapes produce different received waveforms, and certain shapes lend themselves more readily to the Nyquist strategy. Even within the family of acceptable Nyquist pulses some pulse shapes produce less spreading than others and are more forgiving of small timing errors in real-world implementations (see Chapter 2).

The frequency-domain spreading of a single pulse looks very similar to the time-domain representation: The frequency spectrum of the received pulse appears as a peak of energy centered on the desired frequency, with various harmonics showing up as smaller peaks spaced at intervals above and below the desired frequency. Just as two insufficiently coordinated pulses in the time domain will lead to ISI, the analogous phenomenon where two pulses interfere with each other in the frequency domain has been labeled ICI. However, just as ISI can be avoided by proper coordination in the time domain, ICI can be minimized by proper coordination in the frequency domain. Two pulses on close adjacent frequencies can be similarly positioned such that the energy from the adjacent pulse is near zero at the precise frequency where the energy of the desired pulse is at its maximum. By the careful coordination of pulse shape and frequency spacing, an OFDM system can be constructed at the transmitter such that the ICI between a large set of subcarriers is minimized. Then these carriers are combined into a single complex signal by means of the IFFT engine, and we can be confident that this signal is relatively free of embedded ICI. By analogy, we call the proper frequency spacing the Nyquist frequency. The waveforms from two carriers separated by the Nyquist frequency will have zero cross-correlation,[29] zero ICI, and this is the root of the concept of orthogonality as it is applied in OFDM [33].

Of course, this carefully structured signal is going to undergo severe challenges in the physical channel. We know that over a wide frequency

29. "The OFDM carriers exhibit orthogonality on a symbol interval if synthesized such that they are spaced in frequency exactly at the reciprocal of the symbol interval" [34, p. 2908].

band, divided into, say, 512 subcarriers, some of these carriers will be hit by severe frequency-selective fades. We also know that Doppler frequency offsets and other channel effects may disrupt the carefully pinioned Nyquist spacing of the OFDM subcarriers, destroy the strict orthogonality of the OFDM signal [35], and ICI may be reintroduced by the channel (just as the channel reintroduces ISI in the time domain).[30] To counter this, a receiver-centric solution like frequency-domain equalization can be employed (as described in Chapter 7) to clean up the ICI (just as time-domain equalization is used against ISI). A more powerful strategy is to construct the signal at the transmitter so as to minimize or avoid the effects of channel distortions. We know that these distortions are nonuniform across the entire channel. Multipath fading will impact some subcarriers severely while sparing others. One approach to signal construction for such a channel is to ensure that the data message is appropriately coded (for error correction), and interleaved across the many individual carriers, so that the effects of losing most or all of the data on a few carriers are averaged out. In short, the signal hardening strategies discussed in Chapter 5 can be used to enhance the integrity of the message[31] (Figure 8.14). This strategy does not take full advantage of the power of the OFDM architecture. It is not clear yet why the multicarrier signal structure would yield particular benefits for a simple signal hardening strategy based on removing the structure in the interference. Direct-sequence techniques, and indeed many narrowband techniques, follow the same principles, without all of the complexity of the FFT/IFFT processing. We have not yet penetrated the real rationale for the multicarrier structure. To grasp the true significance of the OFDM idea, we need to develop a different kind of signal-construction strategy.

8.3.2 Tailoring the OFDM Signal to Fit the "Shape" of the Channel Transfer Function

Let us now switch our point of view to the "why" of OFDM.

---

30. "When the normalized Doppler frequency is high, the power of ICI cannot be ignored. ... Where normalized Doppler frequency is high, there is an irreducible error floor even if all the data symbols are pilot symbols, since the pilot symbols are themselves corrupted by the ICI ... " [36, pp. 1377, 1379]. Also: "One of the principal disadvantages of OFDM is sensitivity to frequency offset in the channel ... caused by tuning oscillator inaccuracies and Doppler shift" [34, p. 2909]. See also [37].

31. "In wireless channels, good link performance can be achieved by OFDM when combined with diversity, interleaving, and coding. OFDM inherently provides frequency

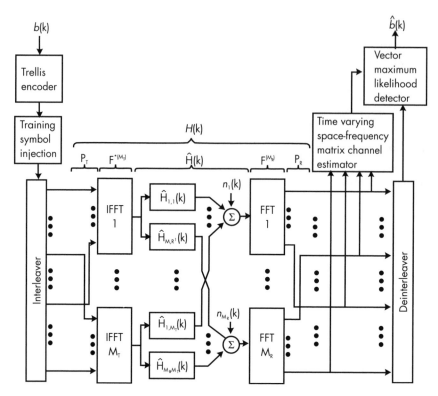

FIGURE 8.14 *Interleaving the OFDM signal. (From: [39]. © 1999 IEEE. Reprinted with permission.)*

The answer, in a single sentence, is this: Multicarrier techniques allow us the opportunity to match the shape of the signal (understood as the power/bandwidth spectrum) to the shape of the channel transfer function, so as to support an optimally efficient transmission strategy. To develop the logic of this approach, let us begin with the simplest case of a multicarrier signal.

The Split-Band Channel

Starting with a wideband channel, let us assume that the entire available bandwidth is divided into two bands (Figure 8.15). This has been referred to

diversity over subchannels, which introduces an opportunity for interleaving in the frequency domain" [38, p. 80].

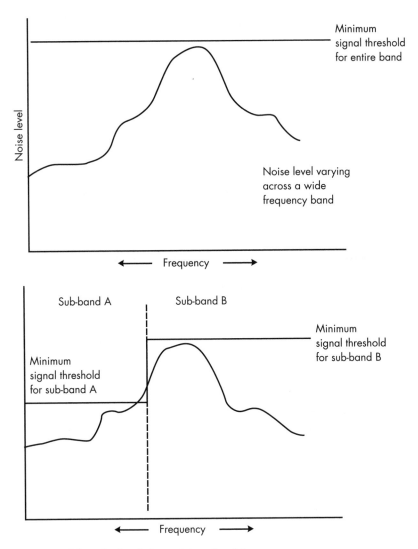

FIGURE 8.15 *The split-band channel (two bands).*

as "a classic problem in discrete-time signal processing" [24, p. 108]. The channel transfer function, which varies across the entire band, may still vary over each sub-band, but the average noise level, for example, for each sub-band can be evaluated independently. This creates an opportunity. Instead of one signal occupying the entire band, we construct two signals—one for each sub-band, each adapted to the power levels required for that band.

Kalet has provided an analytical framework and empirical measures of performance gains:

> The results of the two-tone channel are compared to those of a single-tone transmission over the entire channel bandwidth using ... equalization. The comparison shows that the optimized two-tone system may show significant improvements as compared to the single-tone channel....
>
> As a numerical example ... for a 600 foot cable we find that the maximum bit rate is 49.7 Mbits/s [whereas] we find that the maximum bit rate for a single-tone system is 34.8 Mbits/s. Using multitone it may be possible to achieve an improvement of more than 40 percent as compared to single-tone [signaling]. [32]

Subdividing the Band to Achieve Flat Fading

The idea of splitting the band can be extended: Each sub-band can be further subdivided, and individual noise figure averages for each sub-band can be recalculated. Eventually, the second major advantage of the multicarrier approach begins to emerge: The fading or channel transfer functions for the small individual sub-bands start to become much flatter. That is, we reach a point where we can regard the noise level within the sub-band as a constant across that sub-band (Figure 8.16).

Optimizing the Transmitted Signal in Each Sub-Band: "Water Pouring"

Once we have subdivided the channel down to the point where each sub-band shows flat fading characteristics, we can start to adapt the transmission process more aggressively. For example, for a given level of available transmitter power, we now know how much information can be transmitted in each band. Some bands have a good SNR and can handle higher rates (have higher capacity), while some have poorer SNR and can only accommodate a slower rate of information transfer. Some sub-bands may be unusable and can be avoided. This means that the losses due to nonoptimal matching of the signal power level in the band to the actual noise level in each band are minimized. The signal information is nearly optimally distributed across the entire band (Figure 8.17).

The conceptual touchstone for this idea was again provided by Shannon himself, and is usually referred to as the *water-pouring principle*.[32]

---

32. This arguably misconstrues the detail of the water-pouring argument, but it does point to a similar spirit of adapting the signal structure to the variable characteristics of the channel. "The algorithm is clearly not water-pouring in the classical sense, but

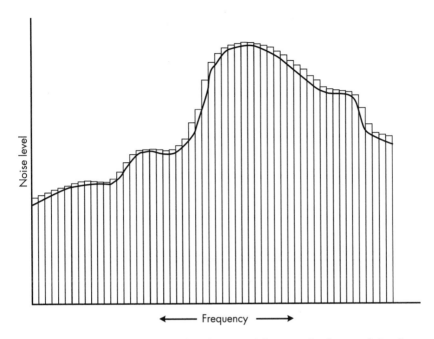

FIGURE 8.16 *Many-banded channel with essential flat noise levels in each band.*

Indeed, Shannon's description of this principle is virtually a prescription for the construction of an OFDM signal:

> We can divide the band [i.e., the channel viewed in the frequency domain] into a large number of small bands, with $N(f)$ [i.e., noise power] approximately constant in each. ... For each elementary band, the white-noise result applies. [40]

Shannon's diagram shows a wide channel that has a varying level of noise across the entire band (Figure 8.18). By dividing the band into sub-bands we accomplish two things: (1) The noise level in each sub-band can be viewed as flat or white, and (2) the transmitter power can be adapted to the noise level in each band.

---

since it puts every increment of power where it will be most effective, it appears to be optimum for multi-carrier transmission ..." [19, p. 8]. See also [14].

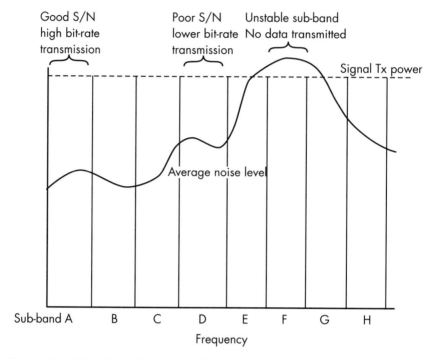

Good S/N   Poor S/N   Unstable sub-band
high bit-rate   lower bit-rate   No data transmitted
transmission   transmission

Signal Tx power

Average noise level

Sub-band A   B   C   D   E   F   G   H

Frequency

FIGURE 8.17 *Many-banded channel with transmitter power optimized for each band.*

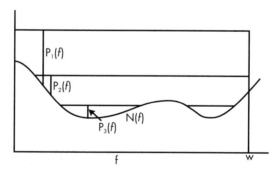

$P_1(f)$

$P_2(f)$

$N(f)$

$P_3(f)$

f   w

FIGURE 8.18 *Shannon's version of the split channel. (From: [40]. © IRE 1949.)*

For frequencies where the noise power is low, the signal power should be high, and vice versa, as we would expect.... With low values of *P* [transmitter power] some of the frequencies will not be used at all. [40, p. 170]

The water-pouring metaphor is suggested by the idea that a perfectly adaptive transmitter would only pour just enough power into each sub-band to cover the noise and reach the desired level.

(In passing, let us take notice once more of the preternatural fecundity of Shannon's original work. In a few paragraphs written in 1947—that is to say, written before the invention of the transistor, the electronic computer, and so much else—Shannon tossed off a brief, but clear and essentially complete description of a communications strategy that is only now, some 50 years later, beginning to achieve commercial success. The phrase "post-Shannon" should not be understood as connoting anything less than a profound respect for the originality of Shannon's insights.)

The true power of the OFDM idea emerges from this adaptive strategy.[33] By dividing the channel into many small subchannels, the frequency-domain channel effects can, in principle, be isolated to a subset of the total number of subchannels. These impacted subchannels can either be taken out of service for the time being, or if the impairment is not fatal, the signal structure for the affected subchannels can be modified.[34] For example, the signaling rate can be reduced, along with the modulation level.[35] Or, to view it from the half-full perspective, "subchannels that suffer less attenuation will carry more information" [24, p. 113].

Receiver-Centric Adaptations to Sub-Band Characteristics

The conversion of a wideband channel with nonuniform and time-varying fading characteristics into a set of sub-bands each with a flat and non-time-varying[36] channel transfer function makes the receiver's job much easier (in principle—leaving aside the FFT complexities for the moment). The ISI

---

33. It has been argued, in fact, that the blind or nonadaptive strategy can end up yielding worse results than a single-carrier implementation. "If the [channel transfer function] varies significantly across the band and a fixed loading is used, the error rate in the too-heavily-loaded sub-bands may be very high, and the overall rate may be greater than for a single-carrier signal" [19, p. 7].

34. "The basic idea behind DMT is to utilize the unevenness of the channel response in order to maximize the total achievable bit rate" [24, p. 112].

35. "The symbol alphabet size for a given subchannel is equal to the maximum the sub-channel can support (with an acceptable symbol error rate), based on channel measurements made during an initialization or training period" [33].

36. The stability of the sub-band channel transfer function over time is, of course, a matter of definition, since over a *long* duration the channel transfer function does vary in a mobile channel. By *non-time-varying*, we mean that it is acceptably stable over the window for detecting the symbol.

distortions in each sub-band can now be corrected, in principle, with a much simpler equalizer. Adaptation *by the receiver* to the varying channel characteristics is another source of performance improvements in a multi-carrier system. For example, we can extend the equalization function from the time domain to the frequency domain, as described by Van Acker et al.:

> A general disadvantage of [time-domain equalization, or *TEQ*] is that TEQ equalizes all tones in a combined fashion and as a result limits the performance of the system. We aim at improving on TEQ performance by changing the receiver structure [so that] equalization is done for each tone separately after the FFT-demodulation....
>
> This enables us to implement true signal-to-noise ratio optimization per tone, because the equalization of one carrier is totally independent of the equalization of other carriers.... Equalization effort can be concentrated on the most affected tones by increasing the number of equalization filter coefficients for these tones. No effort is wasted to equalize unused carriers because there the number of taps can be set to zero. [41, pp. 109, 115][37]

Adaptive Definition of the Sub-Bands Themselves

Finally, the definition of the sub-bands themselves can be parameterized. In most current implementations of OFDM, the entire channel is *uniformly* subdivided into subchannels of equal bandwidth. In principle, a more optimal design can be achieved by allowing a *nonuniform partitioning* of the channel (Figure 8.19):

> The concept of optimal subchannel structuring ... first examines the spectral properties of a given channel. It recommends the best subchannel structure for that given channel....
>
> This concept can be viewed as the *tiling of the channel* by variable-shaped *carrier tiles* [emphasis in original]. This scheme does not blindly split a given channel into its subchannels or orthogonal carriers. The unevenness and power levels of the channel are jointly evaluated for the structuring of its subchannels. Therefore the subchannels (carrier functions) are well tailored to the spectral properties of the channel. [24, p. 113]

---

37. See also Bingham [19]: "An MCM signal can be processed in a receiver without the enhancement (by as much as 8 dB in some media) of noise or interference that is caused by linear equalization of a single-carrier signal" [19, p. 5].

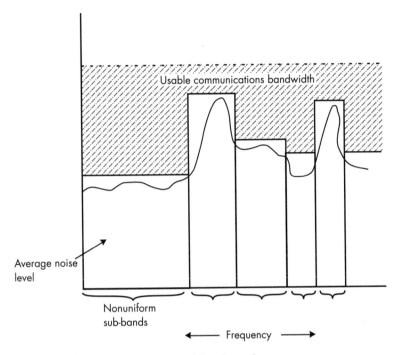

FIGURE 8.19 *Nonuniform partitioning of the channel.*

This might well be the ultimate OFDM—in which not only the code and modulation properties of the signal are designed adaptively for each subcarrier, but the bandwidth of the subcarrier itself is also adaptively defined.

There are thus two different architectural versions of the OFDM signal:

1. A *blind version*, in which the transmitter does not make use per se of information about the specific channel conditions in the individual subcarriers; the signal is coded according to the principles discussed in earlier chapters (e.g., Section 5.4), such that it is effectively decorrelated with, or averaged over, the frequency-selective, time-selective, or spatially selective effects of the fading channel[38];

---

38. If I read them correctly, this is an approach discussed by Pollet and Peeters [42]: "A simplified version of DMT called OFDM ... [in which] the constellation sizes are the same for all modulated carriers."

2. An *adaptive version*, where the transmitter and receiver acquire and use information about the noise levels, fading, or other characteristics of the subcarriers, to adjust various parameters of the transmitted signals on each subcarrier (such as the strength of the error correction, and/or the data rate) in response to this channel state information.

The net result of either form of OFDM, viewed physically, is a signal that completely occupies a wide frequency band, and suffers distortions in both the time and frequency domains (Figure 8.20). The signal contains its information embedded in the convolution of a large number of subcarriers, and the true underlying user payload has been appropriately hardened and spread across the whole frequency domain so that hits on individual subcarriers can be effectively averaged out, and/or the signal can be adjusted adaptively by the transmitter for each subcarrier.

Simplification of Time-Domain Processing

We have pointed out that the intentional multiplication of the signal in the frequency domain (by subdividing the channel into many sub-bands) forces the signal to spread in the time domain. What exactly are the time-domain effects?

It is here that a major benefit of OFDM appears.[39] Because the signaling rate on each individual subcarrier can be slowed down in inverse proportion to the total number of carriers, the actual data rate on each subcarrier becomes quite attenuated. If there are 512 subcarriers, each subcarrier

---

39. The benefit of slowing down the signaling rate is characterized variously. Sandberg and Tzannes [33] highlight the "superior immunity to impulse noise when compared with single carrier systems" which they identify as a "consequence of longer symbol times" [33, p. 1571]. Harada and Prasad [37] stress the reduced equalization requirements: "Converting the data into several slow parallel data streams reduces the effect of multipath fading and suppresses the delay time to within one symbol" [37, p. 1367]. Hara and Prasad [43] point to "the combination of OFDM signaling and CDMA ... has one major advantage that it can lower the symbol rate in each subcarrier so that a longer symbol duration makes it easier to quasi-synchronize the transmissions" [43, p. 127]. My own view is that while the easing of the constraints on time-domain processing is undoubtedly important, it is basically something that affects the ease of implementation and not the fundamental system performance limits, and as such, those trade-offs will tend to shift over time and most likely in favor of applying more computational power. The fundamental advantage of OFDM derives from the adaptive strategy that it enables to tailor the power, bit rate, bandwidth, and other signal parameters to the "shape" of the channel transfer function.

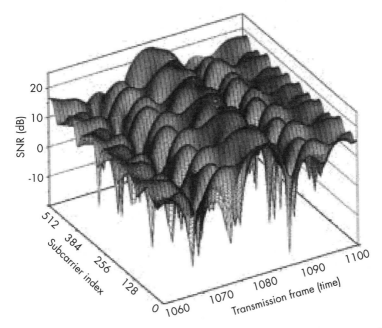

FIGURE 8.20 *Two-dimensional representation of the ODFM signal in time and frequency. (From: [44]. © 2000 IEEE. Reprinted with permission.)*

transmits in principle at 1/512th of the overall signaling rate [24, p. 108]. Each data symbol occupies a longer time period, and in effect incorporates some measure of time diversity in its own physical structure.[40] This allows for a radical simplification of time-domain processing. It is sometimes argued that it eliminates the need for time-domain equalization entirely, because the individual symbol elements on each subcarrier may now substantially exceed the delay spread of the channel, and eliminate ISI.[41] Often,

---

40. "The time-varying channel is a main obstacle to data detection since it destroys the orthogonality of the multicarrier signals. Surprisingly, however, we will show that the time-varying nature of the channel can be exploited as a provider of time diversity ..." [36, p. 1376].

41. "One of the principal advantages of OFDM is its utility for transmission at very nearly optimum performance in unequalized channels and in multipath channels.... Intersymbol interference can be entirely eliminated by the simple expedient of inserting between symbols a small time interval known as a guard interval. The length of the guard interval is mad equal to or greater than the time spread of the channel" [34, p. 2908].

additional guard symbols are inserted to ensure that data symbols do not overlap [35]. Of course, this is a form of overhead, a bandwidth penalty, but it is argued that it is preferable to equalization because it does not increase the processing load on the receiver. On the other hand, many OFDM structures do employ equalization, to trade off the bandwidth penalty against the computational burden [45]. There is also typically some requirement for guardbands in the frequency domain—in the form of unused subcarriers. The total time and frequency overhead penalty (Figure 8.21) varies as a function of the number of subchannels and the required SNR. Nevertheless, it is likely that the overhead is modest by comparison with conventional FDM or TDM systems. Faulkner found that for a large number of subchannels (512) and a modest SNR target, the capacity lost to overhead was less than 3% [46].

The picture of OFDM that emerges is of a signaling strategy in which the spreading function is forced through the radical restructuring of the frequency domain and is manifest in expansion of the time domain. The OFDM system intensively controls the frequency characteristics of the signal, which we see in the use of FFT, IFFT, Nyquist spectrum shaping, frequency-domain equalization, and so forth. On the other hand, OFDM gets a free ride—so to speak—in the time domain (or so it is claimed). The drastic slowdown in the signaling rate and the corresponding elongation of each symbol in time, means that the receiver does not need as much special

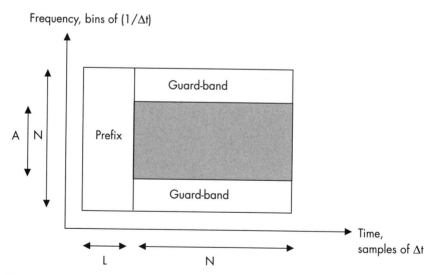

FIGURE 8.21 OFDM overhead penalty. (From: [46]. © 2000 IEEE. Reprinted with permission.)

processing to handle time-domain channel effects. The signal may be viewed as inherently robust in the time domain.[42]

The presentation of OFDM brings into focus again the strange way in which time and frequency mirror one another, as dimensions of reality. Although it is nowadays typically presented as simply another engineering fact for students to absorb and apply, the interlaced phenomenology of time-domain and frequency-domain signal processing should still inspire some wonder, I think. Why, exactly, should limiting a signal in time cause it to spread in frequency, or vice versa? How are we to understand the notion that spreading in both domains is *infinite*, and *bidirectional*, which implies, for example, that the temporal impulse response actually begins "well *before* the impulse is applied to the system!" [47, p. 42]. Which in turn leads the authors of one fairly sober engineering text to warn their readers that "thinking too hard about the physical meaning of continuous distributions of sinusoidal components can sometimes be counterproductive" [47, p. 33].

In any case, we may view OFDM and direct-sequence spread spectrum as mirror images—that is, as quasi-identical, but *inverted*, signal construction programs. Both seek to spread the data signal over a much larger physical domain. The two respective spreading functions are conceptual transforms of each other. As noted earlier, OFDM forces the spreading by multiplying the signal in the frequency domain, which has the effect of expanding the signal in the time domain. The processing complexity is lodged in the frequency domain; the spreading in the time domain makes that side easier to handle. On the other hand, direct-sequence spread spectrum forces the spreading to occur by multiplying the signal in the time domain, which has the effect of expanding the signal in the frequency domain. The time-domain processing requirement for direct sequence is intense and largely drives the receiver architecture, whereas the frequency domain is viewed as inherently robust and does not require much additional processing. By concentrating the complexity in the frequency domain, OFDM avoids or reduces time-domain ISI and the need for equalization,

---

42. Faulkner [46] provides a nice summary of the practical advantages and disadvantages of OFDM signaling. *The pros:* "The major advantages include: an ability to dynamically optimize the data rate to channel characteristics; a resistance to impulse noise; a flexible spectrum occupancy (i.e., an ability to create spectral nulls in the transmission band for interference avoidance); a low cost; and a high tolerance to delay spread.... OFDM [also] eliminates the need for a delay equalizer on dispersive channels and ... the lengthened symbol gives a bigger target for the receiver synchronizer to hit, reducing synchronization accuracy requirements" [46, p. 1877].

but ICI becomes a problem. By concentrating its complexity in the time domain, direct-sequence spread spectrum enjoys robustness in the F-domain, but requires a high-octane time-domain engine.

Given this parallelism between direct-sequence spread spectrum and OFDM, it should not be surprising that attempts to compare the theoretical capacity and link performance (e.g., BER performance) of these two approaches have tended to conclude that—details of hardware implementation aside—the time-division and frequency-division methods should yield approximately the same results [48]. My guess, personally, is that the adaptive strategies that OFDM perhaps more readily supports (compared to direct sequence) may ultimately lead to superior realizable performance.[43]

8.3.4 Multicarrier CDMA

It may have already occurred to the reader that OFDM and direct-sequence spread spectrum can be combined. The spreading codes used by a direct-sequence signal can be applied in the frequency domain rather than the time domain [43]. That is, the direct sequence can be modulated all at once across a large set of different frequencies, instead of serially over a single frequency (Figure 8.22). There are several variations on this approach, but the most common name for this signal structure (albeit yet another misnomer) is *multicarrier CDMA* (MC-CDMA).[44]

The same basic structure is used in both OFDM and MC-CDMA. The difference is that in OFDM an individual data symbol is carried on a

---

43. One recent study asserts rather boldly that multicarrier techniques are superior to direct-sequence spread spectrum: "When perfectly orthogonal code sequences are transmitted over slow, flat fading channels with perfect synchronization, the performance of DS-CDMA and MC-CDMA is equivalent, as the orthogonal multi-user interference vanishes immediately. However, in reality, wide-band CDMA signals sent over multipath channels experience more severe channel distortions and the resulting channel dispersion (i.e., frequency selectivity) erodes the orthogonality of CDMA signals. In such cases, it turns out to be far more beneficial to harness the signal energy in the frequency domain (as in MC-CDMA) than in the time domain (as in DS-CDMA)" [49, p. 1344].

44. Hara and Prasad [43] offer a survey of three different MC-CDMA techniques, which they refer to as MC-CDMA, multicarrier DS-CDMA, and multitone (MT-) CDMA. To my mind, the use of the term CDMA is a misnomer, for what is really involved here is a signal structuring technique and not a multiple access architecture per se. See also the useful introduction to Linnartz's work [50].

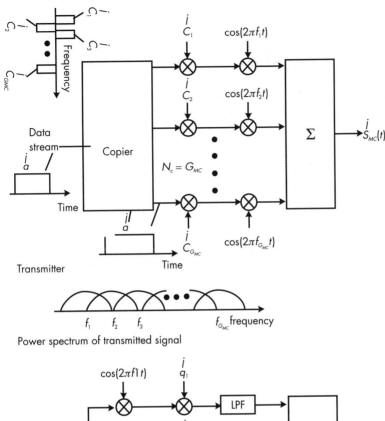

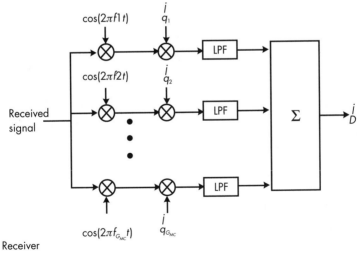

FIGURE 8.22 *Variations on the multicarrier architecture: (a) multicarrier CDMA, (b) multicarrier direct-sequence CDMA, and (c) multitone CDMA. (From: [43]. © 1997 IEEE. Reprinted with permission.)*

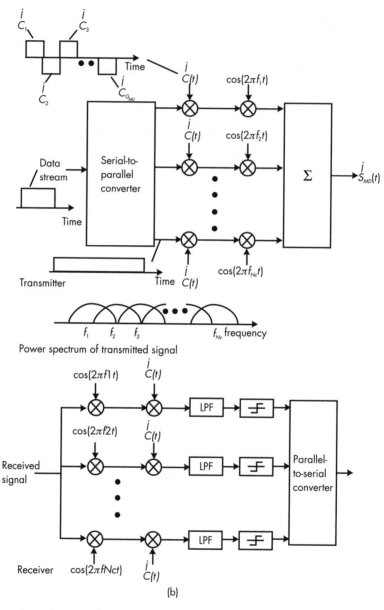

FIGURE 8.22 *Continued.*

single subcarrier, whereas in MC-CDMA each individual data symbol is spread across the entire set of subcarriers. [36, p. 1376]

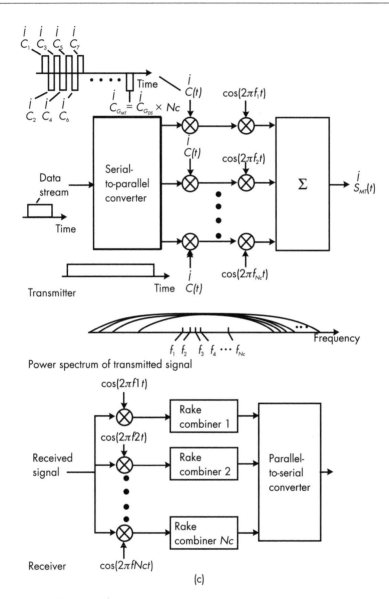

FIGURE 8.22 *Continued.*

In conventional DS[direct sequence]-CDMA, each user bit is transmitted in the form of many sequential chips, each of which is of short duration, thus having a wide bandwidth. In contrast to this, due to the FFT transform associated with OFDM, MC-CDMA chips are long in time duration, but narrow in bandwidth. Multiple chips are not sequential, but transmitted in parallel on different subcarriers. [50, p. 1375]

MC-CDMA is attractive, on the horizon, because it may allow a simplification of some of the receiver processes. It has been argued that "MC modulation ... eliminates multipath and achieves frequency diversity without complex RAKE systems or interleavers" [51, p. 1501]. Whether the time-domain processes (RAKE and so forth) are more or less complex than the corresponding frequency-domain processes used by MCM techniques (e.g., FFT and so forth) is probably a shifting equilibrium point, even in theory. So far, most of these bridges remain unmet and uncrossed, and relatively unilluminated by practical experience (as opposed to simulations).

Generalizing the Multicarrier Concept

Subdividing the signal into smaller frequency bands, and manipulating them in response (i.e., adaptively) to changing channel conditions observed in the frequency domain is a very general and powerful strategy. The wideband channel will show a nonuniform channel transfer function across the band, and if we can tailor the signal so that more of the available signal power is allocated to frequencies where the noise is lower or interference is higher, we can realize significant gains in performance over single-tone strategies that treat the entire band as one signal-to-noise calculation. Striving for generality, some authors look to the concept of *multirate signal processing* to provide a more unified framework for studying OFDM or MCM type signals [15, 52].

Another pathway toward generality is the relatively new discipline of *wavelet analysis* [24, 53]. Whereas Fourier analysis is a pure frequency-domain decomposition, wavelets extend the same general analytical powers into a two-dimensional matrix with both time and frequency factors. Fourier analysis breaks down a complex signal into its frequency atoms; wavelets "decompose a signal into ... *time–frequency atoms*" [54, p. 523].

It is reasonable to speak of time and frequency rather than position and momentum [in the Heisenberg frame of reference].... We will say then that [an event] is *well localized in both time and frequency* if the product of its time and frequency uncertainties is small. A musical note is an example of a time–frequency atom. It may be assigned two parameters, duration and pitch, which correspond to time uncertainty and frequency. [54, p. 524]

Hess-Nielsen and Wickerhauser [54] have developed this metaphor (as I assume we must view it) in a very interesting way, to generalize over the various ways in which the time–frequency plane can be subdivided, or *tiled* as they call it. The framework proceeds from the simpler tilings, such as the Fourier basis (optimal frequency location but no time localization), to more flexible wavelet-based tilings (Figure 8.23). The mathematics are forbidding for the nonspecialist, but the concept is admirably simple: The time–frequency plane of the signal can be carved up, or decomposed, in a variety of ways. If we view the channel transfer function and its distortions as another such plane (albeit a distorted one), the ability to tune the grid by which we divide the total plane into useful sub-bands (in time and frequency) is clearly advantageous.

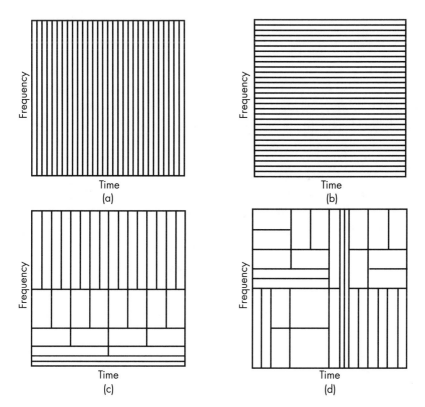

FIGURE 8.23 *Tilings of the time–frequency plane: (a) Dirac function tiling—optimizing in the time domain only; (b) Fourier tiling—optimizing in the frequency domain only; (c) wavelet-based tiling—using time and frequency domains; and (d) arbitrary tiling in time and frequency. (From: [54]. © 1996 IEEE. Reprinted with permission.)*

There is a suggestive parallel, I think, between the noise versus interference argument (discussed in Section 4.5.2) and the comparison between classical direct-sequence signals and adaptive OFDM-type signals. In Chapter 4 we saw how some researchers try to remove all of the structure from the interference in the channel and convert it to a flat noise-like phenomenon, while others argue that the structure should be maintained and leveraged to remove interference deterministically.

Both sides are right and the difference is one of practical philosophy. If the communications system is smart enough to detect and isolate the structure of the interference, then we can achieve optimal results by not homogenizing everything to a noise-like slurry. On the other hand, it may be practically infeasible to handle all the interference events in their structural detail (the interference demon is too complex to build), and using diversity techniques to convert it all to noise may prove the better practical strategy.

The comparison here is between direct-sequence spread spectrum, which follows the noise-conversion philosophy, and OFDM, which looks for the structure in the channel transfer function. If we recall the pothole analogy from Section 8.2, the direct-sequence spread spectrum approach is essentially like creating a big fat tire capable of rolling over the holes in the channel without sinking into them. There is still a bump of sorts, but the tire is wide enough to keep on moving. Of course, such a system doesn't need to try to foresee the occurrence of the potholes. It boasts inherent robustness. Wornell [14] offers a reasonably clear formulation of the noise conversion philosophy underlying even the more sophisticated versions of direct-sequence spread spectrum:

> Using such [spread] signature sets in multipath fading environments has the effect of transforming the collection of channels seen by the individual symbol streams from a collection of coupled Rayleigh fading channels into an uncorrelated collection of identical nonfading simple white marginally Gaussian additive noise channels. In essence, spread-signature CDMA [and indeed all DS-CDMA] converts various degradations due to fading, co-channel interference, and receiver noise into a single, comparatively more benign form of uncorrelated additive noise that is white and quasi-Gaussian. [14, p. 1418]

By comparison, the adaptive OFDM approach (generalized to whatever degree seems appropriate) is more like an agile bicycle weaving rapidly around the potholes which it can see clearly. It is smart—it uses its

Signal Expansion Strategies: Beyond Orthogonality   435

knowledge of the state of the roadway (channel information) to design an evasive strategy. On good roads it can speed up; on bad ones, it will slow down to enable better avoidance of the holes. Oh well, the analogy only carries us so far, but I think the point is clear.

In the long run, I would bet on the smart system (OFDM) to eventually prove superior, but the long run is often very long, and we may well be using dumb direct-sequence spread spectrum signals for quite some time (along with our antiquated internal combustion engines).

8.4 Forced Spatial Spreading: Creating Artificial Multipath

In the mid-1990s, a series of startling announcements from Bell Labs began to outline what may be the most powerful signal construction technique of all. Claims of enormous capacity gains seemed to shatter the limits of orthodox information theoretic calculations (although orthodoxy is resilient, by definition). The underlying concepts are so new that they have yet to penetrate very deeply into the commercial realm. Relatively few working engineers, perhaps, have yet absorbed these ideas, and for all its fascination with finding "the next big thing," even Wall Street has yet to really register these new developments. Nevertheless, research interest is exploding and it seems only a matter of time before we are all learning to speak the language of yet another new wireless revolution.

Let us return to the analogy developed in Section 8.1: the acoustic channel. There, the signal expansion is accomplished by creating new multipath components, both at the receiver (the ear) and especially at the transmitter where multiple speakers are used to enrich the transmit signal. The extra transmitters operate simultaneously, and occupy the same frequency band, but each transmitter is subtly shifted to produce an information-enriched signal at the receiver. The analogy suggests a more direct approach to signal spreading in a wireless system—why not use multiple transmitters to create an even denser multipath channel than nature itself provides?

Is insufficient multipath a real problem? Yes, indeed. Some channels are multipath impoverished, and push the limits of practical signal processing beyond what is feasible or acceptable:

> The indoor wireless channel shows very slow Rayleigh fading with short delay spread and *multipath is hardly available* if the system bandwidth is about 10 MHz. If fading is slow compared to the data rate, large interleaving size is required to provide sufficient coding gain.... If the delay

spread is short and only correlated paths are available ... artificial multipath may be required ... [41, p. 1501]

The science of creating multipath—like the science of good acoustic design—is complex. Whereas we have hundreds of years experience in creating good concert halls, the art of building a spatially enriched RF signal is still in the embryo stage.

8.4.1 Creating Multipath: The Physical Aspect of Spatial Spreading—Transmitter Diversity and Multiple-Input/Multiple Output (MIMO) Channels

The earliest perspective on this new technique is derived as an extension of the antenna diversity concept. *Receive diversity* is a well-known technique for improving the link by averaging fading over multiple independently spaced receiver antenna elements. Receiver diversity is aimed at correcting for the effects of multipath fading that are naturally introduced by the channel. However, our goal here is not to eliminate multipath, but to enhance it. *Transmitter diversity* is the first link in the conceptual chain.[45] By using multiple transmit antenna elements, the signal can be spread in the spatial domain—"creating an artificial multipath distortion" [56, p. 1452].

The unifying principle underlying this translation of the diversity concept from the passive side to the active side is this: The channel itself creates multiple copies of the signal which—if the antenna elements are separated sufficiently—fade independently. These copies can thus be seen as *orthogonal*, in the new sense that we have been developing that here. Two receiving antennas separated by a certain distance will show uncorrelated multipath fading patterns. The signals derived from the two antennas are different, even though they sample the same time/frequency bin. These two uncorrelated copies are produced at two receiver antennas, even though the signal was transmitted from just one transmitting antenna. Now, let us turn this channel around. Assume two transmitting antennas, similarly separated, and we can see how they will produce two uncorrelated images of the same signal, at least in terms of their multipath fading signatures, at a *single* receiver antenna [57]. True, they are superimposed—but in principle they are orthogonal, in the same sense that two superimposed direct-sequence signals can be said to be orthogonal.

---

45. "Transmitter diversity, the achievable performance limits of which are ... less well understood" [55].

The next step is to envision a system with multiple transmit antennas *and* multiple receive antennas. This is known as an MIMO system [58] (Figure 8.24). In a MIMO architecture, *all* transmitters are active in the same frequency band, at the same time; there is no physical orthogonality.

> The essence of MIMO communications is: (1) using multiple transmitters to transmit multiple signals over the same carrier simultaneously as if each signal would have occupied the channel exclusively; and (2) using some signal processing technique to separate individual transmitted signals from the received mixtures out of a receiving antenna array. [59, p. 1]

MIMO systems create huge amounts of extra multipath (much like multispeaker acoustic sound systems). Viewed in information theoretic terms, the capacity of the MIMO channel is increased relative to the conventional *single-input/single-output* (SISO) channel.

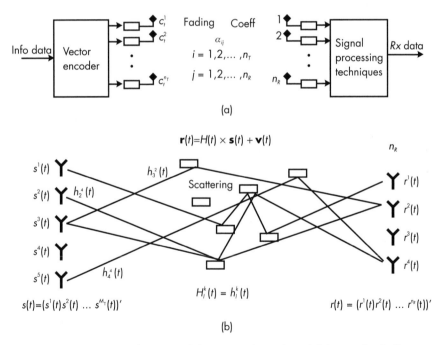

FIGURE 8.24 *MIMO architecture: (a) MIMO channel model (From: [60]. © 2001 Ran Gozali. Reprinted with permission.); (b) channel model with scattering environment and multiple transmit and receive antenna arrays (From: [61]. © 2001 Telenor. Reprinted with permission.); and (c) model of digital communication system with multiple transmitting and receiving antennas (From: [62]. © IEEE. Reprinted with permission.).*

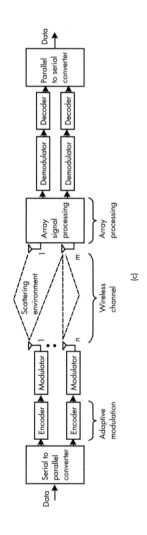

FIGURE 8.24 *Continued.*

Multipath substantially improves capacity for the MIMO case. Specifically, if the number of multipath components exceeds a certain value, then the channel capacity slope in bits per decibel of power increase can be proportional to the number of antennas located at both the input and output of the channel. [63, p. 357]

This is a remarkable result: Channel capacity can be increased simply by adding appropriately spaced antenna elements at the transmitter and receiver, and the gains are proportional to the number of these elements.[46] In must be stressed, however, that it is not just the antenna elements themselves, but their interaction with the reflecting and scattering objects in the channel that create capacity. If the channel is a direct path in free space without significant multipath, the gains cannot be realized: "The capacity advantage of MIMO channel structures can grow without bound as SNR increases, *provided that multipath is present* [emphasis added]" [63, p. 360]. The worse it is, the better it is:

If the multipath is severe, the high SNR capacity can essentially be multiplied by adding antennas to both sides of the radio link. This capacity improvement occurs with no penalty in ... bandwidth ... [63, p. 360]

The significance of this last point is profound: The information-bearing capacity of the signal is expanded but the frequency bandwidth is not. Because of the assumptions inherited from classical spread spectrum and even from the (mis)understanding of the noise–bandwidth trade-off arguments highlighted by Shannon, it may seem counterintuitive—if not just wrong—to suggest that a signal can in any sense be expanded if its occupied spectrum is not increased. Both OFDM and direct-sequence spread spectrum create signals with greatly expanded frequency spectra, but an MIMO-type signal accomplishes the expansion in the spatial domain.

Band-limited wireless channels are narrow pipes that do not accommodate rapid flow of data. Deploying multiple transmit and receive antennas broadens this data pipe. [64, p. 744][47]

---

46. "Theoretically, the channel capacity (bits per second per Hz) for a MIMO communications system is roughly proportional to the number of transmitters" [59, p. 1].

47. See also [58].

The scheme requires no bandwidth expansion, as redundancy is applied in space across multiple antennas, not in time or frequency. [56, p. 1452]

8.4.2 The Coding Aspect of Spatial Spreading: Space–Time Codes

Enriching the transmit signal with extra multipath is all well and good, but how is the receiver supposed to cope with it, let alone turn it to advantage? The receiver now sees not just multiple images of a single transmitted signal, but a "noisy superposition of the faded versions of N transmitted signals" from N different antennas [65, p. 1461]. How can the receiver discriminate the signals from different transmitting antennas?

This question leads us to consider how the different signals are coded. Lacking in traditional physical orthogonality, the signals for each antenna element should be designed to incorporate some form of logical orthogonality akin to the orthogonality of spreading codes in direct-sequence spread spectrum. Indeed, the solutions for orthogonal MIMO coding may be developed in very similar ways. Broadly speaking, the simplest coding strategy for an MIMO signal involves applying either a time or frequency offset to the signals from the two antennas, to create a space–time code or a space–frequency code [56, p. 1453]. A simple example illustrates this strategy: A two-element MIMO transmitter sends the same signal over both antennas, but delayed by one symbol time on antenna 2 (Figure 8.25). This creates a partial logical orthogonality (i.e., decorrelation) between the two signals [64].[48]

A simple repetition code with an offset is a relatively weak orthogonalization function. It is possible to encode the various transmit streams much more effectively to generate greater logical orthogonality that the receiver can use to discriminate the different signal components. Pilot sequences and synchronization sequences can be inserted into data streams transmitted simultaneously from different antennas, to create additional code structure supporting greater logical orthogonality[49] (Figure 8.26). Another approach

---

48. See also Chuang and Sollenberger [38], where they discuss a combination of OFDM and MIMO ideas: "Simplified transmitter diversity can be achieved by transmitting the same OFDM symbols on multiple antennas with delayed transmission times" [38, p. 80].

49. "In general, with N transmit antennas we will have N different synchronization sequences ... and N different pilot sequences.... Since signals at the receiver antennas will be linear superpositions of all transmitted signals, we choose the training sequences ... to be orthogonal sequences" [65, p. 1465].

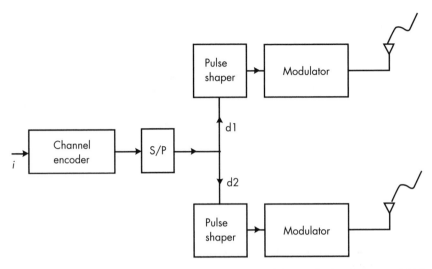

FIGURE 8.25 *Delay coding in an MIMO system. (From: [64]. © 1998 IEEE. Reprinted with permission.)*

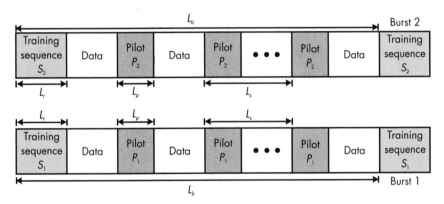

FIGURE 8.26 *More complex MIMO code structures. (From: [65]. © 1998 IEEE. Reprinted with permission.)*

involves the creation of so-called *diagonal* signals, where there is no cross-correlation of between signals transmitted from different antennas [59].

The term that has emerged for the coding strategy that is typically used with multiantenna transmission systems is *space–time coding* [65, 66]. The basic approach is to divide the input data stream (i.e., the source data, plus whatever error correction may be applied) into multiple substreams, often misleadingly referred to as a serial-to-parallel conversion, each of which is fed to a different transmit antenna element. As each substream is mapped to a given antenna, various signal processing operations can be performed to further differentiate the transmitted signal components. The spatial formatting operations [66] (Figure 8.27) used to differentiate two transmission components can include the following:

1. *Time delay:* As noted, a very simple differentiator is used to introduce a delay in the transmission from the second antenna; even if no other form of orthogonality is employed, two signals may be discriminated on this basis.

2. *Frequency offset:* The second stream can be transmitted on a different carrier (invoking the power of some of the OFDM architecture).

3. *Spreading code:* The two streams can be modulated with different orthogonal spreading codes.

4. *Orthogonal pilot sequences:* The substreams can be tagged with different training sequences.

5. *Modulation format:* Each stream can be modulated with a different symbol waveform [67].

Explicit combination of OFDM [or *discrete multitone* (DMT)] with the MIMO channel structure has been described, leading to transmitter–receiver architectures that encompass both space–time trellis coding as well as FFT/IFFT engines [68]. The motivation is again to simplify the processing burden at certain stages (especially the time domain), or to shift it to more tractable portions of the signal processing chain:

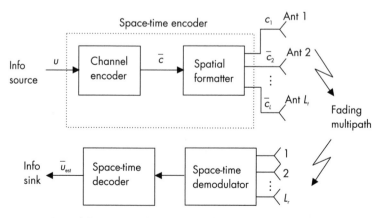

FIGURE 8.27 *Spatial formatting. (From: [66]. © 2000 IEEE. Reprinted with permission.)*

The main disadvantage with space–time ... coding ... is the associated computational complexity....

Complexity can be reduced by using a coding structure similar to the discrete multitone (DMT) solution ... [63, p. 361]

The complexity reduction occurs because ISI is eliminated from each discrete multitone subchannel. [39, p. 856]

The result is appropriately referred to as space–frequency coding, although it also incorporates space–time code structures:

This new space–frequency coding structure results in a matrix of transmission and reception vector solutions for each discrete Fourier transform (DFT) index ... a space–frequency channel matrix ... [63, p. 361]

Frequency- (and possibly time-) domain interleaving are applied to distribute consecutive spatial-vector code segments among well-separated frequency subchannels. Interleaving allows the system to exploit the frequency (and possibly time) diversity of the channel while spatial coding is a form of spatial diversity. [39, p. 858]

The payoff is a gain in capacity. This is still being mapped out, and the indications for capacity increases are best seen as preliminary as of this writing, but they are dramatic. For example, in one simulation, a standard single-antenna system (that is, one antenna at the transmitter and one at the receiver) with a particular pulse-shaping and signal structure was able to achieve 0.72 bps/Hz, and a three-element MIMO system (three transmit and three receive antennas) yielded 2.03 bps/Hz—a gain of almost three times without any frequency-domain bandwidth expansion [39, p. 862].

This may be just the beginning.

8.4.3 Space–Time Architectures: Ultimate RF?

Building a spatially spread signal is the basic stepping-stone to the creation of new multiple access architectures. Though multiple access is beyond the true scope of this volume, we touch on it here in considering the astonishing capacity calculations that are being held out by advocates of this new spreading philosophy. The industry has seen other capacity revolutions announced, and later discounted, but this may well turn out to be the real thing.

The first attempt to organize a comprehensive system architecture based on spatial spreading principles is the BLAST concept, associated with Gerard Foschini and his colleagues at Bell Labs [69]. BLAST is referred to as a layered space–time architecture, in which the data stream is divided into substreams interlaced by complex coding structures to create redundancy and inherent signal diversity [70]. In the initial version proposed by Foschini, the code threads were woven in a diagonal fashion through the space–time matrix. Other, simpler structures have since been proposed [71].

The structure of the BLAST signal is remarkable. All transmitters in the system transmit simultaneously and continuously. The signal is spread in space, but not in frequency—*it remains a narrowband signal*. Perhaps it is more appropriate to say that it is a superimposition of many narrowband signals, but in any case there is no inherent spectral spreading.

> Unlike code-division or other spread spectrum multiple access techniques, the total channel bandwidth utilized in a BLAST system is only a small fraction in excess of the symbol rate [i.e., it is narrowband!] similar to the excess bandwidth required by a conventional QAM system. Second, unlike FDMA, each transmitted signal occupies the entire system bandwidth. Finally, unlike TDMA, the entire system bandwidth is used simultaneously by all of the transmitters all of the time. [71]

The system exploits the logic of MIMO spreading:

> No explicit orthogonalization of the transmitted signals is imposed by the transmit structure at all. Instead, the propagation environment itself, which is assumed to exhibit significant multipath, is exploited to achieve the signal decorrelation necessary to separate the co-channel signals. [71]

The gains are phenomenal [57, 72, 73]. Wolniansky et al. [71] report results from an indoor RF system operating in a standard office environment, with relatively impoverished multipath ("the delay spread is negligible, the fading rates are low"), with 8 transmitting antennas and 12 receiving antennas. Using a simpler version of the BLAST coding, and 16-QAM modulation, they achieved a raw spectral efficiency of almost 26 bps/Hz. This is compared to the classical expectation of 1 bps/Hz for binary signaling, and perhaps 4 bps/Hz for 16-level signaling (like 16-QAM). At an SNR of 34 dB, which is high but achievable, they were able to achieve throughputs of as much as 40 bps/Hz.

We believe these spectral efficiencies are unprecedented for the wireless channel. It is worthwhile to point out that spectral efficiencies of these magnitudes are essentially impossible to obtain using ... a single transmitter ... simply because the required constellation loadings would be immense. For example, to obtain the equivalent of 32 bits per ... symbol ... using a single transmitter, would require a constellation with 2^{32} or more than a billion points, which seems well outside the realm of practicality, *regardless of SNR* [emphasis added]. [71]

Even simple configurations also show "exceedingly large capacities" [72, p. 313], especially compared to nonspatially spread cases. Foschini and Gans [72] report that for two-thread systems (that is, with just two transmitting and two receiving antennas), "the capacity exceeds 7 bits/cycle." For four-thread systems, the gain is 19 bps/Hz: "whereas if [there is only one thread] there is only about 1 bit/cycle" [72, p. 322]. These configurations are certainly within the realm of current commercial feasibility from a hardware standpoint.

How far can this logic be extended? A 16-thread diagonally coded BLAST system is reported to show a capacity of 71 bps/Hz [74].

With a 32-thread system—32 transmitting and 32 receiving antennas—the capacity reaches more than 180 bits/symbol [72, p. 324]!

This—if the results hold—may be something approaching the ultimate RF technology. Consider how far we have come. Physical orthogonality is completely eliminated. Foschini et al. [74] go so far as to suggest abandoning the classical reuse-oriented cellular system, even for narrowband signals.[50] The signal is spread so aggressively that its information-carrying capacity expands beyond expectation, and almost (we might say) beyond all reason. Like the abundant acoustic environment created in a well-designed concert hall, the RF channel has been amplified, reduplicated, and transformed by the seemingly simple expedient of adding well-positioned transmitters. Interference is now so embedded in the structure that we must take pains to make sure there is enough of it.

The other question that must be addressed, sooner or later, is this: Where does this leave us, in terms of the critical review of Shannon theory? The idea of a *narrowband* signal that—*even in theory*—could carry billions of times the information load that anyone thought possible just a few years ago poses a profound challenge to our understanding of the communications process.

---

50. "Three sectors in a conventional cellular system can be combined to form one 'edge-excited' (inward-facing) cell to enhance capacity" [73, p. 176].

Academic information theorists are struggling to prove that Shannon's theory still embraces these results, but that seems hard to swallow. Shannon's simple channel model, the communication engineer's "coat of arms" that we started with in Chapter 2, is woefully inadequate to portray this sort of channel environment. Whatever the notion of capacity meant in Shannon's calculations, it seems that we have blown it apart. As the example of the billion-point constellation illustrates, it is not even clear that we can continue to think of signal and noise in the conventional terms. *"Regardless of SNR ..."* The phrase rings with the irrelevance of old concepts and fated paradigms. If we are achieving results that cannot even be properly expressed in the old terminology, it is a clear enough sign that we are approaching a new era in the history of the field.

REFERENCES

[1]   Calhoun, G., *Digital Cellular Radio,* Norwood, MA: Artech House, 1988.

[2]   Calhoun, G., *Wireless Access and the Local Telephone Network,* Norwood, MA: Artech House, 1992.

[3]   Kendall, G., "A 3D Sound Primer: Directional Hearing and Stereo Reproduction," http://www.northwestern.edu/musicschool/classes/3D/pages/sndPrmGK.html, reprinted from *Computer Music Journal,* Vol. 19, No. 4, Winter 1995, pp. 23–46.

[4]   Blauert, J., "Localization and the Law of the First Wavefront in the Median Plane," *J. of the Acoustic Society of America,* Vol. 50, 1971, pp. 466–470.

[5]   Lord Rayleigh, "On Our Perception of Sound Direction," *Philosophy Magazine,* Vol. 13, 1907, pp. 214–223.

[6]   Kuhn, G. F., "Physical Acoustics and Measurements Pertaining to Directional Hearing," in Yost, W. A., and G. Gourevitch, (Eds.), *Directional Hearing,* New York: Springer-Verlag, 1987.

[7]   Rasch, R. A., and R. Plomp, "The Listener and the Acoustic Environment," in D. Deutsch, (Ed.), *The Psychology of Music,* New York: Academic Press, 1982.

[8]   Carr, C., "Sounds, Signals and Space Maps," *Nature,* Vol. 415, January 3, 2002.

[9]   Hyde, P., and E. Knudsen, "The Optic Tectum Controls Visually Guided Adaptive Plasticity in the Owl's Auditory Space Map," *Nature,* Vol. 415, January 3, 2002, pp. 73–76.

[10]   Warren, R., "The Sound and the Hurry," *Chicago Tribune,* January 21, 2002, Sec. 6, p. 1.

[11] Kendall, G., "A 3D Sound Primer," http://www.northwestern.edu/musicschool/classes/3D/pages/envPsyAcst.html., p. 1.

[12] Bissell, C. C., and D. A. Chapman, *Digital Signal Transmission,* New York: Cambridge University Press, 1992.

[13] Wornell, G. W., "Spread-Signature CDMA: Efficient Multiuser Communication in the Presence of Fading," *IEEE Trans. on Information Theory,* Vol. 41, No. 9, September 1995, pp. 1418–1438.

[14] Wornell, G. W., "Emerging Applications of Multirate Signal Processing and Wavelets in Digital Communications," *Proc. IEEE,* Vol. 84, No. 4, April 1996, pp. 586–603.

[15] Dixon, R. C., *Spread Spectrum Systems,* New York: Wiley, 1984.

[16] Bhashyam, S., A. M. Sayeed, and B. Aazhang, "Time-Selective Signaling and Reception for Communication over Multipath Fading Channels," *IEEE Trans. on Communications,* Vol. 48, No. 1, January 2000, pp. 83–94.

[17] Sayeed, A. M., A. Sendonaris, and B. Aazhang, "Multiuser Detection in Fast Fading Multipath Environments," *IEEE J. on Selected Areas in Communications,* Vol. 16, No. 9, December 1998, pp. 1691–1701.

[18] Kammler, D. W., *A First Course in Fourier Analysis,* Upper Saddle River, NJ: Prentice-Hall, 2000.

[19] Bingham, J. A. C., "Multicarrier Modulation for Data Transmission: An Idea Whose Time Has Come," *IEEE Communications Magazine,* May 1990, pp. 5–14.

[20] Ifeachor, E. C., and B. W. Jervis, *Digital Signal Processing,* Reading, MA: Addison-Wesley, 1993.

[21] Cooley, J. W., and J. W. Tukey, "An Algorithm for the Machine Computation of Complex Fourier Series," *Math. Comp.,* Vol. 19, 1965, pp. 297–301.

[22] Johnson, J. R., and R. W. Johnson, "Challenges of Computing the Fast Fourier Transform," *Proc. of Optimized Portable Application Libraries (OPAL) Workshop,* Kansas City, MO, June 2–3, 1997.

[23] Rockmore, D. N., "The FFT: An Algorithm the Whole Family Can Use," *Computing in Science and Engineering,* Vol. 2, No. 1, January–February 2000, pp. 60–64.

[24] Akansu, A. N., et al., "Wavelet and Subband Transforms: Fundamentals and Communication Applications," *IEEE Communications Magazine,* December 1997, pp. 104–115.

[25] Weinstein, S. B., and P. M. Ebert, "Data Transmission by Frequency-Division Multiplexing Using the Discrete Fourier Transform," *IEEE Trans. on Communications,* Vol. 19, No. 5, October 1971, pp. 628–634.

[26] Chang, R. W., "High-Speed Multichannel Data Transmission with Bandlimited Orthogonal Signals," *Bell System Technical J.,* Vol. 45, pp. 1775–1796.

[27] Saltzburg, B. R., "Performance of an Efficient Parallel Data Transmission System," *IEEE Trans. on Communications Technology*, Vol. 15, December 1967, pp. 805–811

[28] Darlington, S., "On Digital Single-Sideband Modulators," *IEEE Trans. on Circuit Theory*, Vol. CT-17, August 1970, pp. 409–414.

[29] Ruiz, A., J. Cioffi, and S. Kasturia, "Discrete Multiple Tone Modulation with Coset Coding for the Spectrally Shaped Channel," *IEEE Trans. on Communications*, Vol. 40, No. 6, June 1992, pp. 1012–1029.

[30] Doelz, M. L., E. T. Heald, and D. L. Martin, "Binary Data Transmission Techniques for Linear Systems," *Proc. IRE*, Vol. 45, May 1957, pp. 656–661.

[31] Pollet, T., et al., "Equalization for DMT-Based Broadband Modems," *IEEE Communications Magazine*, May 2000, pp. 106–113.

[32] Kalet, I., "The Multitone Channel," *IEEE Trans. on Communications*, Vol. 37, No. 2, February 1989, pp. 119–124.

[33] Sandberg, S. D., and M. A. Tzannes, "Overlapped Discrete Multitone Modulation for High Speed Copper Wire Communications," *IEEE J. on Selected Areas in Communications*, Vol. 13, No. 9, December 1995, pp. 1571–1585.

[34] Moose, P., "A Technique for Orthogonal Frequency Division Multiplexing Frequency Offset Correction," *IEEE Trans. on Communications*, Vol. 42, No. 10, October 1994, pp. 2908–2914.

[35] Arslan, G., B. L. Evans, and S. Kiaei, "Equalization for Discrete Multitone Transceivers to Maximize Bit Rate," *IEEE Trans. on Signal Processing*, Vol. 49, No. 12, December 2001, pp. 3123–3135.

[36] Choi, Y.-S., P. J. Voltz, and F. Cassara, "On Channel Estimation and Detection for Multicarrier Signals in Fast and Selective Rayleigh Fading Channels," *IEEE Trans. on Communications*, Vol. 49, No. 8, August 2001, pp. 1375–1387.

[37] Harada, H., and R. Prasad, "An OFDM-Based Wireless ATM Transmission System Assisted by a Cyclically Extended PN Sequence for Future Broad-Band Mobile Multimedia Communications," *IEEE Trans. on Vehicular Technology*, Vol. 50, No. 6, November 2001, pp. 1366–1374.

[38] Chuang, J., and N. Sollenberger, "Beyond 3G: Wideband Wireless Data Access Based on OFDM and Dynamic Packet Assignment," *IEEE Communications Magazine*, Vol. 38, No. 7, July 2000, pp. 78–87.

[39] Raleigh, G. G., and V. K. Jones, "Multivariate Modulation and Coding for Wireless Communication," *IEEE J. on Selected Areas in Communications*, Vol. 17, No. 5, May 1999.

[40] Shannon, C. E., "Communications in the Presence of Noise," *Proc. IRE*, Vol. 37, 1949, pp. 10–21. Reprinted in Sloane, N. J. A., and A. D. Wyner, (Eds.), *Claude Elwood Shannon: Collected Papers*, Piscataway, NJ: IEEE Press, 1993, pp. 160–172.

[41]   Van Acker, K., et al., "Per Tone Equalization for DMT-Based Systems," *IEEE Trans. on Communications*, Vol. 49, No. 1, January 2001, pp. 109–119.

[42]   Pollet, T., and M. Peeters, "Synchronization with DMT Modulation," *IEEE Communications Magazine*, April 1999, pp. 80–86.

[43]   Hara, S., and R. Prasad, "Overview of Multicarrier CDMA," *IEEE Communications Magazine*, Vol. 35, No. 12, December 1997, pp. 126–133.

[44]   Keller, T., and L. Hanzo, "Adaptive Multicarrier Modulation: A Convenient Framework for Time-Frequency Processing in Wireless Communications," *Proc. IEEE*, Vol. 88, No. 5, May 2000, pp. 611–640.

[45]   Al-Dhahir, N., and J. Cioffi, "Optimum Finite-Length Equalization for Multicarrier Transceivers," *IEEE Trans. on Communications*, Vol. 44, No. 1, January 1996, pp. 56–64.

[46]   Faulkner, M., "The Effect of Filtering on the Performance of OFDM Systems," *IEEE Trans. on Vehicular Technology*, Vol. 49, September 2000, pp. 1877–1884.

[47]   Bissell, C. C., and D. A. Chapman, *Digital Signal Transmission*, New York: Cambridge University Press, 1992, p. 42.

[48]   Papathanassiou, A., A. K. Salkintzis, and P. T. Mathiopoulos, "A Comparison Study of the Uplink Performance of W-CDMA and OFDM for Mobile Multimedia Communications Via LEO Satellites," *IEEE Personal Communications*, June 2001, pp. 35–43.

[49]   Natarajan, B., et al., "High-Performance MC-CDMA Via Carrier Interferometry Codes," *IEEE Trans. on Vehicular Technology*, Vol. 50, No. 6, November 2001, pp. 1344–1353.

[50]   Linnartz, J.-P., "Performance Analysis of Synchronous MC-CDMA in Mobile Rayleigh Channel with Both Delay and Doppler Spreads," *IEEE Trans. on Vehicular Technology*, Vol. 50, No. 6, November 2001, pp. 1375–1387.

[51]   Sanada, Y., and M. Nakagawa, "A Multiuser Interference Cancellation Technique Utilizing Convolutional Codes and Orthogonal Multicarrier Modulation for Wireless Indoor Communications," *IEEE J. on Selected Areas in Communications*, Vol. 14, No. 8, October 1996, pp. 1500–1509.

[52]   Vaidyanathan, P. P., *Multirate Systems and Filter Banks*, Englewood Cliffs, NJ: Prentice-Hall, 1993.

[53]   Daubechies, I., "Where Do Wavelets Come From?—A Personal Point of View," *Proc. IEEE*, Vol. 84, No. 4, April 1996, pp. 510–513.

[54]   Hess-Nielsen, N., and M. V. Wickerhauser, "Wavelets and Time–Frequency Analysis," *Proc. IEEE*, Vol. 84, No. 4, April 1996, pp. 523–540.

[55]   Narula, A., M. D. Trott, and G. W. Wornell, "Performance Limits of Coded Diversity Methods for Transmitter Antenna Arrays," *IEEE Trans. on Information Theory*, Vol. 45, No. 7, November 1999, pp. 2418–2433.

[56] Alamouti, S. M., "A Simple Transmit Diversity Technique for Wireless Commu-
      nications," *IEEE J. on Select Areas in Communications*, Vol. 16, No. 8, October
      1998, pp. 1451–1458.

[57] Shiu, D.-S., et al., "Fading Correlation and Its Effect on the Capacity of Multiele-
      ment Antenna Systems," *IEEE Trans. on Communications*, Vol. 48, No. 3, March
      2000, pp. 502–513.

[58] Tarokh, V., et al., "Combined Array Processing and Space–Time Coding," *IEEE
      Trans. on Information Theory*, Vol. 45, No. 4, May 1999, pp. 1121–1128.

[59] Luo, H., et al., "The Autocorrelation Matching Method for Distributed MIMO
      Communications over Unknown FIR Channels," http:/www.research.att.com/
      ~macsbug/Signal_Processing/ICASSP2001.pdf.

[60] Gozali, R., "A Tutorial on Space–Time Coding and MIMO Channels,"
      http:/www.mprg.ee.vt.edu/people/rgozali/tutorial.pdf.

[61] Schneider, C., "MIMO Communications Channels," Telenor White Paper, 2000.

[62] Catreux, S., et al., "Data Throughputs Using Multiple-Input Multiple Output
      (MIMO) Techniques in a Noise-Limited Cellular Environment," *IEEE Trans. on
      Wireless Communications*, Vol. 1, No. 2, April 2002, pp. 226–240.

[63] Raleigh, G. G., and J. M. Cioffi, "Spatio-Temporal Coding for Wireless Commu-
      nication," *IEEE Trans. on Communications*, Vol. 46, No. 3, March 1998,
      pp. 357–366.

[64] Tarokh, V., N. Seshadri, and A. R. Calderbank, "Space–Time Codes for High
      Data Rate Wireless Communication: Performance Criterion and Code Construc-
      tion," *IEEE Trans. on Information Theory*, Vol. 44, No. 2, March 1998,
      pp. 744–765.

[65] Naguib, A. F., et al., "A Space–Time Coding Modem for High-Data-Rate Wire-
      less Communications," *IEEE J. on Selected Areas in Communications*, Vol. 16,
      No. 8, October 1998, 1459–1478.

[66] Hammons, A. R., and H. El Gamal, "On the Theory of Space–Time Codes for
      PSK Modulation," *IEEE Trans. on Information Theory*, Vol. 46, No. 2, March
      2000, pp. 524–542.

[67] Paulraj, A. J., and B. C. Ng, "Space–Time Modems for Wireless Personal Com-
      munications," *IEEE Personal Communications*, February 1998, pp. 36–48.

[68] Lu, B., X. Wang, and K. R. Narayanan, "LDPC-Based Space–Time Coded
      OFDM Systems over Correlated Fading Channels: Performance Analysis and
      Receiver Design," *IEEE Trans. on Communications*, Vol. 50, No. 1, January 2002,
      pp. 74–88.

[69] Foschini, G. J., "Layered Space–Time Architecture for Wireless Communication
      in a Fading Environment When Using Multiple Antennas," *Bell Laboratories
      Technical J.*, Vol. 1, No. 2, Autumn 1996, pp. 41–59.

[70]   Lozano, A., and C. Papadias, "Layered Space–Time Receivers for Frequency-Selective Wireless Channels," *IEEE Trans. on Communications*, Vol. 50, No. 1, January 2002, pp. 65–73.

[71]   Wolniansky, P. W., et al, "V-Blast: An Architecture for Realizing Very High Data Rates over the Rich-Scattering Wireless Channel," *Proc. 1998 URSI Int. Symp. on Signals, Systems, and Electronics*, Pisa, Italy, September 29–October 2, 1998, pp. 295–300.

[72]   Foschini, G. J., and M. J. Gans, "On Limits of Wireless Communications in a Fading Environment with Using Multiple Antennas," *Wireless Personal Communications*, Vol. 6, 1998, pp. 311–335.

[73]   Driessen, P. F., and G. J. Foschini, "On the Capacity Formula for Multiple Input—Multiple Output Wireless Channels: A Geometric Interpretation," *IEEE Trans. on Communications*, Vol. 47, No. 2, February 1999, pp. 173–176.

[74]   Foschini, G. J., et al., "Simplified Processing for High Spectral Efficiency Wireless Communication Employing Multi-Element Arrays," *IEEE J. on Selected Areas in Communications*, Vol. 17, No. 11, November 1999, pp. 1841–1852.

9

Epilogue:
The Red Queen and the Kitten

"And as for you," she went on, turning fiercely on the Red Queen ...

She took her off the table as she spoke, and shook her backwards and forwards with all her might. The Red Queen made no resistance whatever; only her face grew very small, and her eyes got large and green: and still, as Alice went on shaking her, she kept on growing shorter—and fatter—and softer—and rounder—and—and it really was a kitten, after all.

So Alice hunted among the chessmen on the table till she had found the Red Queen: then she went down on her knees on the hearthrug, and put the Kitten and the Queen to look at each other. "Now, Kitty!" she cried, clapping her hands triumphantly. "You've got to confess that that was what you turned into!" ("But it wouldn't look at it," she said, when she was explaining the thing afterwards to her sister: "it turned away its head, and pretended not to see it: but it looked a little ashamed of itself, so I think it must have been the Red Queen.")

—*Alice's Adventures in Wonderland*, Lewis Carroll (1832–1898)

If this were a screenplay, we would be writing "Fade to black" at this point. The dramatic arc is complete. We began with the pre-Shannon framework, which was all about (or mostly about) transmitter power. We traced the familiar tenets of the Shannon revolution, and followed the consequences out through the evolution of complex second-generation wireless architectures in which classical Shannonesque technologies like error coding, compression, modulation, diversity processing, interleaving, and signal shaping were assayed. We pushed on into the strange new world of post-Shannon physical-layer techniques, such as the signal-construction strategies described in the preceding chapter. As we reached the end of the arc, we found ourselves in a looking-glass world where the traditional engineering values are reversed. Interference is no longer simply something to be suppressed, but may be intentionally created to strengthen the transmitted signal. Aggressive spreading of the signal in the time, space, and frequency domains has been found to be a way of vastly increasing the information-carrying capacity of a channel.

Above all, we have reached the end of physical orthogonality, it would seem. The philosophy of keeping different users or different signals separated from one another by physical buffers, which had been the overarching design principle of all wireless communications systems since Marconi and De Forest, has in the post-Shannon era been turned on its head, refuted, and replaced with new architectural principles based on eliminating physical orthogonality of the old sort entirely. Spread spectrum, OFDM, MIMO—all of these signal structures transcend physical orthogonality, eliminate the buffers and the guardbands and the reuse patterns, and put their faith in powerful signal processing to extract the information content of the desired signal from a composite channel that is fully shared physically with many other users.

This should be the basis for a satisfying sense of closure. The story of scientific progress is often told according to this template: At one time, children, we really did believe X, but now we *know* that not-X is really the case. The world *is* round and dinosaurs really *were* warm-blooded. Aren't we lucky to live in an enlightened era? Close the book and go to sleep. If we have really proved that the idea of physically separating different wireless signals was a wasteful and inefficient strategy, we can see the triumph of nonorthogonal techniques, like spread spectrum and the rest, as a triumph of truth.

9.1 Countertrends

But what if the story does not end there, or ends in another way? What if we find ourselves, like Alice, shaking the Red Queen and suddenly having it turn back into a kitten?

The chessboard is a world of full-information. Nothing is concealed from us. The only limit is the amount of processing power at our disposal. It is a model for the information theoretic paradise. It is a world where software triumphs over physics. The Red Queen perhaps may symbolize the hubris of this vision.

All too often, however, a little shaking from the outside will dissolve this world into something more complex, organic, unpredictable. A little interference, representing the intrusion of physical reality, and the software breaks down. Alice complains that "It's a very inconvenient habit of kittens that, whatever you say to them, they *always* purr. If they would only purr for 'yes' and mew for 'no,' or any rule of that sort so that one could keep up a conversation! But how can you talk with a person if they always say the same thing?" A paradigm of the detection problem, perhaps, in a noisy environment?[1]

Whether or not we may safely fade to black at the end of Chapter 8, there are indications that the story is not finished and that the pendulum swing may one day reverse back toward "reorthogonalization." Consider the following four straws in the backdraft, that come from four articles published sequentially in a recent IEEE publication,[2] authored by four of the most distinguished figures in the modern wireless industry, dealing with four apparently quite separate wireless applications. An interesting pattern will emerge.

9.1.1 Verdú

First, Sergio Verdú—the father of multiuser detection concepts, and a champion of the aggressive adoption of signal processing solutions for interference control—writes in "Wireless Bandwidth in the Making" [1] how fortunate we are to live in an era where we can approach closely the Shannon limits on channel capacity, and proceeds to survey a variety of techniques for boosting capacity and system performance. He gives the nod to his own

---

1. Readers will also note the analogy with some of Shannon's examples used to develop the idea of the entropy of a source.

2. The July 2000 issue of *IEEE Communications Magazine*.

preferred approach of working to extract the structure of the interference, rather than treating it as background noise: "By exploiting that structure, multi-user detection can increase spectral efficiency, receiver sensitivity, and the number of users the system can sustain" [1, p. 55]. This is familiar by now.

He touches further on a number of counterintuitive conclusions regarding the handling of interference, the value of signal spreading, the immense apparent power of signal structures like OFDM and MIMO. Verdú seems all for the post-Shannon future—nonorthogonality and signal spreading. Then he hesitates. He notes briefly that "some CDMA ... systems are starting to move away from the philosophy of *simultaneous* [emphasis added] uncoordinated transmissions" [1, p. 54]. Eventually we reach a strange caveat, couched in vague language:

> However, recent information-theoretic results point out that too much spreading (or, in general, too many degrees of diversity) may actually decrease the spectral-efficiency of time-varying channels because of the need to estimate more and more unknown fading parameters with less and less energy devoted to each. Under those time-varying conditions, it is actually preferable to concentrate the energy of the [signal] in a small region of the time-frequency plane. [1, p. 57]

What is he saying here?

We have spent several chapters now building to the logical climax that signal spreading is beneficial, and the more diversity we can implement, the better. Verdú seems to say that there are limits to this swing of the pendulum, too. What are those limits? He hints at the answer: *processing cost*—"the need to estimate more and more unknown parameters." The implication is that this cost grows faster than the benefit. The term *information-theoretic* hints further at the fact that this cost is not just to be measured in the quantity pricing for the latest digital signal processors. There is something fundamental about this exponentially expanding cost of computation associated with acquiring the side information to process the expanding signal.

It calls to mind the ultimate undoing of Maxwell's demon. We remember that Maxwell invented his demon as a way of poking at a soft spot in theories about the second law of thermodynamics, and the principle that entropy (disorder) in physical systems always tends to increase. Maxwell's demon was a creature of atomic proportions, sitting at a tiny gate or valve connecting two containers filled with gas. The demon watched as individual molecules of the gas approached his gate. If the molecule was traveling fast, the demon opened the gate to let it through to the other container. If the molecule was moving slowly, the demon kept the gate closed. Over time, Maxwell suggested, one container would accumulate the hotter, faster moving molecules. A temperature differential would build up, and this—by the principles of conventional thermodynamics—could drive an engine and perform work. The demon also poses a philosophical problem, by apparently reversing the flow of entropy, and violating the second law.

The conundrum was long-lived. Maxwell conjured his demon in 1872. Well into the twentieth century, some prominent scientists were troubled by the possibility that the demon might be a disproof of the second law. The solution to this puzzle was finally proposed in the 1950s, when it was observed that the demon is actually an information processing device. It costs him energy to acquire the knowledge he needs to separate fast-moving from slow-moving molecules. This processing cost absorbs and offsets the hypothetical benefits in thermodynamic terms and work potential.

We have used Maxwell's demon as an inspiration for the creation of an interference demon, described in Chapter 4. Our demon would reside somewhere near the front end of a radio receiver, and he would be able to detect whether a given "molecule" of signal energy arriving at his gateway belonged to the desired signal or to an interfering signal. The multiuser detector described by Verdú in his work is a first approximation of this demon, but perhaps the implication of the fate of his predecessor, along with the "information-theoretic results" referenced by Verdú, is that the interference demon is also incapable of keeping ahead of his game. As the number of signal dimensions (or degrees of diversity) increases, the demon's operating costs become too great. At some point, it becomes simpler to fall back to the old fashioned strategy of creating a much more narrowly targeted signal—"*concentrating the energy ... in a small region of the time-frequency plane.*"

9.1.2 Abramson

The next article comes from Norman Abramson, who is known as the father of the ALOHA multiple access scheme, which was developed under his leadership by a team at the University of Hawaii in the 1960s, and implemented in the first modern data network called ALOHANET, operating at 9,600 bps throughout the state of Hawaii—the first packet radio system to be commercially deployed.

The ALOHA access scheme was a fundamental breakthrough in systems architecture, with significance far beyond that first application. The essence of the ALOHA idea is that different dispersed users (i.e., mobile units, in a cellular application) listen to the channel and if it is unoccupied, they transmit in bursts or packets. If two users happen to transmit at the same time, this overlap is detected by one of several means and the transmitters know that they have to retransmit the lost packets. An algorithm to control the retransmission attempts so as to minimize the likelihood of repeated mutual interference is applied. The system is simple, does not require central coordination, and can support reasonable throughput as long as the loading is not too great. There is no setup overhead to create a circuit between the two nodes. It is especially well suited for short messages. ALOHA thinking underlies many of the connectionless communications architectures in wide use today, including the Ethernet protocol family, and will be an important topic in the next volume of this work.

In the article, Abramson tackles the problems of "Internet Access Using VSATs" [2] (i.e., very small aperture satellite terminals) and issues associated with adapting access and transmission protocols to operate over delay-constrained satellite channels. Again, as Verdú did, Abramson ranges over a lot of ground with a fairly focused interest in boosting the capacity or bit rate of the VSAT channel. He presents and evaluates a number of multiple access architectures, assesses their strengths and weaknesses, paying particular attention to CDMA with direct sequence spreading as a signal architecture and multiple access technique. He finds that CDMA is hampered by the requirement to mingle and manage many different spreading codes, one for each active user, in the same physical channel, which introduces considerable complexity. So, instead of having multiple spreading codes coexisting simultaneously, in superimposition, Abramson proposes a new access scheme, which he calls *spread ALOHA multiple access* (SAMA). As he describes it:

> In CDMA networks discussed earlier, the use of code division presents a problem in the design of a family of spreading codes for spread spectrum operation.... The problem is solved by choosing the spreading codes at random. In a SAMA network, however, it is not necessary to find a family of spreading codes with a separate code for each transmitter. It is only necessary to find a single spreading code to be shared by all users....
>
> SAMA can be viewed as a version of CDMA which uses a common code for all remote transmitters in the multiple access channel. *In a SAMA channel, different users are separated by a random timing mechanism, as in a conventional ALOHA channel, rather than by different codes* [emphasis added]. [2, p. 67]

Let us be clear about what this is. SAMA is a reorthogonalization of CDMA. It preserves the spread spectrum structure of the physical signal, but it no longer uses a unique direct-sequence spreading code for each user and does not therefore discriminate users on the basis of that code (i.e., it is hardly "a version of CDMA"). Instead, the users are discriminated in time ("a random timing mechanism") and if they overlap, there is a retransmission ("as in a conventional ALOHA channel").

This is not to imply that SAMA is not an attractive architecture. It does take advantage of signal spreading to achieve significant processing gains (an additional 14 dB), and as Abramson points out it leads to a much simpler transmitter structure than a multichannel narrowband system occupying the same overall bandwidth. He claims that it achieves an advantage of 5–7 dB over standard CDMA in the SIR (multiuser interference) for the same number of users (Figure 9.1).

This result means that in the important mutual interference-limited case, a SAMA channel can accommodate five times as many users as the equivalent CDMA channel. [2, p. 68]

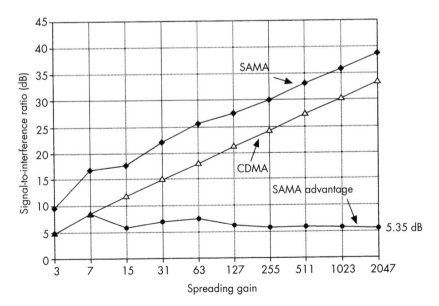

FIGURE 9.1 *Orthogonal versus nonorthogonal CDMA. (From: [2]. © 2000 IEEE. Reprinted with permission.)*

This is a big number, given that CDMA is promoted on the basis of its alleged higher capacity compared to orthogonal systems. Abramson's results, apparently from simulations, would show that the orthogonal version of a direct-sequence spread spectrum signal significantly outperforms the nonorthogonal version, at least in the VSAT application. Is the Red Queen morphing back into the kitten here?

9.1.3 Viterbi et al.

The next in sequence is an article [3] by six Qualcomm authors, including Andrew Viterbi, the originator of the Viterbi decoding algorithm now in wide use, a cofounder of Qualcomm, and another father figure for the CDMA community. This article is a well-reasoned presentation of the rationale and architecture for Qualcomm's data-only air interface specification. This spec is intended to be fully compatible with existing IS-95 CDMA hardware and software, and is referred to here as high-data-rate CDMA, or HDR [3].

Once again, the presenting problem to solve is how to increase capacity, or data speed, to support new 3G-type data services. The article reviews the shortcomings of the current voice-oriented CDMA protocol from an unusually candid perspective.[3] The various sources of lost capacity are highlighted and solutions assessed. However, one problem remains intractable: the mutual interference from base stations sharing the same physical downlink.

> Interference on the reverse link enjoys the advantage of the law of large numbers, whereby the cumulative interference from multiple low-power transmitters tends to be statistically stable. The forward link, on the other hand, suffers interference from a small number of other high-power base stations. This becomes particularly serious at the vertices of the (imaginary) cellular hexagon ... [3, p. 71]

The solution proposed by Viterbi et al. is ... *to restore physical orthogonality.*

With a dedicated RF carrier, the HDR downlink takes on a different form than that of the IS-95 designs. [As shown in Figure 9.2] The

---

3. This is one of the few citations I have found where the capacity penalty of soft handoff is discussed openly. It is amazing how often this is omitted from CDMA capacity calculations, including those provided in the article by Chuang and Sollenberger [4].

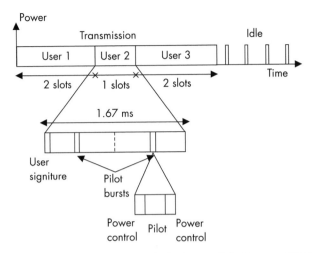

FIGURE 9.2 *Structure of the HDR downlink. (From: [3]. © 2000 IEEE. Reprinted with permission.)*

downlink packet transmissions are *time-multiplexed* and transmitted at the full power available.... Furthermore, when users' queues are empty, the only transmissions ... are short pilot bursts and periodic transmissions of control information, effectively *eliminating interference from idling sectors* [emphasis added]. [3, p. 73]

Let us be clear about what this means. There is no more code division here, as such. Different users no longer share the same spectrum simultaneously on the downlink. They are allocated to different time slots. The philosophy of interference averaging has been abandoned. Each mobile unit directs the base station to transmit from only one sector at any given time. Idle sectors are turned off.

In short, this new and improved form of CDMA (if we may still use "CDMA" as sort of a brand name) is no longer a true CDMA system. In fact, it *is* a true TDMA system. Physical orthogonality has been restored.

9.1.4 Chuang and Sollenberger

The last article, "Beyond 3G ... " is contributed by the Wireless System Research Group at Bell Labs, headed by Nelson Sollenberger [4]. In this article, Chuang and Sollenberger plump for the merits of OFDM (among other things) as a way to again increase capacity and user data rates. Specifically, they promote a system design they call *wideband OFDM*, which they rationalize as somehow simpler than, for example, "using multiple transmit antennas."[4] They suggest that OFDM "promises to overcome ... challenging implementation issues" associated with various other techniques like dynamic channel assignment.

Chuang and Sollenberger do develop an interesting observation, however, regarding the general approach to interference mitigation. Following work by Pottie [5], they characterize two alternative strategies: interference averaging, and interference avoidance. Dynamic channel allocation—in which channel assignment reflects the immediate interference loads within the system—is an example of an *avoidance* strategy. Classical direct-sequence CDMA is a good example of an *averaging* strategy.

I am not sure this is an exhaustive classification. It seems to me that there is a third, even more interesting strategy that involves the active creation of interference, or the manipulation of found interference for signal processing ends. Nevertheless, the rough dichotomy between averaging strategies (which would include most applications of diversity, interleaving, and many coding schemes) and avoidance strategies (including many of the adaptive techniques we have discussed in the last few chapters that modify the signal or the receiver parameters based on specific characteristics of the channel at a particular instant) is conceptually useful. For example, it captures the distinction between OFDM (fundamentally an avoidance technique) and DS/SS (an averaging technique), which we discussed at the end of Chapter 8. Moreover, it defines a watershed in performance: "Interference avoidance techniques can outperform interference averaging techniques by a factor of 2–3 in spectrum efficiency" [4, p. 81]. This makes sense. If there is a way to keep interference *out* of the signal, it stands to reason that this should give better results than *any* countermeasures applied after the interference has crept in. In purely information theoretic terms, it may be possible to prove that structured interference can be perfectly diagnosed and subtracted, in principle. Implementation is always imperfect, and one way to restate the assertion just quoted is to say that averaging solutions usually involve a 3- to 6-dB implementation loss. Better by far if we can avoid contaminating the signal in the first place, it would seem.

---

4. Presumably, of course, Chuang and Sollenberger are well aware of the other major post-Shannon contribution from Bell Labs: the BLAST MIMO-type architectures proposed by Foschini and his collaborators. Whether the "Beyond 3G" article and its so-called "wideband OFDM" are intended as a counterstroke to BLAST is not clear to an outsider.

9.2 Predictions

In the wireless industry today, we are in the high-momentum phase of a pendulum swing toward the abandonment of physical buffers between signals, and the use of coding and other signal processing devices to create new kinds of orthogonality that do not rely on gross physical differentiators like time and frequency. Along with this, we find a similar momentum toward the use of highly *spread* signals, capable of carrying much larger amounts of information than the traditional narrowband signals that have been the mainstay of wireless communications for 50 years or more.

The four articles we have just reviewed indicate, I think, that the question is not as settled as it might seem. The pendulum is tuned by friction, so to speak. Technology trends usually begin with simulations and laboratory results or controlled field results that suggest bold new targets for system performance. In practical implementation, many factors ("friction") intervene to cause us to fall short of those targets. The history of claims for capacity increases surrounding the introduction of CDMA for cellular communications—the first post-Shannon system to achieve commercial acceptance—is instructive.[5] The tuning point, the momentum, and even the direction of wireless technology research are likely to be reset again in the future as new implementation bottlenecks emerge. These may set limits that block the achievement of more idealized targets suggested by some of the new architectures.

For example, it is my guess that when all is said and done it will be too difficult to build the interference demon, even in the somewhat limited form envisioned by the optimistic proponents of multiuser detection techniques. I think multiuser detection will be used to attack much more limited targets, such as eliminating the single strongest interferer, and the complexity of implementation may limit it to uplink applications (where it can be installed at the base station).

Second, I would predict that, in general, *unplanned* interference—that is, interference from sources external to the system—is going to set some

---

5. The controversy is by now largely passé, because arguments other than capacity are more in favor of promoting CDMA, but we may recall that original claims for capacity gains for CDMA relative to analog or TDMA were very large, up to 40 times or more. (See Chapter 6 of [6].) In the end, there is still uncertainty, but the relative advantage claimed in practice for IS-95 (CDMA) over IS-136 (TDMA) is quite modest and perhaps nonexistent. This is another subject that will be addressed in much more detail in the forthcoming Volume 2 of this work.

unfortunate constraints on what some of the new architectures can achieve. Some of the new systems being discussed in the literature these days strike me as conceptually overoptimized. They are like advanced jet engines that work wonderfully until they swallow an errant goose and explode. OFDM-type architectures have been made to work in the stable transmission environment of copper wire, although the complexity is significant and xDSL is still in the early phases of commercial rollout. I personally doubt that some of the more ambitious OFDM architectures can be made to work easily in the challenging mobile channel. It may take longer than some expect.

Third, I would forecast a general reevaluation of the benefits and easy gains to be had from reintroducing some degree of physical orthogonality (separating signals more in space, time, and frequency) into the nonorthogonal architectures that currently predominate. In some cases, the restoration of a modicum of physical orthogonality is simple and may not even be recognized as such. For example, the move to more fully *synchronized* CDMA systems is an easy way to restore some important physical orthogonality (analogous to the use of slotted ALOHA,[6] which has twice the capacity of unslotted ALOHA). Indeed, almost anywhere that we find the word *synchronization* in a discussion of a wireless system architecture, we can be sure that physical orthogonality in the time domain is being strengthened.

Whether this mini-trend to reorthogonalization might even reverse the current intellectual polarity of the field, as is suggested by the HDR architecture outlined in the Qualcomm article, will bear watching. Personally, I think that spread signals are here to stay. Wideband signals in the frequency domain will support high-rate data more readily than bundling many narrowband signals, and without duplication of expensive transmitter and receiver hardware. There is no inherent or necessary link between spreading the signal and the use of nonorthogonal multiple access architectures like CDMA.

As shown by both the Abramson article and the Qualcomm article, it is perfectly possible to use a direct-sequence spread spectrum signal without mixing users in the same spectrum simultaneously. I suspect that orthogonality will remain one of the tools in the toolkit for wireless engineers and system architects. As satisfying as it would seem for the storyline to end

---

6. Slotting an ALOHA system means creating an overall structure in the time domain, defining specific time slots, and requiring mobile units to transmit only within specific time slots. This reduces the chances of overlap and interference between packets, but it requires that the entire system—base station and mobile units—maintain tight synchronization.

with the full triumph of software over physics, the real world continues to intrude on our dreams of managing the airwaves with chess-like precision.

9.3 Whither Shannon?

Throughout this book, we have plucked from time to time at the thread of a profound argument—too profound to pursue in depth in a work of this kind. The question is whether Shannon theory is truly adequate for all conceivable sorts of communications systems. Does it establish immutable "laws" and impassable "limits" (as it is often claimed), or is it more like a vigorous first approximation? Something like Newtonian physics that holds true for the simpler cases but which we now recognize is inadequate for many others? If so, may not the laws and limits we propound today be subject to revision tomorrow?

The field of information theory is strongly resistant to any such critique of Shannon theory. A great deal of energy is devoted in the academic literature to providing reassurances that indeed Shannon theory is fully compatible with, and has even anticipated, each startling new development. I have never really seen any sort of critical treatment of the theory, although it is certainly wide open to the charge that on certain important points it is simply wrong[7] and on others far too simplistic.[8] The simple model of the communications process that we started with calls to mind the famous two-body problem (e.g., the motions of the Earth and the Moon, viewed as a closed system), which Newton's theory of gravity was able to explain. Right behind that, of course, is the three-body problem, which is intractable. In information theory, pretty much anything beyond the Gaussian single-user channel is also intractable.

My real view is perhaps less friendly than the Newtonian analogy would suggest. I am inclined to see Shannon theory as a gigantic "what-if" proposition. What if we could assume for a moment that the communications process involves a single user, transmitting *digital symbols only*, in a well-defined channel, with a background of stable, structureless noise as the only problem to contend with? What if all communications problems could be reduced to the model of the telegraph signal?

---

7. The assertion that "semantics" has no role in information theory is a clear example.

8. The *independence of source and channel coding* theorem is either wrong, or grossly simplistic, as we have remarked repeatedly throughout this book.

It may be objected that Shannon dealt with situations where the noise was not white, Gaussian, or so on (his water-pouring analogy cited in Chapter 8 as an ur-insight behind OFDM). He also tried to address analog signals through the expedient of the rate distortion concept. He also discussed—and certainly later authors have also discussed—the case of two or more users sharing the channel.

The epilogue of one book is not the place to initiate another, but let us up take the last counterexample, briefly. Yes, indeed, information theory has examined the case where two users share the same channel, even in a fully nonorthogonal manner (like CDMA). What is the capacity of that channel, in information theoretic terms? We are told that it is *exactly the same as the capacity for a single user!* By now, the conceptual basis of this argument should actually be quite clear to us, and per se unobjectionable. The two signals each have structure, and if the receiver can decode one, it can decode the other, and by simply decoding both and subtracting the interfering information from the mix, the receiver can yield up the desired signal *regardless of the interference* [7].

However, although we can understand the logic, we must admit that in any practical sense the result is absurd. It is absurd to say that interference has, even in principle, no impact on capacity. It is like saying that if we could only communicate with each other in a perfect vacuum, or inside the center of a black hole, we would enjoy perfect communications no matter how much interference. The Shannon theory pretends to be a theory of communications in our world, and not in some looking-glass world. Yet this result would suggest that the number of users who may share a CDMA channel is infinite, because multiple users have no effect on capacity. The artificiality of the concept of capacity is one of the weak links in the intellectual architecture.

Interference, as an issue, may be the clue. The Shannon theory has trouble in general accounting for interference effects, in my opinion. It reduces external degradations to two idealized cases, based on predictability of the interfering signal. In the case where the interfering signal is completely unpredictable, random, and structureless, we call it noise. Because by definition we cannot anticipate the effects of noise, we are forced to retain it inside the communications process. This is why many people refer to noise as the worst form of interference, because by definition it cannot be removed from the received signal no matter how much processing power we employ. The other idealized case is what Shannon in his papers called *distortion*. Distortion is produced by a deterministic process, and if we can discover the rules governing this process, we can reverse them and completely eliminate the distortion. In the idealized world of the Red Queen and her software, we can simply assume the availability of infinite processing power, in which case any source of interference that has structure can be predicted, can be analyzed and subtracted from the received signal, and thus can be considered to carry no weight in the information theoretic calculations of channel capacity whatsoever.

That these two idealized positions are inadequate for characterizing the domain of interest is suggested by the repeated observation of how interference may play a more positive role in the construction of a signal and the creation of an effective communications process. Whether it is the improvement realized by adding noise to a signal (e.g., dithering), or the purposeful introduction of interference between previously discretized signal elements (e.g., partial response modulation), or the strange positive uses of multipath to enhance system performance that we have touched on in Chapters 7 and 8 in particular, it seems clear that there must be other options than either tolerating or eliminating interference in the communications process.

The critique I would offer is even more fundamental. The "what if" of information theory is really the same "what if" that underlies mathematical thinking itself, or at least those parts of mathematics that are derived from arithmetic. It is the "what if" that results from the assumption that the domain of interest (whether mathematical objects like sets or magnitudes, or information theoretic objects like signal elements) *can be quantized.* For it is indeed clear, even from Shannon's first papers, that without quantization there is no ability to speak about information. A purely analog signal has infinite information, cannot be reproduced by any finite process, and *must* be quantized in order to be brought under the theory.

Quantization is in turn related to the fundamental notions of traditional logic, such as the excluded middle (the idea that any proposition must be either true or not true, with no third option like "sort of true" or "true-and-not-true"). It is related to the fundamental concept of linearity and nonlinearity. It is related very much to the problems of language that are the foundation of analytic philosophy. Finally, we may say that we can view quantization as one of the portals by which we can enter the ancient intellectual realm of the mind-world problem, which has to do with how the mind creates and manipulates information out of the smeared, overlapping, continuously shifting phenomena that the world presents to us.

Here we must leave this thread, except for one final pluck. As I have delved into the foundations of information theory, drawn in by the strange results that are increasingly being turned up by researchers in the wireless field and other fields, I am struck by the idea that information theory—its results, its problems, its puzzles—may have value for shedding light on some of these older and presumptively more fundamental problems of science and philosophy. To cite but one example, I find it fascinating that even quantization, the translation of an analog signal into 1's and 0's, does not entirely remove the underlying "analogness" of that signal. As we have seen in Chapter 7, one of the more active areas in source coding (especially for images) involves deciding how to allow some of the analog redundancy to leak through the quantization scheme. This notion, if translated back to some of the older versions of the quantization problem (such as classical logic), would point to some potentially interesting lines of thought. What if the world can somehow sneak through the grid of logic and assert something of its prelogical, prequantized nature? Perhaps it may shed light on how we should view the many conundrums that seem to erupt relentlessly from all closed logical systems, troubling logicians and philosophers for centuries, from the Cretan Liar, to Russell's paradox, to Gödel's theorems. Perhaps as well it would help us to understand the counterintuitive phenomena of quantum physics, which elude the principles of conventional logic.

In any case, it is clear that Shannon's revolution deserves its luster in the history of science. It would be hard to find another work in that entire history that contains as many fertile insights, as compactly argued, in a form that is fundamentally accessible even to a prepared layman. The great works of Newton, Boltzmann, Maxwell, and Einstein all require extensive interpretation in order to be digested by nonspecialists. Very few of us read them in the original. That is not true of Shannon, which I suspect every engineering student reads at some point in his or her career. Shannon's contributions are still remarkable for their clarity and transparency, and their honesty. Indeed, most of the points of the critique we have sketched here are clearly presented, anticipated, and acknowledged in Shannon's work itself.

But precisely because of the inherent honesty of his presentation, I doubt that Shannon himself would have argued for the assumption of *completeness* that is often made on his behalf nowadays. I think he would not be surprised at the idea that nearly 60 years after he laid the conceptual foundations for a decisive right turn in the thinking about communications processes, there might be enough new information by now to warrant a few course corrections.

REFERENCES

[1] Verdú, S., "Wireless Bandwidth in the Making," *IEEE Communications Magazine*, Vol. 38, No. 7, July 2000, pp. 53–58.

[2] Abramson, N., "Internet Access Using VSATs," *IEEE Communications Magazine*, Vol. 38, No. 7, July 2000, pp. 60–68.

[3] Bender, P., et al., "CDMA/HDR: A Bandwidth-Efficient High-Speed Wireless Data Service for Nomadic Users," *IEEE Communications Magazine*, July 2000, pp. 70–77.

[4] Chuang, J., and N. Sollenberger, "Beyond 3G: Wideband Wireless Data Access Based on OFDM and Dynamic Packet Assignment," *IEEE Communications Magazine*, July 2000, pp. 78–87.

[5] Pottie, G. J., "System Design Choices in Personal Communications," *IEEE Personal Communications Magazine*, October 1995, pp. 50–67.

[6] Calhoun, G., *Wireless Access and the Local Telephone Network*, Norwood, MA: Artech House, 1992.

[7] Cover, T., and J. Thomas, *Elements of Information Theory*, New York: Wiley, 1991.

About the Author

George M. Calhoun is the managing director of Davinci Solutions LLC, a consultancy specializing in communications technology and business development. He has been involved in the commercial wireless industry since 1980. He is the author of *Digital Cellular Radio* (1988) and *Wireless Access and the Local Telephone Network* (1992), both published by Artech House. His e-mail address is geo@georgecalhoun.com.

INDEX

CDMA (continued)
  signals overlaid on one another, 404
  soft handoff, 405
  spread signature, 399, 434
  spread spectrum, 287, 352
  synchronized, 464
  TDMA vs., 122–23
  wideband, 366, 378
Codeless error detection, 310–12
Code(s)
  block, 175–77
  concatenated, 207–10
  convolutional, 177–82
  cyclic, 175
  efficient, 169
  error-resilient entropy (EREC), 305
  Golay, 185
  group, 175
  Huffman, 249–50
  linear, 175
  minimum distance of, 171
  noise-like, 398
  overlapping, 402
  PN, 398
  polynomial, 175
  Reed-Solomon, 182–83
  resynchronizing variable-length, 305
  reversible variable-length, 305
  robust, 170
  spreading, 398
  systematic, 175
Code space
  codewords in center of, 171
  diagram, 173
  expansion, 168
  spreading codewords throughout,
    169, 172
  VQ, 262
Codewords
  in center of code space, 171
  distance between, 171
  in efficient code, 169
  Reed-Solomon generation of, 193
  in robust code, 170
  scattered, 172
  synchronizing, 304
  valid, 169, 172

variable-length (VLCs), 304
Coding, 140–213
  BLAST, 444
  channel, 37–38, 101–2, 163–86
  concatenated, 207–10
  correlative, 224
  decoding algorithms, 182–84
  defined, 145
  delay, 441
  designing, 232
  entropy, 248–50
  FEC, 160
  gain, 184–86
  hierarchical structures, 206–13
  image, 303–8
  implicit, 311
  iterative, 207, 210
  nonlinearity and threshold effects,
    153–59
  parallel, 207
  partial response, 224
  perceptual, 254–56
  philosophy, 140–74
  as redundancy construction,
    159–63
  repetition, 162
  run-length, 245–46
  scheme choice, 190
  source, 173
  space-time, 441
  standard view, 144–45
  summary, 173
  techniques comparison, 185
  theory, 160, 163, 165
  trellis, 202–6
  turbo, 210–13
  See also Signal hardening
    techniques
Coherence
  distance, 104, 120, 222
  frequency, 218
  time, 221
Coherence bandwidth
  defined, 120, 218
  frequency, 106
Collision channel, 310
Combining strategies, 341–42

Recent Titles in the Artech House
Mobile Communications Series

John Walker, Series Editor

Advances in 3G Enhanced Technologies for Wireless Communications, Jiangzhou Wang and Tung-Sang Ng, editors

Advances in Mobile Information Systems, John Walker, editor

CDMA for Wireless Personal Communications, Ramjee Prasad

CDMA Mobile Radio Design, John B. Groe and Lawrence E. Larson

CDMA RF System Engineering, Samuel C. Yang

CDMA Systems Engineering Handbook, Jhong S. Lee and Leonard E. Miller

Cell Planning for Wireless Communications, Manuel F. Cátedra and Jesús Pérez-Arriaga

Cellular Communications: Worldwide Market Development, Garry A. Garrard

Cellular Mobile Systems Engineering, Saleh Faruque

The Complete Wireless Communications Professional: A Guide for Engineers and Managers, William Webb

EDGE for Mobile Internet, Emmanuel Seurre, Patrick Savelli, and Pierre-Jean Pietri

Emerging Public Safety Wireless Communication Systems, Robert I. Desourdis, Jr., et al.

The Future of Wireless Communications, William Webb

GPRS: Gateway to Third Generation Mobile Networks, Gunnar Heine and Holger Sagkob

GPRS for Mobile Internet, Emmanuel Seurre, Patrick Savelli, and Pierre-Jean Pietri

For further information on these and other Artech House titles, including previously considered out-of-print books now available through our In-Print-Forever® (IPF®) program, contact:

Artech House
685 Canton Street
Norwood, MA 02062
Phone: 781-769-9750
Fax: 781-769-6334
e-mail: artech@artechhouse.com

Artech House
46 Gillingham Street
London SW1V 1AH UK
Phone: +44 (0)20 7596-8750
Fax: +44 (0)20 7630-0166
e-mail: artech-uk@artechhouse.com

Find us on the World Wide Web at:
www.artechhouse.com